Advances in Comparative and Environmental Physiology 1

AF788023

Volumes already published

Vol. 1: NaCl Transport in Epithelia
Edited by *R. Greger*

Volumes in preparation

Vol. 2: From the Contents: Hibernation – Vapor Absorption
Nutrient Transport – Stable Pullutants

Vol. 3: Molecular and Cellular Basis of Social Behavior in Vertebrates
Edited by *J. Balthazart*

Vol. 4: Animal Adaptation to Cold
Edited by *L. C. H. Wang*

Vol. 5: Vertebrate Gas Exchange
Edited by *R. G. Boutilier*

Advances in

Comparative and Environmental Physiology 1

NaCl Transport in Epithelia

Guest Editor: R. Greger

With Contributions by

R. Bakker · G. A. Gerencser · R. Gilles · R. Greger
J. A. Groot · F. Lang · E. H. Larsen · W. S. Marshall
L. G. Palmer · A. Péqueux · E. Schlatter · M. J. Welsh

With 70 Figures

Springer-Verlag
Berlin Heidelberg New York
London Paris Tokyo

Guest Editor: *Prof. Dr. Rainer Greger*

Physiologisches Institut
der Universität Freiburg
Hermann-Herder-Straße 7
7800 Freiburg

ISBN-13: 978-3-642-73287-4 e-ISBN-13: 978-3-642-73285-0
DOI: 10.1007/978-3-642-73285-0

Library of Congress Cataloging-in-Publication Data. NaCl transport in epithelia/edited by Rainer Greger. P. cm. – (Advances in comparative and environmental physiology; 1) Includes index. ISBN 0-387-18700-6 (U.S.) 1. Salt – Physiological transport. 2. Epithelium – Metabolism. 3. Physiology, Comparative. I. Greger, Rainer. II. Series. QP535.N2N33 1988 591.19'214–dc19 88-12166 CIP

Softcover reprint of the hardcover 1st edition 1988

Typesetting: K+V Fotosatz GmbH, Beerfelden

Foreword to the Series

The aim of the series is to provide comprehensive, integrated reviews giving sound, critical, and provocative summaries of our present knowledge in environmental and comparative physiology, from the molecular to the organismic level.

Living organisms have evolved a widespread range of basic solutions to cope with the different problems, both organismal and environmental, with which they are faced. A clear understanding of these solutions is of course of fundamental interest for all biologists, zoologically or medically oriented. They can be best comprehended in the framework of the environmental and/or comparative approaches. These approaches demand either wide surveys of animal forms or a knowledge of the specific adaptive features of the species considered. This diversity of requirements, both at the conceptual and technological level, together with the fact that physiology and biochemistry have long been mainly devoted to the service of medicine, can account for the fact that these approaches emerged only slowly amongst the other new, more rapidly growing disciplines of the biological sciences.

The field has now gained the international status it deserves and the organization of a series devoted to it appeared timely to me in view of its actual rapid development and of the interest it arouses for a growing number of biologists, physiologists, and biochemists, independently of their basic, major orientation.

Liège, Belgium, Spring 1988 *Raymond Gilles*

List of Editors

Editor-in-Chief

R. Gilles
Laboratory of Animal Physiology
University of Liège
22, quai Van Beneden
B-4020 Liège, Belgium

Editorial Board

E. Florey
Fakultät für Biologie
Universität Konstanz
Universitätsstraße 10
D-7756 Konstanz 1, FRG

Ch. P. Mangum
Department of Biology
The College of William & Mary
Williamsburg, VA 23185, USA

T. J. Shuttleworth
Department of Physiology
University of Rochester Medical Center
601 Elmwood Avenue
Rochester, NY 14642, USA

G. N. Somero
Scripps Institution of Oceanography
University of California
La Jolla, CA 92093, USA

R. E. Weber
Biologisk Institut
Odense Universiteit
Campusvej 55
DK-5230 Odense M, Denmark

Preface

After the pioneering studies by Ussing and co-workers, studies of epithelial NaCl transport have come a long way. The first phase of the phenomenological description of the cell as a black box has been followed by studies of cellular mechanisms, the interplay of the different transport components, and the mechanisms of regulation. A broad spectrum of methods has been applied to many epithelia in a variety of species. For the individual epithelia transport schemes have been proposed, and, at this point I think it is appropriate to take a pause and search for elements common to several epithelia. This aspect triggered the publication of this book, and in fact the various chapters emphasize that the functional components, expressed in the various epithelia, are not infinite in number, but they occur in epithelia which are separated in evolution by several hundred million years.

The authors come both from the field of veterinary and human physiology as well as from biology. In my opinion, the close contacts and collaborations between physiologists and biologists have been essential for the progress in this field. I wish to thank all authors for their contributions, and I hope that the reader will appreciate this collection of up to date reviews on epithelia in nonvertebrates and vertebrates.

Freiburg, May 1988 *Rainer Greger*

Contents

Chapter 4

NaCl Transport in the Kidney
F. Lang (With 6 Figures)

Chapter 5

NaCl Transport in Amphibian Skin
E.H. Larsen (With 16 Figures)

Chapter 6

NaCl Transport in Tracheal Epithelium
M.J. Welsh (With 5 Figures)

Chapter 7

NaCl Transport in Salt Glands
E. Schlatter and *R. Greger* (With 1 Figure)

Chapter 8

Regulation of NaCl Transport in Tight Epithelia
L. G. Palmer (With 1 Figure)

Chapter 1

NaCl Transport in Gills and Related Structures

A. Péqueux[1], *R. Gilles*[1], and *W. S. Marshall*[2]

Contents

Part I: Invertebrates. *A. Péqueux* and *R. Gilles*

Part II: Vertebrates. *W. S. Marshall*

[1] Laboratory of Animal Physiology, University of Liège, 22, Quai Van Beneden, B-4020 Liège, Belgium
[2] Department of Biology, St. Francis Xavier University, Antigonish, Nova Scotia, Canada B2G 1CO

Advances in Comparative and Environmental Physiology
Vol. 1. Edited by R. Greger

Part I: Invertebrates

A. Péqueux and *R. Gilles*

1 Introduction

This first part of the review will survey our actual knowledge on various aspects of NaCl transport and related mechanisms at work in gills of invertebrates.

In his "Introduction to the study of gills", Hughes (1982) recalls "that the view of a gill as an outpushing from the body wall of an animal is a very old one but originally had little to do with respiration function".

This function seems to have been recognized for the first time in the 16th century. All gill-like extensions of the body wall cannot be considered, however, as gills involved in gas exchanges.

Besides their prominent respiratory function, gills have also been recognized as implicated in or basically responsible for other important functions such as feeding, locomotion, and osmoregulation.

A characteristic feature of invertebrate gills is that their structure and shape vary according to the animal group and/or species and its surroundings. Various types of "outpushings of the body surface" can be considered as assuming similar functions as the gills. According to that view, they certainly have to be related to the gill tissue. An example of this are the body wall extensions at the posterior end of mosquito larvae more commonly designed as anal papillae.

Gills of invertebrates may be considered as organized on the same general principles as those of vertebrates. As a general rule, their exchange surface is considerably increased by more or less extended branching of a main axis. Most invertebrate gills consist of a large number of filaments or flattened lamellae arranged along a central stem (Fig. 1).

Their basic structure and their attachment to the axis may vary from one group to another, but their essential tissular organization is fairly common to most invertebrates. Cross-sections across lamellar surfaces of gills reveal an organization into two sheets of epithelia consisting of a single layer of cells, and joined together periodically by trabeculae or trabecular cells (Fig. 1). Blood or hemolymph circulate through the spaces or lacunae delimited by the epithelia and the trabeculae.

A comparative survey of our actual knowledge on the NaCl transport systems at work in all invertebrate gills that are involved in osmoregulation should be an enormous subject. The interest of such a study should be moreover limited since most of these groups would exhibit, in their gills, specialized patches or extended surfaces, so-called salt-transporting regions, revealing a relative uniformity in their structure at the cell level. In salt-transporting regions of invertebrate gills, the structural organization of the epithelial cells seems indeed to be quite similar in the different species. They all present an asymmetric organization with distinct apical and basal sides, both usually showing a complex elaboration of large infoldings. At the basal side, these deep infoldings are moreover closely associated

(Talbot et al. 1972; Foster and Howse 1978), crayfish (Curra 1965; Morse et al. 1970; Bielawski 1971; Fisher 1972; Burggren et al. 1974; Dunel-Erb et al. 1982), and crabs (Copeland 1968; Copeland and Fitzjarrell 1968; Finol and Croghan 1983; Péqueux et al. 1980; Gilles and Péqueux 1981; Barra et al. 1983). The gill of the Chinese crab *Eriocheir sinensis* has appeared to be a most interesting model for studying the relation between the ultrastructure and the ion transepithelial movements related to blood osmoregulation in crustaceans. It will be shown in the following that the *E. sinensis* model of the structural and functional organization of the gill can satisfactorily be applied to other hyperregulatory crustaceans. In the Chinese crab, indeed, a clear distinction can be made between the three most posterior pairs of gills characterized by a thick epithelium and the anterior pairs which are lined by a thin epithelium. An extended study conducted in this laboratory in collaboration with Drs. Barra and Kirsch from Strasbourg (France) has traced the difference between anterior and posterior gills down to the level of their ultrastructure (Péqueux et al. 1980; Gilles and Péqueux 1981; Barra et al. 1983). In *anterior* gills, the epithelium is about 2 – 4 µm thick and the nuclei generally protrude within the hemolymph space. The apical side of the cells just beneath the cuticle is very lightly folded, which makes the extracellular compartment under the cuticle extremely reduced. The amount of intracellular organelles is very limited and the plasma membrane does not exhibit extensive basolateral infoldings (Fig. 2). In that epithelium, the distance between the hemolymph and the external medium is only about 3 – 5 µm, more often less than 2 µm, thus such tissue thinness doubtlessly favors gas passage from the outside towards the hemolymph. Therefore, it seems quite reasonable to consider that the main and essential function of that kind of gill tissue must be respiration. In crabs from dilute media, the thick epithelium of the *posterior* gills (10 – 20 µm) is characterized by a complex and well-developed network of extended, apical, evaginated, and digitated folds and villi managing a large and very characteristic extracellular subcuticular compartment (Figs. 2 and 3). As a general rule, such an organization of the apical membrane to form microvilli, folds, or leaflets is typical of cells involved in active transport processes (Berridge and Oschman 1972; Cioffi 1984). It is also now interesting to note that it is the only region of the cell to exhibit major morphological changes when the transport activity is turned off (see Sect. 2.2).

At the serosal side of the cell, the basolateral plasma membrane is infolded deeply into the cell; often infoldings go up to the apical invagination of the apical membrane (Figs. 2 and 3). Between and along the basolateral folds numerous mitochondria are packed which can exhibit unusual shapes. In the blue crab *Callinectes sapidus,* Copeland has described tremendously flattened and cup-shaped mitochondria with modified cristae, organized in so-called mitochondrial pumps (Copeland and Fitzjarrell 1968). That author defined the mitochondrial pump as "a metabolically linked ion pump located in the cell plasma membranes intimately associated with the mitochondrial membranes at a distance of several hundred Ångströms or less" (Copeland 1967). In the osmoregulatory gills of crustaceans, mitochondria may be extremely abundant, almost completely filling the cytoplasmic space within and outside of these folds.

Although details may vary from one animal to another, such extensive elaboration of the plasma membranes in the form of tubules, crypts, infoldings,

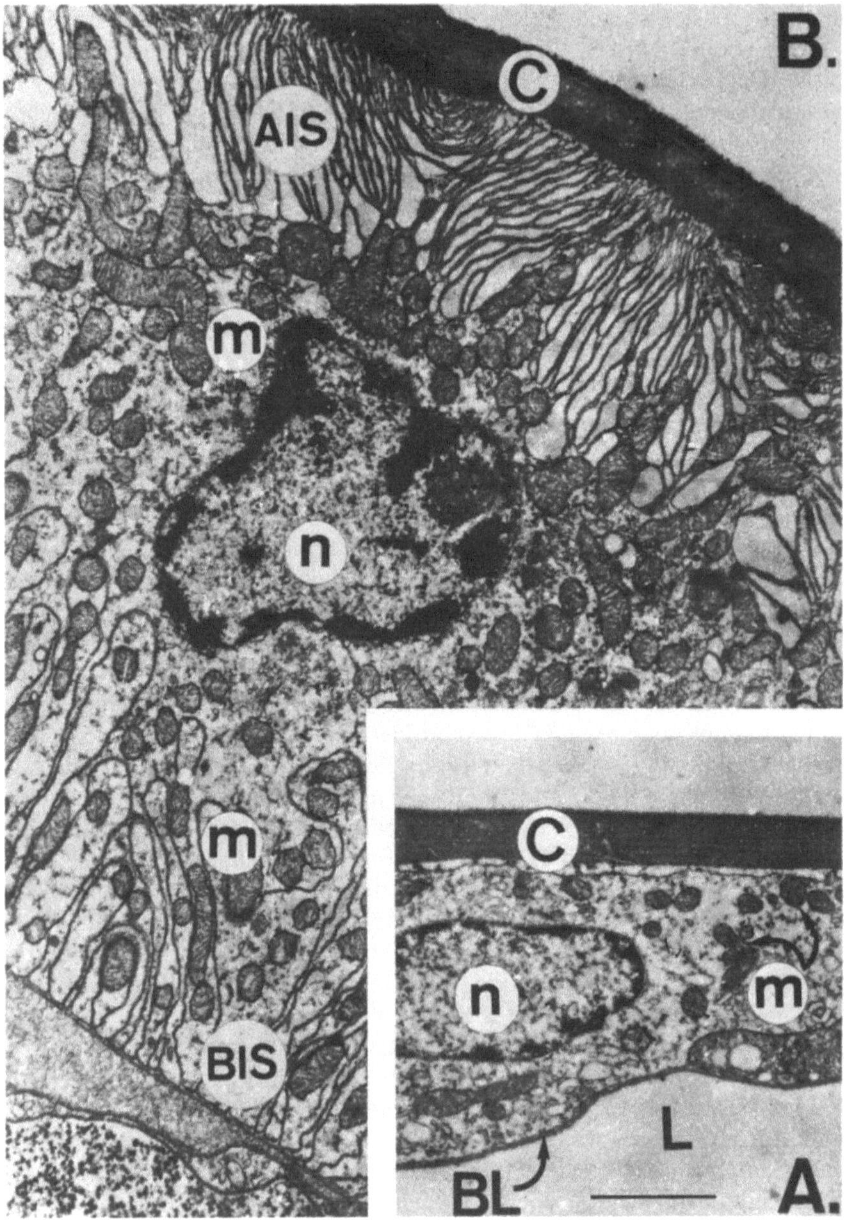

Fig. 2A, B. Ultrastructure of epithelial cells from gill lamella of the FW Chinese crab, *Eriocheir sinensis*. *A* Anterior gills; *B* posterior gills. Bar: 1 μm. Cuticle (*C*); mitochondria (*m*); nucleus (*n*); apical infolding systems (*AIS*); basolateral infolding system (*BIS*) (Gilles and Péqueux 1985; Barra et al. 1983)

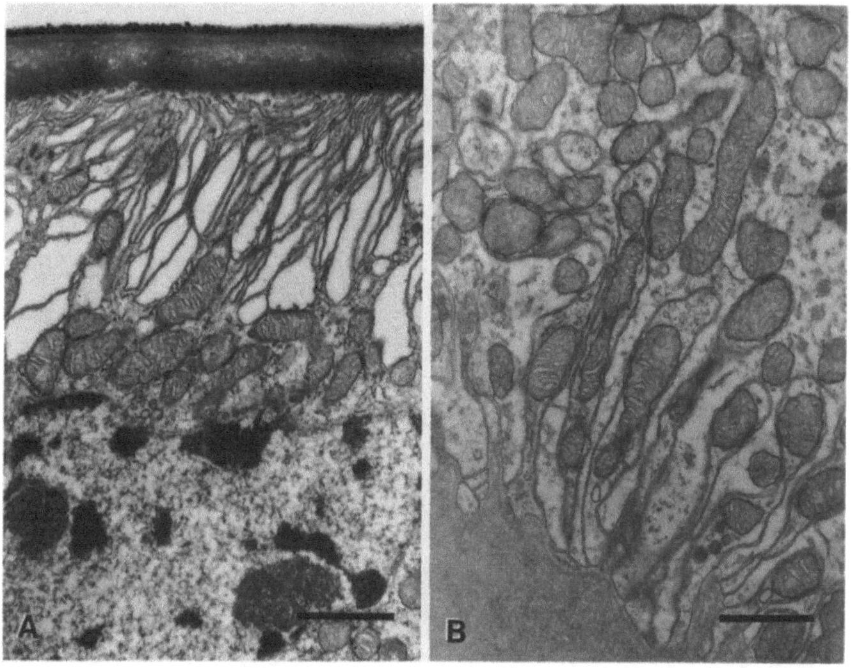

Fig. 3A, B. Ultrastructure of epithelial cells from posterior gill lamella of FW Chinese crabs, *E. sinensis*. *A* Details of the apical infolding system (AIS) and of the subcuticular space; *B* of the basolateral membrane infoldings (BIS). Bar: 1 µm (Barra and Péqueux, unpubl.)

and interdigitations opening to the extracellular space and in close association with mitochondria not only seems to be common to the gill tissue of many other osmoregulator crustaceans but, furthermore, to most of the organs responsible for hydromineral regulation in other animal groups (for review, see Berridge and Oschman 1972; Gilles and Péqueux 1981).

The question which now arises is whether there are changes in ultrastructure that can be correlated with changes in physiological function. In the preceding section, we have examined the fine structure of gills of euryhaline decapods acclimated to dilute media, i.e., to media from which they have to actively transport Na^+ and Cl^- ions. Most of these animals become isosmotic to the external medium when in seawater. One can thus expect major physiological changes under these conditions as the transport activity of the gills must have become minimum.

2.2 Gill Structure and Salinity Acclimation

Using the silver-staining method, Copeland and Fitzjarrell (1968) reported that silver-stained patches of the gills of hyperosmotic regulators were increased in size after acclimation to dilute media. This observation was best explained by considering a significant reduction in the number or size of transporting sites in

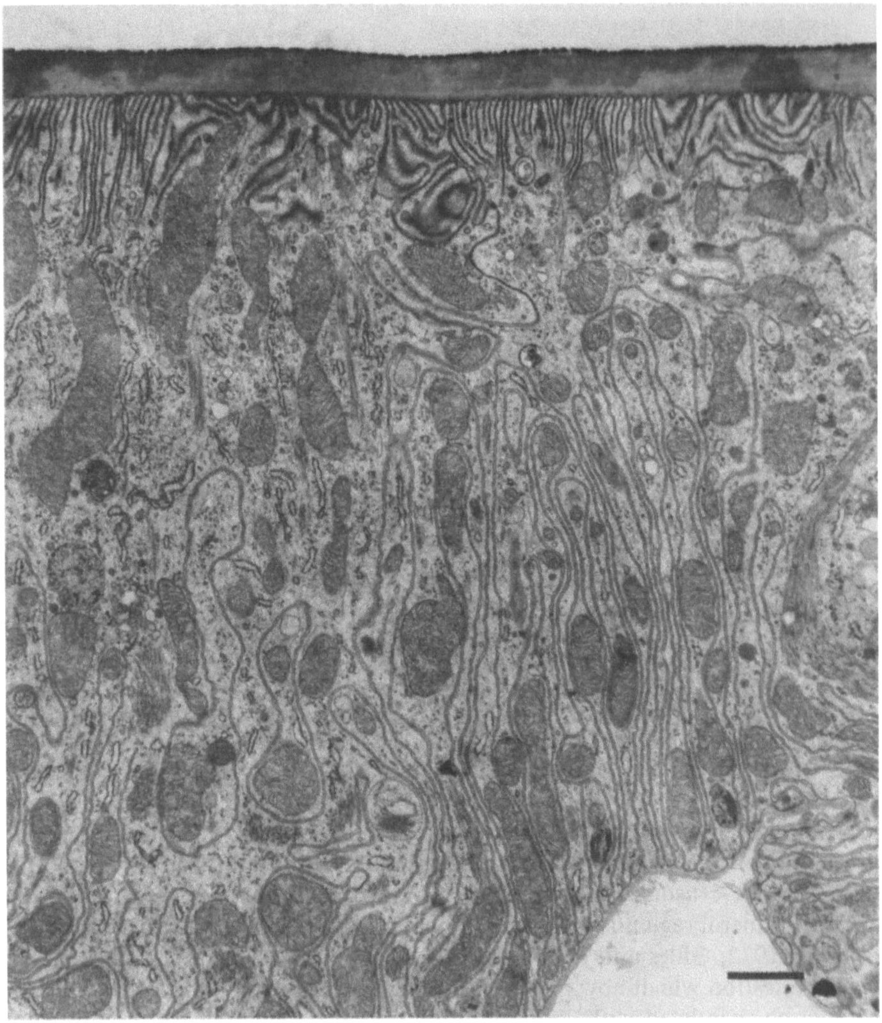

Fig. 4. Ultrastructure of epithelial cells from posterior gill lamella of the Chinese crab *E. sinensis* acclimated to seawater. Bar: 1 µm (Barra and Péqueux, unpubl.)

concentrated media, i.e., when the transport activity is known to be considerably decreased. Up to now, possible salinity effects on branchial fine morphology have been little examined. We recently undertook such a study on the gills of the Chinese crab *Eriocheir sinensis*. This species is of particular interest for this purpose as it can be easily acclimated to concentrated media and since the Na^+ transport activity under these conditions is completely turned off (see Sect. 3.1.3).

In seawater, the ultrastructure of the anterior gill epithelium remains basically similar to the structure observed in a dilute environment. The major structural changes occur in the "salt-transporting" epithelium of the posterior gills. A structural difference which is immediately visible is the considerable regression of

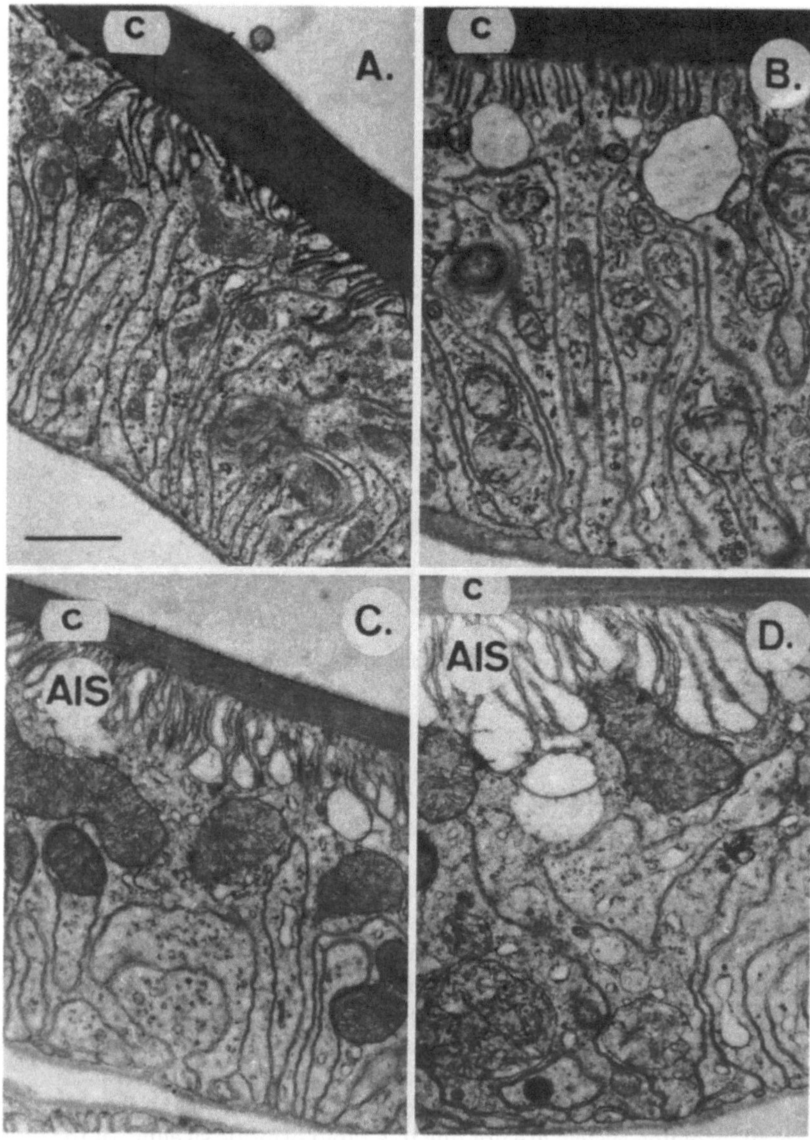

Fig. 5A – D. Change in the ultrastructure of the epithelium posterior gill lamella of the Chinese crab *E. sinensis* during acclimation from SW to FW. *A* 10 h of acclimation; *B* 24 h; *C*, *D* 88 h. Bar: 1 μm. Cuticle (*c*); apical infolding systems (*AIS*) (Gilles and Péqueux 1985)

the apical infolding system (Fig. 4). This leads to a severe decrease of the apical surface area of the cells and to the disappearance of the large subcuticular extracellular space. Similar changes in the development of the apical folding system have also been reported to occur in the gills of the amphipods *Gammarus oceanicus* (Milne and Ellis 1973) and *Gammarus duebeni* (Lockwood and Inman 1973), in the brown shrimp *Penaeus aztecus* (Foster and Howse 1978), and in the

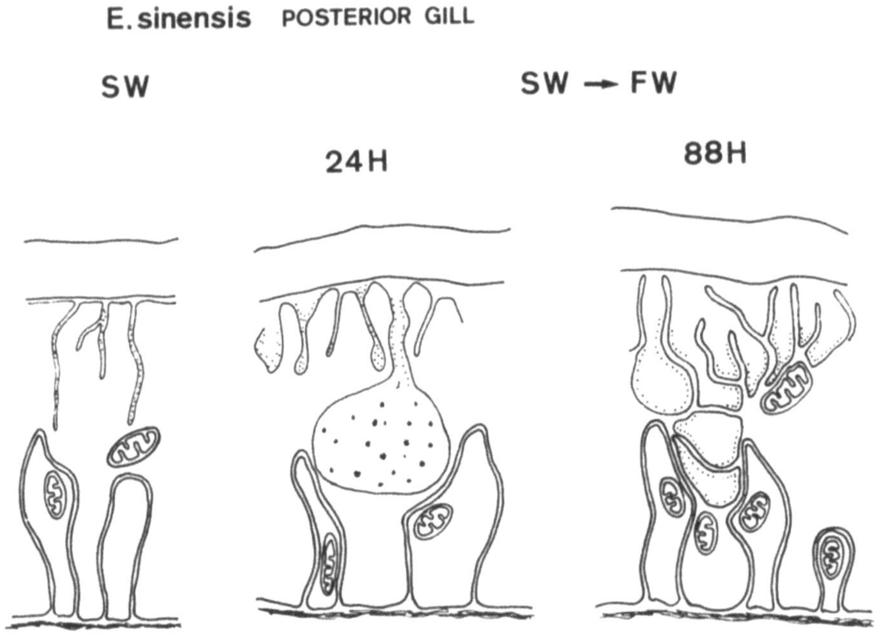

Fig. 6. Changes of the ultrastructure that occur in the posterior gill epithelium of *E. sinensis* during acclimation from SW to FW

crab *Goniopsis cruentata* (Martelo and Zanders 1986). There may also be some differences in the structural organization of the basolateral infoldings and the associated mitochondria, but the situation seems to be much less clear. In the Chinese crab *E. sinensis,* as well as in the brackish-water crab *Uca mordax* (Finol and Croghan 1983), the basal folds system of the transporting epithelium does not seem to be affected by differences in salinity regimes. On the contrary, the extracellular spaces between the folds in the basal cell membrane and the conformation of mitochondria undergo significant changes in the gills of *Gammarus oceanicus* (Milne and Ellis 1973). Further investigation needs to be done before a complete discussion of that problem can be undertaken.

Evidence suggests that the differences observed in the development of the apical infolding system are not due to a rapid osmotic swelling of the intercellular spaces that could occur during transfer from seawater to dilute water. The appearance of the apical infoldings system is indeed very progressive in the course of low salinity acclimation. In the Chinese crab, we have observed that only a few bleb figures appear during the first 24 h following transfer of SW animals to FW, at the basis of the few folds present at the apical side of the salt-transporting cells of the posterior gills (Fig. 5). The number and size of blebs, as well as the number of folds, increase then progressively during the following days, the overall process taking about 4–5 days to evolve to the typical picture observed in animals fully acclimated to FW. The evolution of the structural organization as a function of the acclimation time is diagrammatically summarized in Fig. 6.

3 Insights from Physiological Studies

Krogh (1938, 1939) was the first to propose the independent absorption of Na^+ and Cl^- in crustacean gills. Considering the different rates of anion and cation uptake in the crayfish and in the Chinese crab, he postulated that electroneutrality was maintained by countertransport of NH_4^+ and/or H^+ for Na^+ and of HCO_3^- for Cl^-. Results that could be interpreted as indicative of separate uptake systems were given later for different hyperosmoregulating species, for example, the crayfish *Astacus pallipes* (Shaw 1959a, 1960a, b), the shore crab *Carcinus maenas* (Shaw 1961), and the blue crab *Callinectes sapidus* (Cameron 1978a, b, 1979). Most of these studies have been achieved, however, on whole animals, assuming that the gill was the major site of ion movements. The results obtained following this line of approach can certainly be usefully fitted into possible models of transport systems, but they cannot provide us with a clear characterization of the mechanisms and driving forces at work in the ion transport at the gill level itself. Wrong conclusions concerning ion movements across the gill epithelium may indeed result from a misunderstanding of possible interferences with the transport activity of other tissues or from the interaction with the metabolic activity when fluxes are related to the movements of ionic species such as protons or ammonia. Further, this kind of approach cannot take into consideration the possible differences between gills in their capability to transport ions. Such differences, however, do exist. It will be shown in the following that the differences in the structural organization of the crustacean gills reported in Section 2 can be related to physiological specialization. These differences may be important. In the Chinese crab *E. sinensis,* a net separation between anterior and posterior gills can even be done, based on a clear distinction between respiratory type and salt-transporting type epithelia. The separation is not always so sharp in other crab species, and, in some cases, both types of epithelia can be found in the same gill. This has been demonstrated, for example, in the blue crab *Callinectes sapidus* (Copeland and Fitzjarrell 1968), in the land crab *Gecarcinus lateralis* (Copeland 1968), and in the shore crab *Carcinus maenas* (Compère et al. 1985). In other groups of crustaceans, the gill organization may be different, but, as a general rule, the structural difference with possible functional correlates still exists in hyperregulating species. In the crayfish, for example, two branchial cavities contain a large number of gills which are podobranchiae, arthrobranchiae, and pleurobranchiae (André 1960). It has been suggested that gas exchanges should concern the thin filament epithelia of arthro-, pleuro-, and podobranchiae, while ion exchanges should occur essentially at the level of the thicker lamina epithelium of the podobranchiae (Bock 1925; Fisher 1972; Dunel-Erb et al. 1982). In crabs, the thick epithelium generally restricted to the posterior gills is moreover the only one to show morphological changes during the course of salt adaptation, i.e., when the NaCl transport process has to be activated or even developed. According to that view, the posterior gills of the Chinese crab *E. sinensis* should be the only ones containing the transepithelial salt active uptake machinery involved in the control of the blood ion balance in dilute media. That it is indeed the case has been demonstrated by direct flux measurements with radioactive tracers and by potentiometric studies conducted on perfused gill preparations.

3.1 NaCl Fluxes Across Crustacean Gills

Much of our understanding of the nature of Na^+ and Cl^- movements at the gill level of hyperregulators has been gained in investigations on perfused preparations of isolated gills.

3.1.1 Na^+ Fluxes in Isolated Preparations of Crustacean Gills

Early reports of Na^+ transport across isolated gills in crustaceans were presented by Koch about 30 years ago. Working on isolated but not perfused gill preparations of the euryhaline Chinese crab *E. sinensis,* Koch reported the influx of Na^+ to be larger in the posterior gills than in the anterior ones (Koch 1954; Koch et al. 1954). He suggested already that the anterior and posterior gills of that crab may differ in their overall ability to transport Na^+.

On perfused gill preparations of the blue crab *Callinectes sapidus,* Mantel (1967) showed that gills isolated from crabs acclimated to seawater were able to transport Na^+ inward in the absence of an electrochemical gradient. There was a large variation in the individual measurements (influx values ranging from 135 to 3060 μmol Na^+ h^{-1} per gill) but, in each case, there was a net influx of Na^+.

Very high Na^+ flux data have also been obtained by Wanson and Péqueux (1981) and by Lucu and Siebers (1986) on different species of *Carcinus.* When gills from SW *Carcinus maenas* were perfused and immersed in SW saline, Na^+ in- and effluxes values reached approximately 2500 – 3000 μmol Na^+ g^{-1} h^{-1}

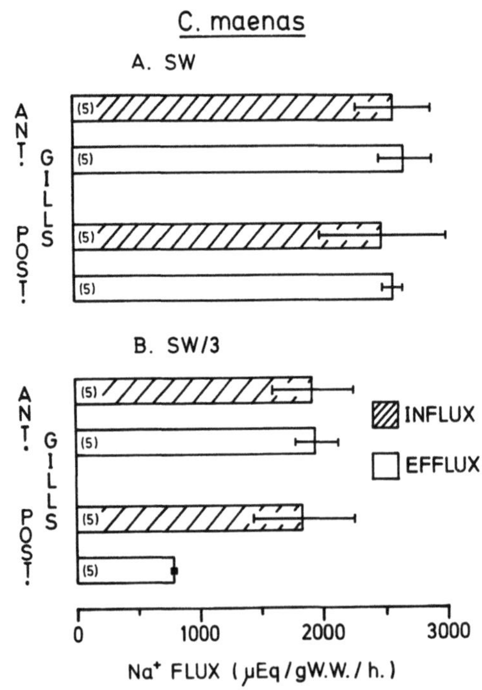

Fig. 7. $^{22}Na^+$ fluxes in isolated, perfused anterior and posterior gills of shore crabs, *Carcinus maenas* acclimated to SW and to 30% SW. Mean results of (n) experiments ± SD. (After Wanson and Péqueux 1981, and unpubl.)

with no significant net flux of Na^+ (Fig. 7) (Wanson and Péqueux 1981 and unpubl. results). Similar results have been obtained on *Carcinus mediterraneus* by Lucu and Siebers (1986) who reported unidirectional flux values of the same order of magnitude (around 4000 ± 1800 µmol Na^+ g^{-1} h^{-1}). These authors also reported a 40 – 50% reduction of the amplitude of Na^+ fluxes when the gills were transferred from external SW to dilute SW. Unidirectional Na^+ fluxes have also been measured across gills of the same crab species acclimated for a few weeks to dilute SW. In *C. maenas,* Na^+ in- and effluxes, respectively, felt to 1834 ± 429 (n = 5) and 770 ± 62 (n = 3) µmol g^{-1} h^{-1} in the three most posterior pairs of gills perfused by and incubated in the same saline (Fig. 7), resulting thus in a net uptake of Na^+ across the gill epithelium (Wanson, unpubl. results). A net uptake of Na^+ also occurs under similar conditions of incubation at the level of the gill pairs Nos. 7 and 8 of the shore crab *C. mediterraneus* (Lucu and Siebers 1986). In the anteriorly located gills of *C. maenas,* Na^+ in- and effluxes do not differ significantly from the fluxes measured in the posterior gills when under SW conditions. In dilute SW, flux values are reduced by about 25% but do not further result in any significant net movement of Na^+ (Wanson unpublished results). There is no information concerning the anterior gills of *C. mediterraneus.*

Na^+ fluxes have been more systematically studied on the Chinese crab *E. sinensis* and the existence of a functional difference between the various pairs of gills suggested by the early works of Koch and by the above data collected on *Carcinus* has found strong experimental support. This study definitively leads to separate considerations of the three most posterior pairs of gills from the anterior ones in the fully euryhaline Chinese crab *E. sinensis.* By comparison with the Na^+ flux data reported on the crab species mentioned above, the Na^+ movements measured across gill preparations of freshwater Chinese crabs remained always of a much lower order of magnitude. In all cases and whatever the gill type considered, Na^+ fluxes never exceeded 400 µmol Na^+ g^{-1} h^{-1} (Péqueux and Gilles 1981). In the anterior gills, the Na^+ influx is lower than the Na^+ efflux when gills are immersed in artificial freshwater and there is thus a net Na^+ loss under such conditions. However, both the Na^+ in- and effluxes appear to be dependent on the external concentration of that ion, in such a way that they approach the same order of magnitude when the external Na^+ exceeds 50 mM (Fig. 8). From considerations based on the comparison between the ratios that can be calculated from Na^+ flux measurements and those calculated according to Ussing's equation, taking into consideration the transepithelial Na^+ concentration gradient and the transepithelial potential difference, it can be established that Na^+ fluxes in the anterior gills are essentially passive. The study of the Na^+ fluxes as a function of the external Na^+ level, however, reveals saturation kinetics indicating that the movements of that ion, though passive, are carrier-mediated and not the result of a simple diffusional process. Worth noting is the fact that the Na^+ fluxes decrease as the external Na^+ concentration decreases, in spite of the fact that the Na^+ gradient across the epithelium is increasing. Reduction of the Na^+ influx following dilution of the external medium has also been reported in isolated gill preparations of *Carcinus maenas* (King and Schoffeniels 1969; Wanson, Péqueux and Gilles, unpubl.) and, more recently, of *Carcinus mediterraneus* (Lucu and Siebers 1986). Of particular interest is the Na^+ efflux

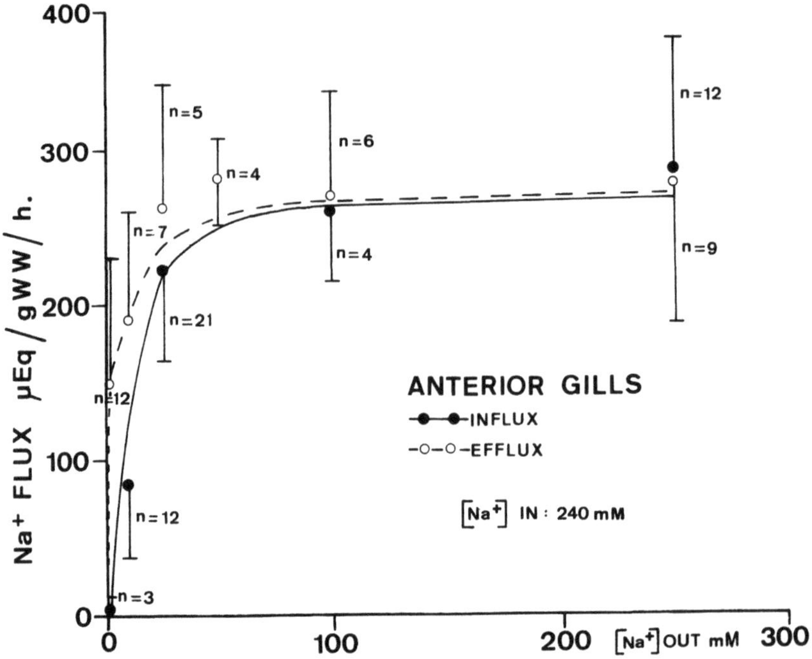

Fig. 8. $^{22}Na^+$ fluxes in isolated, perfused anterior gills of FW Chinese crabs, *E. sinensis*. Mean results of (n) experiments ± SD (Péqueux and Gilles 1981)

decrease observed in *E. sinensis* anterior gills. This can be best explained by considering that the permeability of these gills to Na^+ is dependent on the level of that ion in the external medium, thus decreasing at low external Na^+ levels. A decrease in Na^+ permeability with decreasing concentration of the ion in the environment has already been proposed to occur in fish (Maetz 1971, 1974; Kirschner 1979; Evans 1980). Changes in permeability to ions in crustaceans seem more difficult to assess from the available literature. As stated by Mantel and Farmer in their recent review (1983), little if any change in permeability can be shown in osmoconforming and moderately regulating crabs but they consider that, in strong regulators, the rate of salt loss across the body surface should be reduced upon exposure to dilute media. The results obtained with the isolated anterior gills of *E. sinensis* bring experimental support to that idea, as do the results of Baldwin and Kirschner (1976) on the fiddler crab *Uca*. These authors also suggested indeed that the reduction of Na^+ efflux they observed in Cl^--free water indicated a reduction in ionic permeability of the gill of *Uca*. Such an immediate decrease in permeability is, of course, of great physiological interest, especially for strong regulators which often experience very dilute media. This will indeed decrease the importance of the salt loss, thus making the maintenance of the blood osmotic balance easier. As will be seen in the following, besides this "immediate" adaptability of the Na^+ permeability, a slower "long-term" decrease in permeability is also occurring in the Chinese crab *E. sinensis* and can be considered as another feature of the anterior gills facilitating acclimation to dilute media by limitation of salt loss. An alternative explanation of such efflux reduction was proposed by Cameron (1978a, 1979). It is based on the idea that a reduction in the efflux of a particular ion, when the external concentration of that ion is reduced, can be interpreted as evidence for exchange diffusion, provided that no change in electrical potential takes place that would account for the efflux reduction. Accordingly, the translocation of an ion from the inside to the outside is inhibited because there is a shortage of similar ions on the outside to be simultaneously translocated to the inside. The results obtained by Cameron (1978a, 1979) on the blue crab *Callinectes sapidus* are therefore consistent with the independent exchange diffusion hypothesis, since a reduction of Na^+ efflux occurring in Na^+-free water is not accompanied by any change in Cl^- efflux, nor is there a change in transepithelial potential difference.

Whatever the exact nature of their components, the unidirectional movements of Na^+ in the anterior gills of the Chinese crab are strictly passive. In contrast, the Na^+ flux in the posterior gills can be ascribed essentially to active processes. Whatever the magnitude of the Na^+ gradient applied across the epithelium, a significant, even large, Na^+ influx can always be measured in perfused preparations of posterior gills, while there is never any significant efflux of that ion (Péqueux and Gilles 1981). The permeability to Na^+ of the posterior gills of Chinese crabs acclimated to freshwater is thus probably extremely low and the large Na^+ influx that occurs at that level must thus be entirely accounted for by active processes. The active influx of Na^+ of the Chinese crab posterior gills has been shown to be dependent on the external concentration of the ion, revealing saturation kinetics with maximum influx value for external Na^+ levels of about 50 mM (Fig. 9). By analogy with enzyme kinetics, apparent Km and vmax values for the transepithelial Na^+ carrying system have been calculated (Péqueux and Gilles 1981). The Km value of 13.7 mM obtained accordingly for the Chinese crab gill is rather close to the one (20 mM) reported by Shaw (1961) when studying Na^+ uptake in whole specimens of another euryhaline decapod, *Carcinus maenas*, but is high when compared to those available for freshwater crustaceans

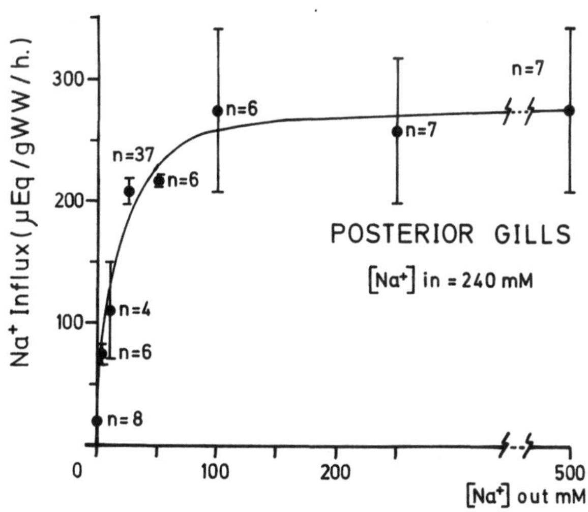

Fig. 9. Influx of $^{22}Na^+$ in isolated, perfused posterior gills of the FW Chinese crab, *E. sinensis*. Mean results of (n) experiments ± SD. (After Péqueux and Gilles 1981; Gilles and Péqueux 1985)

like *Astacus fluviatilis* (0.2 mM, Shaw 1959a), *Potamon niloticus* (0.1 mM, Shaw 1959b), or *Gammarus pulex* (0.15 mM, Shaw and Sutcliffe 1961). It must, however, be considered that the vmax that can be calculated for these marine euryhaline decapods is higher than the one that can be measured in freshwater species (331 μmol g^{-1} wet wt h^{-1} in *E. sinensis,* Péqueux and Gilles 1981; 10 mmol kg^{-1} h^{-1} in *C. maenas,* Shaw 1961; and 0.15 mmol kg^{-1} h^{-1} in *Astacus fluviatilis,* Shaw 1959a). A possible explanation can be that the rather low affinity of the carrier for Na^+ found in the marine euryhaline custaceans is, at least partly, compensated by a high number of pumping sites.

The question is now which mechanisms are implicated in the active uptake of Na^+. This question still remains a matter of controversy. As early as 1938, a component of the unidirectional fluxes of Na^+ was proposed by Krogh to consist of an electroneutral exchange for NH_4^+ and/or H^+. These exchanges have been fairly well documented in fish (Maetz 1973, 1974; Maetz and Garcia-Romeu 1964; Evans 1977; Payan 1978) and evidence has been provided for crustaceans consisting of a demonstration of a Na^+ effect on net H^+ excretion (Kirschner et al. 1973). In the crayfish, results on ammonia production in relation to the salinity of the acclimation medium have been explained as indicating a correlation between Na^+ influx and NH_4^+ movement in the opposite direction (Shaw 1960b).

As discussed earlier (Gilles 1980a, b; Gilles and Péqueux 1981), such increased ammonia excretion is found in many crustaceans acclimating to dilute media. During this process, the animals, however, also show increased tissue ammonia production as well as increased ammonia blood levels. These increases are, in fact, due to the increased tissue deamination activity related to cell volume control. They thus cannot be directly ascribed to the activity of a Na^+/NH_4^+ antiport system. More recently, ammonia efflux as well as Na^+ influx from living blue crabs *Callinectes sapidus* have been demonstrated to be sensitive to external amiloride (Pressley et al. 1981). This mutual sensitivity to amiloride, a well-known inhibitor of Na^+ transport has been considered by these authors as further evidence for a Na^+/NH_4^+ exchange. At variance with these results, other studies have shown that amiloride does not affect the ammonia efflux from intact blue crabs (Kormanik and Cameron 1981). The interest of this finding is unfortunately weakened by the fact that these experiments were carried out on seawater crabs, i.e., crabs in which the activity of the Na^+ active transport is minimum, if even existent. They are, however, in agreement with the results of experiments showing that amiloride has only a small effect on ammonia excretion in the crayfish at concentrations completely blocking Na^+ active uptake (Kirschner et al. 1973). In vitro, Lucu and Siebers (1986) moreover demonstrated that addition of 2.5×10^{-4} M amiloride to the external medium of isolated perfused gills of the shore crab *Carcinus mediterraneus* inhibits both the Na^+ influx and efflux, indicating that the inhibitor must act at another level than the only Na^+ uptake process. It appears from these investigations that the effect of amiloride on crustaceans in vivo remains controversial and that amiloride sensitivity alone may not be a critical indicator for Na^+/NH_4^+ exchange. Several other studies have shown indeed that amiloride can inhibit apical Na^+ conductive cellular or paracellular leak pathways (Balaban et al. 1979) and even basolateral

Na^+/K^+ activated ATPase (Soltoff and Mandel 1983). Moreover, as pointed out earlier (Gilles and Péqueux 1985), amiloride inhibition of ammonia efflux could also be explained by an indirect effect on the NH_3 equilibrium distribution, as it is indeed an effective blocker of the apical Na^+/H^+ exchange system.

In each of these in vivo experiments, an alternative explanation based on acid-base disturbance and consequent effect on ammonia equilibrium distribution cannot be ruled out. The question of the gill ammonia transport and of its possible implication in the movements of Na^+ is thus far from being clear and the interest reader is referred to the recent review of Evans and Cameron (1986) on that subject.

The data obtained in vitro with perfused preparations of isolated *E. sinensis* gills clearly show, however, that in this species at least, a Na^+/NH_4^+ exchange transport can only account for a very small part, if any, of the total Na^+ influx occurring in the posterior gills. If there is a dependency of the ammonia efflux on the external level of Na^+ in these gills, comparison of the flux data for Na^+ and NH_4^+ makes, however, the hypothesis of a coupling between the movements of both ions unlikely; the influx of Na^+ is indeed much larger than the efflux of ammonia (Péqueux and Gilles 1978a). It is thus clear that the part of the total Na^+ influx that could be accounted for by an Na^+/NH_4^+ exchange in the posterior gills of *E. sinensis* is probably extremely small. This conclusion is further substantiated by the fact that increasing the concentration of NH_4Cl in the perfusion saline fails to induce any significant change in the Na^+ influx (Péqueux and Gilles 1981). It appears thus more reasonable to assume that a large part of the active Na^+ uptake is not coupled to NH_4^+ excretion.

This conclusion leaves us with the question as to which mechanisms are implicated in the major part of the active Na^+ uptake. Whether it is due to another kind of neutral antitransport mechanism or to an electrogenic system is still essentially a matter of speculation.

The idea of a Na^+/K^+-coupled, ouabain-sensitive component of the transepithelial transport of Na^+ in the posterior gills of *E. sinensis* has been considered, but did not lead to a clear picture of the situation (Péqueux and Gilles 1981). Indirect evidence, however, suggested the possible existence of such a component of the transepithelial transport of Na^+ or, at least, of a correlation between them. In the shore crab, *C. maenas,* Siebers et al. (1985) reported that the transepithelial electrical potential difference that they considered as essentially produced by the activity of the Na^+ pump (see Sect. 3.2), is inhibited when K^+ is omitted or ouabain is applied. Similar results were obtained on posterior gill preparations of FW Chinese crabs *E. sinensis* (Péqueux et al., unpubl.). This suggestion has found experimental support in a very recent paper in which Siebers et al. (1986) reported a decrease in Na^+ influxes from 500 – 1000 μmol $g^{-1}h^{-1}$ to 250 – 400 μmol $g^{-1}h^{-1}$ upon application of 5 mM ouabain in the perfusion medium of isolated posterior gills of *C. maenas.* Prominent also must be the part played by an ouabain-sensitive component in the context of the regulation of the intracellular level of both K^+ and Na^+ ions which must involve control of the movements of these ions between hemolymph and intracellular fluid. Another good, perhaps better, candidate could be a Na^+/H^+-coupled system. As early as 1960, Shaw (1960b) suggested that H^+ might indeed serve as a counterion in Na^+ uptake in *Astacus* since H^+ excretion varied as a function of the external Na^+ concentration and that high external H^+ concentrations inhibit Na^+ uptake. In that context, the results obtained by Ehrenfeld (1974) on the crayfish *Astacus leptodactylus* are also interesting. That author indeed showed that injection of acetazolamide, a carbonic anhydrase inhibitor, inhibits Na^+ uptake as well as acid efflux, ammonia efflux, and base extrusion, indicating that hydration of CO_2 to form H^+ and HCO_3^- is an integral component of the Na^+ uptake mechanism in this species. Similar results were obtained in salt-depleted *Cambarus* (Kirschner et al. 1973) and in freshwater blue crabs *Callinectes sapidus* (Cameron 1979). One should also keep in mind that amiloride is an effective blocker of the Na^+/H^+ exchange system too and, in that context, all the reports on an inhibitory effect of amiloride on transepithelial Na^+ transport could be considered as evidence for a possible involvement of a Na^+/H^+ antitransport in Na^+ unidirectional fluxes (Cameron 1979; Pressley et al 1981; Lucu and Siebers 1986). It is clear that further experiments are needed in order to obtain more insight to these questions.

3.1.2 Cl^- Fluxes in Isolated Preparations of Crustacean Gills

The intense staining of gills due to precipitation of silver chloride when the animals are placed in a solution of silver nitrate may be considered as an early indication that Cl^- ions penetrate through the gill tissue (see Sect. 2). On the other hand, the negative electrical potential of the hemolymph in relation to the external medium in several decapods species either in intact animals (Shaw 1960c; Mantel 1967; Baldwin and Kirschner 1976; Spaargaren 1977; Cameron 1978a; Pressley et al. 1981) or in an isolated preparation of gills (Lockwood 1962; Croghan et al. 1965; Mantel 1967; Smith and Linton 1971; Péqueux and Gilles 1978b) has also been considered as an indication that the gill epithelium is the major site of active transport of chloride (see also Sect. 3.2). According to Bielawski (1964), the rate of Cl^- transport could even be a rough measure of the whole ion transport in the gills. Surprisingly, Cl^- transport across the gill epithelium has been little investigated in vitro using perfused preparations and informations in that field are still very scanty. Early direct evidence for a system transporting Cl^- against the concentration gradient has been presented by Bielawski on isolated gill preparations of the crayfish *Astacus leptodactylus* and *Astacus astacus* (Bielawski 1964).

Working with isolated perfused gills of *Callinectes sapidus* acclimated to seawater, Mantel reported briefly that there is not necessarily a net influx of Cl^- accompanying the net Na^+ influx measured in the absence of an electrochemical gradient (Mantel 1967). That finding could perhaps be explained by the fact that these experiments were done on seawater crabs, i.e., on crabs in which the activity of the transepithelial transport systems is minimum, if even existent. More recently, Lucu and Siebers (1986) measured transepithelial in- and efflux of Cl^- in isolated perfused gills 7 and 8 of *Carcinus mediterraneus* acclimated to dilute seawater. With identical media on both sides of the gills, the influx and efflux of Cl^- they obtained ranged respectively, between 726 ± 249 ($n = 4$) and 536 ± 198

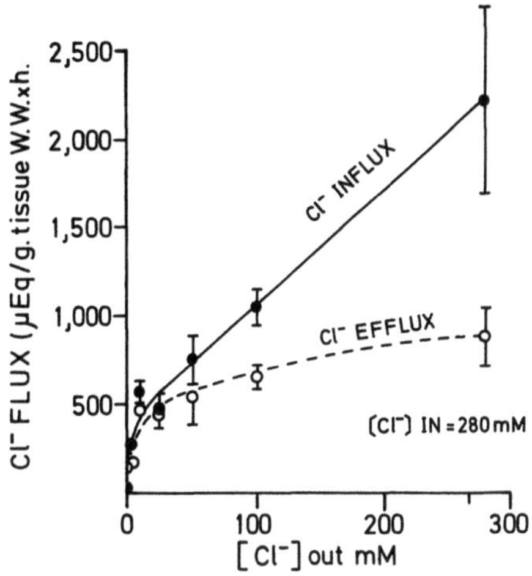

Fig. 10. $^{36}Cl^-$ fluxes in isolated, perfused posterior gills of FW chinese crabs, *E. sinensis*. (After Gocha et al. 1987)

(n = 4) μmol g^{-1} h^{-1}. According to considerations based on the comparison between calculated and measured flux ratios according to Ussing's equation, these authors concluded that an active transport process is implicated in the inward movement of Cl^-.

In these works, reference is never made to the possible existence of differences in Cl^- transport activity between the various pairs of gills.

Experiments conducted in our laboratory on isolated perfused gills of freshwater Chinese crabs *E. sinensis* establish that such differences may be very important (Gocha et al. 1987). In the anterior gills, there is no Cl^- influx nor efflux regardless of the Cl^- gradient applied across the epithelium. Cl^- flux data of very low amplitude have only been obtained when both faces of the gills were bathed with the same saline, i.e., in the absence of any concentration gradient, but their real significance may be questioned. If the anterior gills of the Chinese crab cannot be in any way involved in any transepithelial Cl^- active uptake then it is worth noting that their permeability to Cl^- ions must also be extremely low.

In contrast, the posterior gills epithelium seems to be permeable to Cl^- and the large net Cl^- influx which can be measured at that level must be of an active nature. As shown in Fig. 10, both the Cl^- influx and efflux in the posterior gills are dependent on the external concentration of Cl^- and result in a net passage of Cl^- from the external medium to the hemolymph. The Cl^- efflux overcomes the influx only when the outside Cl^- concentration falls to 1 mM or less. Of particular interest is the decrease of the Cl^- efflux magnitude as a function of the dilution of the external medium, in spite of the fact that the transepithelial Cl^- gradient is increasing. As it has already been suggested when considering the movements of Na^+ ions, this supports the idea that the permeability of the posterior gills to Cl^- may be adjusted to minimize the salt loss. Important also are the differences that exist between the different pairs of gills in their ability to transport Cl^- ions in the shore crab *Carcinus meanas* (Wanson et al. unpublished). Wanson has indeed demonstrated that, at variance with the efflux, the influx of Cl^- is dependent on its concentration in the external medium in both

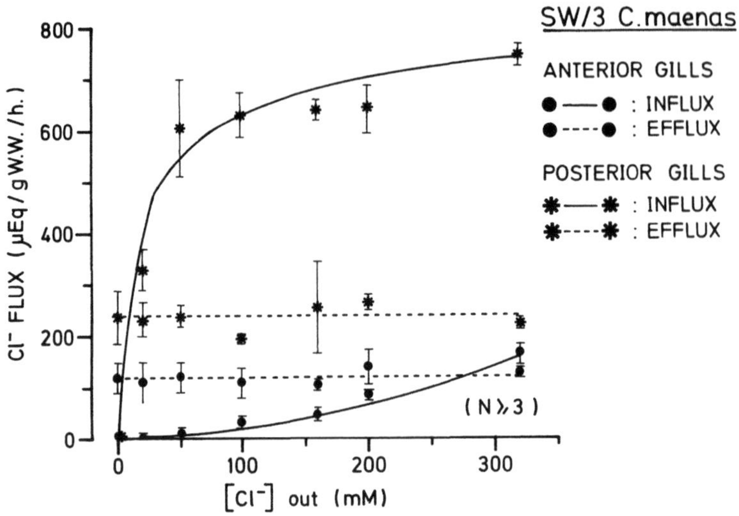

Fig. 11. $^{36}Cl^-$ fluxes in isolated, perfused anterior and posterior gills of shore crabs; *C. maenas* acclimated to 30% SW. Mean results of (n) experiments ±SD. (Wanson et al., unpubl. results)

the anterior and posterior gills of *C. maenas* while of a much larger magnitude in these latter gills (Fig. 11). This results also in a net uptake of Cl^- restricted to the posterior gills only.

As all these results are, in some cases, essentially preliminary, they still do not bring much insight as to the mechanisms implicated in the active uptake of Cl^-, i.e., the various components of the unidirectional fluxes. As stated earlier, most experiments done up to now have been carried out on whole animals and have been fitted into the Krogh model implicating independent transport processes for Na^+ and Cl^-, Cl^- being exchanged for HCO_3^-. There are numerous lines of evidence in support of this hypothesis, including a role for carbonic anhydrase in providing HCO_3^- for the exchange (see also Sect. 4.2). Disruption of hemolymph Cl^- regulation has been demonstrated in decapod crustaceans maintained in a dilute environment and treated with acetazolamide, a well-known inhibitor of carbonic anhydrase (Burnett et al. 1981; Henry and Cameron 1983). The effects of acetazolamide have also been studied on unidirectional ion fluxes in crayfish (Ehrenfeld 1974) and in freshwater blue crabs (Cameron 1979). In both cases, these authors observed an increase of Cl^- influx accompanied by a greater increase in Cl^- efflux. Ehrenfeld (1974) explained these results by suggesting that acetazolamide could cause Cl^-/HCO_3^- exchange to become predominantly a Cl^-/Cl^- exchange. It has also been shown that changes in blood pH brought about by external hypercapnia cause, in the blue crab *Callinectes sapidus,* changes in Na^+ and Cl^- flux rates that could be explained on the basis of H^+ and HCO_3^- movements to regulate pH (Cameron 1978b). If electroneutral exchange of Cl^- for HCO_3^- may thus be considered as a possible component of the unidirectional flux of Cl^-, Cameron (1979) assessed that it must be only a very small part of the Cl^- fluxes. A further component of the unidirectional flux of

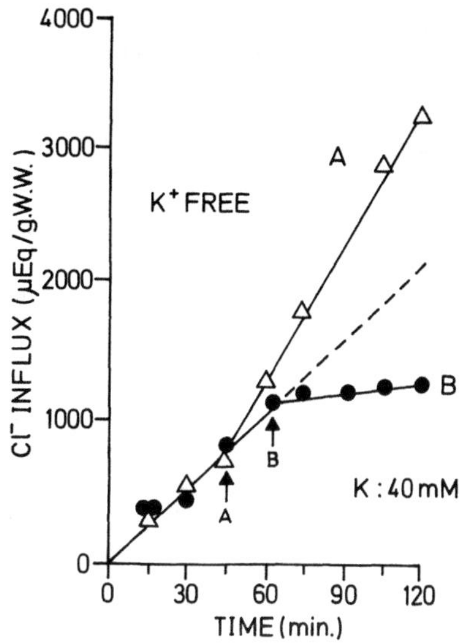

Fig. 12. Effect of the K^+ concentration of the perfusion saline on the $^{36}Cl^-$ influx of isolated, perfused posterior gills of FW Chinese crabs, *E. sinensis.* Altered K^+ saline is applied at *arrows: A* K^+-free saline; *B* 40 mM K^+ saline. (After Gocha et al. 1987)

Cl^- meets indeed the usual criteria of exchange diffusion (Cameron 1978a, 1979). From the preceding, it appears that the problem is still far from being clearly solved and certainly needs further investigations. Very recently, evidence has been presented that Cl^- influx was dependent on the K^+ concentration of the medium bathing the serosal side of the gill epithelium (Gocha et al. 1987). In a perfused preparation of isolated posterior gills from freshwater *E. sinensis,* the Cl^- influx was either enhanced or decreased by respectively decreasing or enhancing the K^+ concentration of the perfusion medium (Fig. 12). This indicates that the K^+ concentration gradient at the serosal face of the epithelium generates an electrical potential difference which could act as a force driving Cl^- ions from inside the cells to the blood.

3.1.3 Ion Fluxes and Salinity Acclimation

Although the various patterns of blood osmotic regulation as a function of the environmental salinity are very well documented in intact specimens of euryhaline crustaceans (see for example the extensive review of Mantel and Farmer 1983), still little is known as to the possible changes in ion fluxes activity that occur at the gill level.

While this was not the actual subject dealt with in the recent paper of Lucu and Siebers (1986), evidence can be found suggesting that the magnitude of Na^+ movements across the epithelium of gills 7 and 8 of *Carcinus mediterraneus* is considerably larger in crabs kept in 38‰ salinity seawater than it is in crabs acclimated to 14.5‰ salinity dilute seawater. Under symmetrical incubation conditions (almost the same solution on both sides of the epithelium), these authors reported that influx and efflux were respectively, 3835 ± 1813 (7) and 4493 ± 1545 (6) μmol Na^+ g^{-1} h^{-1} in seawater crabs, while these flux data fall respec-

Table 1. Na^+ fluxes in isolated, perfused gill preparations of *Eriocheir sinensis* acclimated to freshwater (FW) or to seawater (SW). (After Péqueux and Gilles 1981; Gilles and Péqueux 1985)

Acclimation medium	Na^+ in saline ($mEq\,l^{-1}$)		Na^+ fluxes[a] ($\mu Eq\,g^{-1}$ w.w. h^{-1})	
	Out	In	Influx	Efflux
Anterior gill				
FW	500	480	625 ± 195(4)	628 ± 175(4)
SW	500	480	3448 ± 960(3)	2679 ± 632(8)
Posterior gill				
FW	500	480	Undetectable	Undetectable
	25	480	4.3 ± 3.8(8)	Undetectable
SW	500	480	Undetectable	Undetectable
	100	480	8.4 ± 8.8(11)	–
	25	480	5.9 ± 13.2(18)	7.2 ± 4.6(10)
	0.96	480	0.03 + 0.08(12)	–
	100	240	5.5 ± 8.3(12)	–
	25	240	10.9 ± 13.3(11)	9.4 ± 8.3(12)
	0.96	240	0.8 + 1.0(12)	–

[a] Not determined. (n) number of experiments.

tively to 758 ± 124(6) and 512 ± 106(6) μmol Na^+ g^{-1} h^{-1} in dilute seawater. In *Carcinus mediterraneus,* it thus appears that acclimation to a dilute environment has resulted in a considerable decrease in the magnitude of Na^+ fluxes.

These data are of the same order of magnitude as the flux results we obtained with the anterior gills of Chinese crabs *E. sinensis* acclimated to freshwater or to seawater (Table 1). The results shown in Table 1 deal with Na^+ fluxes measured in isolated perfused gills (anterior "respiratory" gills and posterior "salt-transporting" gills) of Chinese crabs kept in freshwater or acclimated to seawater and submitted to changes in the Na^+ content of the perfusion saline as if the crabs were transferred from one medium to the other. In *anterior* gills of seawater acclimated crabs, the Na^+ fluxes are of a particularly large magnitude. It is interesting to note that Na^+ fluxes obtained with gills from freshwater crabs remain mostly lower when bathed with the same seawater saline. This indicates that the permeability to Na^+ of the anterior gills must be significantly lower in animals acclimated for a long period of time to a dilute medium. This "long-term adaptation" of the permeability has to be considered differently from the "immediate" permeability change that occurs following a sudden change in the external Na^+ concentration (see previous Sect. 3.1.1). Both mechanisms, however, act together in limiting the importance of the salt loss with which the animals have to cope in dilute media. They both are of the utmost physiological interest for animals which experience fluctuating salinities.

In *posterior* gills of seawater crabs, the Na^+ fluxes fall to undetectable values. Moreover, the results reported in Table 1 show that Na^+ influx becomes also undetectable regardless of the Na^+ concentration of the incubation medium, when gills from a freshwater crab are perfused with a "seawater" saline. If the same gills were perfused with a freshwater saline, a large active Na^+ influx would occur. These observations thus strongly suggest that the pumping activity of the posterior gills of the Chinese crab can actually be turned on and off according to the Na^+ concentration of the hemolymph medium.

Interesting also is the fact that there is no detectable transport activity in these "seawater" gills regardless of the Na^+ level in the external medium and regardless of the nature of either the freshwater or seawater of the perfusion saline. This means that acclimation to salt concentrated media leads to the "inactivation" of pumping sites and also to a progressive loss of the capability of transporting actively. Accordingly, the main trigger system modulating the transport activity would be the Na^+ level in the serosal medium. Conversely, upon transfer of seawater crabs to a salt dilute environment, the pumping activity is progressively restored within 3–8 days of acclimation (quoted in Gilles and Péqueux 1981, 1983, 1985). This time-course evolution fits perfectly the time course of the change in the ultrastructure of the apical membrane network of salt-transporting epithelium (see previous Sect. 2.2) and of the change in membrane ATPase (see Sect. 4.1.1).

3.2 Potentiometric Studies

In many vertebrates tissues and epithelia, much insight into the mechanisms controlling the ion fluxes has been gained by studying the nature and the modalities of the transepithelial electrical potential difference. The use of potentiometric methods on invertebrate material has been extremely scanty up to now.

This section will consider the few studies giving measurements of potential difference in crustaceans, either in vivo, i.e., measured between the pericardial blood space and the external medium, or in vitro, i.e., measured between the two faces of perfused preparations of isolated gills.

3.2.1 In Vivo Potential Difference Measurements

Studies giving measurements of electrical potential difference (PD) in decapods in vivo (Table 2) show that the hemolymph of the animal is usually slightly negative with respect to the outside (Bryan 1960; Potts and Parry 1964a; Parry and Potts 1965; Mantel 1967; Smith and Linton 1971; Baldwin and Kirschner 1976; Greenaway 1972, 1976; Spaargaren 1977; Cameron 1978a, b, 1979; Robinson and Potts 1979; Zanders 1980; Pressley et al. 1981; Winkler 1986). In all these studies, the magnitude of the PD is not largely affected by, or is even completely independent of, the salinity of the external acclimation medium.

From his work on the shore crab *C. maenas,* Zanders (1980) concluded tentatively that the trans body-wall PD in dilute media is probably due to diffusion potentials, the small size of these reflecting the low ion selectivity of the surfaces involved in the exchanges between crabs and medium. A similar conclusion as to the diffusional nature of the PD in Dungeness crabs *Cancer magister* has also been proposed by Robinson and Potts (1979). On the FW blue crab *C. sapidus,*

Table 2. Trans body-wall electrical potential difference (PD) of intact crustaceans in media of various salinities

Animal species	External medium	PD (mV)[a]	Reference
Artemia salina	SW	$+23.4 \pm 0.6(36)$[b]	Smith (1969)
Astacus fluviatilis	Tap water	$-6.6 \pm 4.8(10)$[c]	Bryan (1960)
Austropotamobius pallipes	Tap water	$-17.7 \pm 1.5(9)$[b]	Greenaway (1972)
Callinectes sapidus	SW	$-1.9 \pm 0.2(4)$[b]	Mantel (1967)
	SW	$-0.3 \pm 0.4(3)$[b]	Pressley et al. (1981)
	50% SW	$-3.9 \pm 0.9(4)$[b]	Mantel (1967)
	50% SW	$-1.4 \pm 0.9(4)$[b]	Pressley et al. (1981)
	FW	-2 to -4	Cameron (1978a)
	FW	$-3.4 \pm 1.4(4)$[c]	Cameron (1979)
Cancer magister	SW	$-0.65 \pm 0.46(13)$[c]	Robinson and Potts (1979)
Carcinus maenas	SW	-1.5	Mantel (1967)
	SW	$-2.6 \pm 1.3(8)$[c]	Winkler (1986)
	60% SW	$-0.8 \pm 0.1(15)$[b]	Zanders (1980)
	50% SW	-3.0	Mantel (1967)
	30% SW	$-3.3 \pm 1.3(4)$[c]	Winkler (1986)
	25% SW	$-0.8 \pm 0.7(18)$[b]	Zanders (1980)
Eriocheir sinensis	SW	$-0.83 \pm 0.43(3)$[c]	Péqueux (unpubl. data)
	FW	$-79.8 \pm 3.9(7)$[c]	Péqueux (unpubl. data)
Libinia emarginata	80–120% SW	No potential	Mantel (1967)

[a] Sign refers to inside (hemolymph).
[b] Mean results ±SE of (n) measurements.
[c] Mean results ±SD of (n) measurements.

Cameron (1978a) demonstrated that the PD response to changes in external Na^+, K^+, and Cl^- is negligible, while values as large as −30 mV were observed in experiments consisting in a progressive dilution of tap water with distilled water. These effects were shown to be almost entirely due to the lack of Ca^{2+} in the external medium (Cameron 1978a).

In response to hypercapnia, the same author, however, reported a slight drop (2.6 mV) of the PD (Cameron 1978b), while neither external SCN^-, external amiloride, nor injected acetazolamide affected in any way the PD of crabs resting in normally aerated water (Cameron 1979). Working on SW blue crabs, Pressley et al. (1981) have established that substitution of Na^+-free choline SW for normal SW causes significant hyperpolarization in crabs acclimated either to 17‰ salinity or to full 35‰ SW. Addition of amiloride to SW, however, did not produce any change in PD, while the inhibitor caused a slight depolarization when applied to the 17‰ SW crabs (Pressley et al. 1981). According to these authors, a Na^+ diffusion potential might make a large contribution to the overall PD under Na^+-free SW conditions, at least larger than in normal SW. This idea was moreover in agreement with the observed effects of amiloride, as an amiloride-induced decrease in the permeability of cell membranes or paracellular pathways would have a larger effect on the PD under Na^+-free conditions than under of a smaller Na^+ gradient.

According to Bryan (1960), the PD across the body surface of the crayfish *Astacus fluviatilis* indicates that the high blood Na^+ concentration is maintained by active uptake of Na^+.

It is interesting to note in Table 2 that P.G. Smith (1969) reported a blood positive PD in a crustacean of another group, the brine shrimp *Artemia salina*. Smith explained his results as an indication that Cl^- is subject to active transport out of the animal, while K^+ is pumped in and Na^+ is approximately in equilibrium (Smith 1969). His observations on the PD and electrical resistance further suggested that the gills were more permeable to Na^+ ions than to Cl^- ions and the resulting diffusion potential modulated the Na^+ flux.

It appears that these in vivo studies certainly do not lead to a clear picture of the real nature of the electrical potential difference existing between the external medium and the hemolymph of the species considered. These studies do not give conclusive information as to the processes participating in the establishment of the PD, nor as to what is going on at the level of the different pairs of gills.

Let us now consider the insights obtained by experiments carried out on isolated preparations of gills, either perfused or not.

3.2.2 Potential Difference Across Isolated Gill Epithelium

Transepithelial potential differences have been measured across isolated gills of the crabs *Callinectes sapidus* (Mantel 1967; Smith and Linton 1971), *Carcinus maenas* (King and Schoffeniels 1969; Wanson and Péqueux 1981; Siebers et al. 1985; Winkler 1986), *Carcinus mediterraneus* (Lucu and Siebers 1986), *Eriocheir sinensis* (Péqueux and Gilles 1978b, 1981, 1988; Gilles and Péqueux 1985, 1986), *Maia squinado* (King and Schoffeniels 1969), and the crayfish *Austropotamobius pallipes* (Croghan et al. 1965).

From experiments carried out on excised gills of FW acclimated blue crabs *Callinectes sapidus*, Smith and Linton (1971) concluded that the negative potential (blood with respect to the medium) that they measured between the hemolymph space and the external dilute medium was the result of differential permeability and active transport processes. These authors moreover proposed that the whole gill was preferentially permeable to Na^+ with respect to Cl^-.

Contradictory results were obtained on the same material at isotonicity, i.e., when perfusion and incubation were carried out with the same saline. The fact that at isotonicity a strong negative potential was displayed by metabolically active gills has been considered by Smith and Linton (1971) as an indication of an electrogenic inward transport of a negative ion. At variance with these results, Mantel (1967) measured a small positive potential that she nevertheless considered as too small to be accounted for solely on the basis of a net Na^+ influx.

From the studies mentioned at the beginning of this section, it appears that the origin and the modalities of the transepithelial potential difference across

Table 3. Electrical potential difference (PD) across the perfused gills of FW Chinese crabs *E. sinensis.* [a] (After Pèqueux and Gilles 1988)

Gill No.	4	5	6	7	8
Identified as	Anterior gills		Posterior gills		
PD (mV)	+45.2	+24.8	−26.6	−25.5	−16.3
±	±	±	±	±	±
SD	10.3	6.8	14.2	15.3	9.2
(n)	(8)	(16)	(10)	(13)	(6)

[a] Both faces of the epithelium are bathed with the same saline. Sign refers to the "inside" (perfusion medium).

isolated crustacean gills are still a matter of controversy. Most of these works, however, agree on the fact that the PD, or at least one of its components, must be of an active nature. Siebers et al. (1985) even considered that it represents the driving force for the active uptake of Na^+ in crabs that hyperregulate their body fluids when in reduced salinities.

These studies generally do not take into consideration the possibility that the different gills could have different functional roles and therefore could display different types of ion movements. Scanty reports of differences in potentiometric properties between the various pairs of gills are, however, found in the works of Croghan et al. (1965) on the crayfish *Austropotamobius pallipes,* Winkler (1986) on the shore crab *Carcinus maenas,* and Pèqueux and Gilles (1978b) on the Chinese crab *Eriocheir sinensis.* In fact, Croghan et al. (1965) reported that in the crayfish, there is a continuous gradation of gill types from a predominantly cation-permeable type towards a more chloride-permeable type. This question has been dealt with more extensively in a recent work carried out on the Chinese crab *E. sinensis* (Pèqueux and Gilles 1988). It has been shown (Table 3) that the three more posterior pairs of gills are not electrically equivalent to the anterior pairs in crabs from freshwater, i.e., in crabs that hyperregulate their body fluids. Their polarity is indeed of opposite sign when they are bathed with the same saline on both sides of the epithelium. Moreover, their PD varies as a function of the NaCl level in the incubation medium (Table 4). In the absence of an ion concentration gradient across the epithelium, the PD in the *anterior* gills is always inside positive, varying from 6 to 60 mV. From considerations based on the analysis of PD changes induced by specific ionic substitution, it has been accounted for by several diffusive components for Na^+, K^+, and Cl^-. At the serosal side of the epithelium, there are diffusion potentials for K^+ and Cl^- which are partly short-circuited by an Na^+ diffusion potential. At the apical face, the PD has been demonstrated to be essentially generated by Na^+ diffusion. Besides these cellular components, a transepithelial diffusion of Na^+, likely to occur through paracellular pathways, must also be taken into consideration when Na^+ gradients are applied to the gill epithelium.

The *posterior* gills, when bathed with the same saline on both sides to avoid transepithelial gradients of NaCl, show a PD varying from 5 to 50 mV and which is always inside negative. This PD has been explained considering diffusive

Table 4. Effects of external NaCl concentration on the transepithelial PD of FW *E. sinensis* perfused gills.[a] (After Péqueux and Gilles 1978b, 1988)

External [NaCl] (mM)	PD (mV) in	
	Anterior gills	Posterior gills
1	$-52.5 \pm 2.4(4)$	$-11.1 \pm 5.2(4)$
10	$-38.3 \pm 1.4(6)$	$-25.1 \pm 8.7(4)$
25	$-11.1 \pm 6.6(7)$	$-23.9 \pm 4.4(4)$
100	$-1.7 \pm 4.8(7)$	$-27.0 \pm 2.3(4)$
240	$+26.0 \pm 10.7(22)$	$-20.9 \pm 5.5(16)$

[a] Sign refers to the "inside" (perfusion medium). Mean results of (n) experiments ±SD. Perfusion medium: FW crab saline (240 mEq $Na^+ l^{-1}$). Incubation medium: artificial FW + NaCl up to concentrations quoted in the first column.

movements of Cl^- at the apical side and diffusive movements of K^+ short-circuited by Cl^- outward movements at the serosal side (Péqueux and Gilles 1988). However, the posterior gills of *E. sinensis* are known to be the site of independent, active transport processes of both Na^+ and Cl^-, the activity of which could interfere with PD establishment (Péqueux and Gilles 1981; Gocha et al. 1987). Application of transport inhibitors to perfused gill preparations during PD measurements gave some insight into the mechanisms implicated in the processes. The fact that SITS (4-acetamido-4′ isothiocyanatostilben-2-2′ disulfonic acid, disodium salt) induces a depolarization essentially when added in the external medium has been considered as an argument supporting the idea that a Cl^-/HCO_3^- antiporter is implicated in the inward movement of Cl^- at the apical side of the epithelium asides the diffusive movements previously reported (Fig. 13). If we now consider the active movement of Na^+, the results of Fig. 14 show that amiloride produces an immediate drop of the PD induced by gluconate substitution of Cl^-, i.e., of the PD related to cation movements at the serosal side of the epithelium which is sensitive to the movement of Na^+ at the apical side (Péqueux and Gilles 1988). These results thus indicate that the movement of Na^+ at the apical side is largely dependent on an amiloride-sensitive transport system. They are in good agreement with the work of Lucu and Siebers (1986) who reported amiloride-sensitive fluxes of Na^+ in isolated posterior gills of *Carcinus mediterraneus*. Since the possibility of an important Na^+/Na^+ exchange component must be excluded in the posterior gills of *E. sinensis* (there is indeed no significant efflux of Na^+) and, further, as a Na^+/NH_4^+ exchange does not play a significant role in the Na^+ uptake, the essential part of the Na^+ active entrance at the apical side of *E. sinensis* posterior gills is probably related essentially to an amiloride-sensitive Na^+/H^+ antiporter.

Interesting to consider in this context is the early conclusion of Smith and Linton (1971) according to which the preferential permeability to Na^+ with respect to Cl^- of the blue crab gills is primarily a function of the cuticle. Until recently, little, or almost nothing, was known about the physical properties of the cuticle and its permeability. Recent works, however, establish that gill

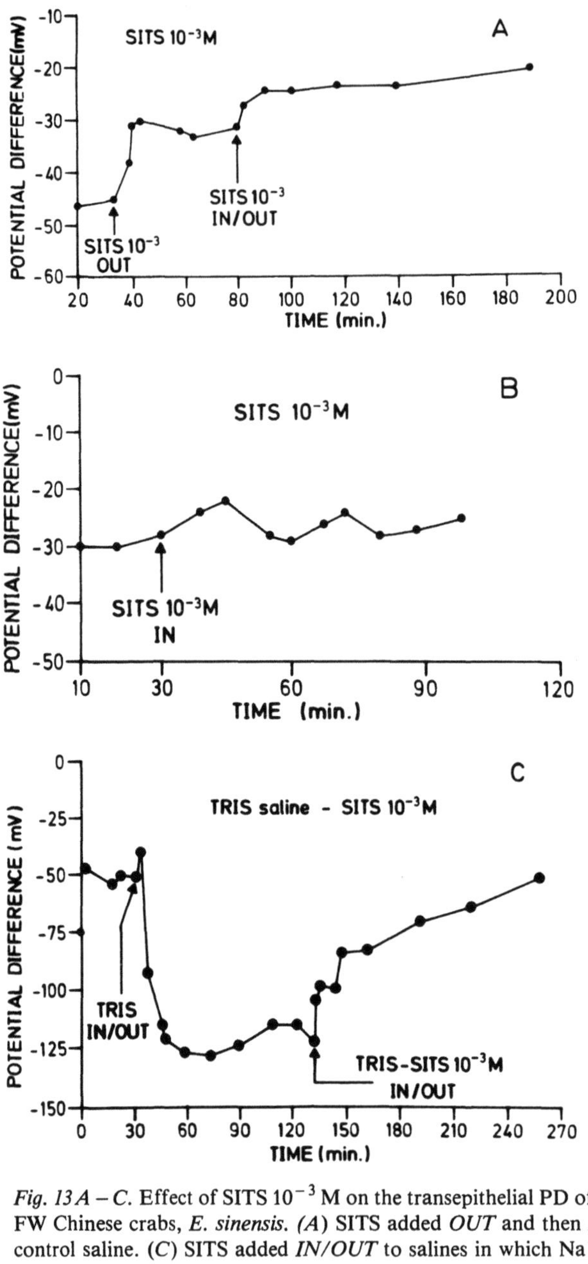

Fig. 13A – C. Effect of SITS 10^{-3} M on the transepithelial PD of isolated, perfused posterior gills of FW Chinese crabs, *E. sinensis*. (*A*) SITS added *OUT* and then *IN/OUT* or (*B*) SITS added *IN*, to control saline. (*C*) SITS added *IN/OUT* to salines in which Na^+ has been replaced by Tris. (After Gilles and Péqueux 1985, 1986; Péqueux and Gilles 1988)

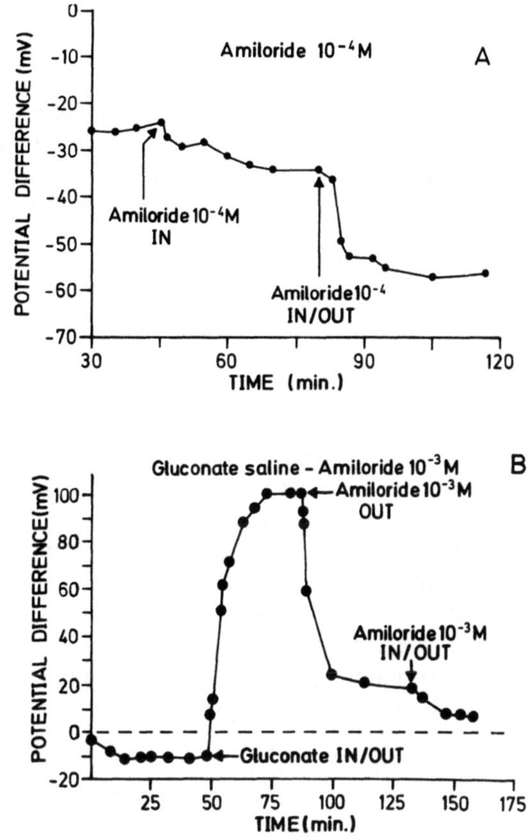

Fig. 14A, B. Effect of amiloride on the transepithelial PD of isolated, perfused posterior gills of FW Chinese crabs, *E. sinensis*. *A* Amiloride 10^{-4} M added *IN* and then *IN/OUT*, to control saline. *B* Amiloride 10^{-3} M added *OUT* and then *IN/OUT* to salines in which Cl^- has been replaced by gluconate. (After Gilles and Péqueux 1985, 1986; Péqueux and Gilles 1988)

cuticles may exhibit selective ionic permeabilities, rectifying properties, and an asymmetrical behavior. As these characteristics do not concern all invertebrates, they will not be considered in detail in this chapter, but the interested reader is referred to the recent review on the topic by Lignon (1987).

3.3 A Model of NaCl Transport Across the Gill of Crustaceans

In this section, an attempt will be made to fit the different information reported in the above sections into a model for NaCl transport across the gill(s) of crustaceans.

First, we will try to fit the different results discussed above into a model for Na^+ and Cl^- transport across the anterior and the posterior gills of freshwater Chinese crabs, *Eriocheir sinensis*.

We will then consider the possibility of applying the *E. sinensis* model to other crustaceans.

3.3.1 Model of NaCl Transport in E. sinensis Gills

From the various investigations conducted on *E. sinensis* gills, either structural, physiological, or biochemical, it appears that the anterior gills do not "transport" NaCl even if unidirectional ion movements occur. They will, however, be considered in this section dealing with modeling of transepithelial NaCl transport. In the gills of most of the other hyperregulators studied so far, such a clear distinction as in *E. sinensis* between the gills exhibiting two distinct types of epithelia cannot be done. The gill properties of these species are thus more likely resulting from the properties of each of their tissular components, the individual knowlegde of which being then of utmost interest.

To summarize, the results obtained on the *anterior* gills can be fitted into a functional model, as represented diagramatically in Fig. 15, in which the PD is accounted for by several diffusive components for K^+, Na^+, and Cl^-. In the absence of a transepithelial concentration gradient of NaCl, K^+ and Cl^- diffusion potentials are generated at the serosal side of the epithelium. They are partly short-circuited by the potential resulting from the inward diffusive movement of Na^+ at the same level. Under these "symmetrical" conditions of incubation, a Na^+ diffusion into the cells, at the apical side, must also be considered. Besides these cellular diffusive components, transepithelial movements of Na^+ further occur through paracellular pathways (Fig. 15).

The situation appears a bit more complex in the *posterior* gills. The different results discussed in the previous sections have been summarized in the tentative model presented in Fig. 16. This functional model of posterior gill epithelium agrees with the different data obtained when both sides of the preparation are bathed with the same saline, i.e., in the absence of a transepithelial concentration gradient. It is, however, worth noting that in this crab species living in freshwater, the intracellular level of Na^+ and Cl^- in the gill epithelium is higher than in the external medium and lower than in the serosal/hemolymph one. Therefore, the movements of these ions from the external medium to the intracellular and hemolymph compartments cannot occur by diffusion. Active transport processes at both sides have thus to be taken into consideration. The following characteristics must also be considered: (1) Na^+ and Cl^- are transported independently; (2) the permeability of the epithelium to Na^+ is extremely low; (3) the epithelium, however, is quite permeable to Cl^-; (4) a diffusion potential for Cl^- ions at the apical side is one of the overall PD components; (5) the PD is dependent on the level of K^+ in the perfusion medium and is sensitive to addition of ouabain; (6) Cl^- movements at the serosal side can be, at least partly, passive and driven by the diffusion potential generated by the passive outward movement of K^+ down its concentration gradient; (7) Na^+ inward movement at the apical side is amiloride-sensitive and seems to be mostly related to an Na^+/H^+ exchange process; (8) a SITS-sensitive Cl^-/HCO_3^- exchange system is involved in the transport of Cl^- at the apical side of the epithelium.

Although this model agrees with the different data available up to now on *E. sinensis,* it remains certainly basically tentative. For example, nothing is yet known about possible electrogenic components in the transport of Na^+ and Cl^-, or about the role of ion channels. Nevertheless, it shows that the transport

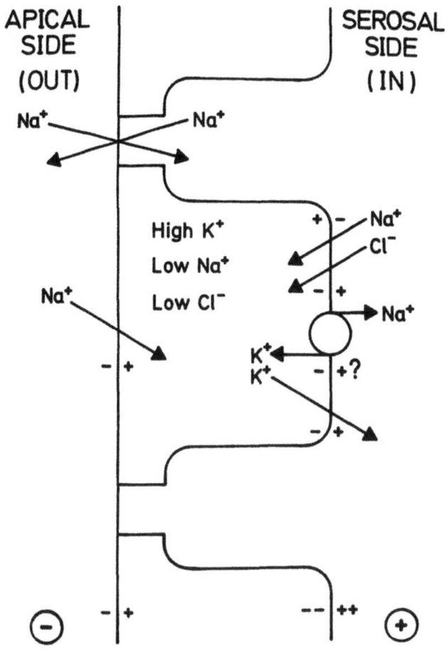

Fig. 15

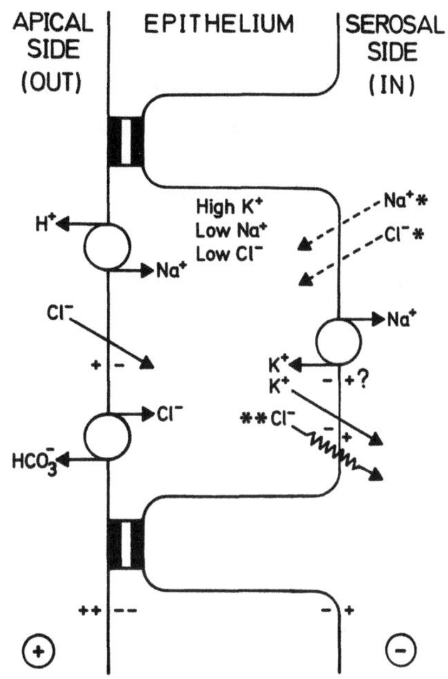

Fig. 16

Fig. 15. Functional model of anterior gills of FW Chinese crabs, *E. sinensis*. Both sides of the epithelium are bathed with the same "FW bloodlike" saline (Péqueux and Gilles 1988)

Fig. 16. Functional model of posterior gills of FW Chinese crabs, *E. sinensis*. Both sides of the epithelium are bathed with the same "FW bloodlike" saline (Péqueux and Gilles 1988)
* The inward diffusive movements of Na^+ and Cl^- at the serosal side do not play a significant part in the establishment of the PD
** The outward movement of Cl^- at the serosal side is driven by the K^+ diffusion potential

systems at work in gills of crustaceans are more complicated than generally thought and that mechanisms other than those considered up to now must be implicated.

3.3.2 *Applicability of the* E. sinensis *Model to Other Crustaceans*

The question which now arises is that of whether the *E. sinensis* model of gill transport can be applied to other crustaceans. This model is by no means exclusive of others. However, it could be quite satisfactorily applied to other euryhaline crabs by adding pathways for Na^+ outward movements or by considering that the transport characteristics of their gills are a mixture of the characteristics described separately for the anterior and posterior gills of the Chinese crab.

The relative importance of these processes and/or of these specific characteristics could vary greatly from one species to another, thus determining their ability to hyperosmoregulate in dilute environments.

In a recent study, the ultrastructure of the gills of an euryhaline osmoconformer *Cancer pagurus* has been compared to that of the weak hyperosmoregulator *Carcinus maenas* and that of the strong regulator *Eriocheir sinensis* (Péqueux et al. 1984a; see also Compère et al. 1985).

C. pagurus gills have essentially a respiratorylike epithelium. In *C. maenas,* the anterior gills are also mostly lined with the same kind of epithelium, while both a respiratory and a salt-transporting epithelium are found in the posterior gills. The salt-transporting epithelium, however, does not exceed 30% of the whole lamellar surface of these posterior gills in *C. maenas* (Compère et al. 1985).

The large unidirectional Na^+ fluxes that have been measured in *C. maenas* gills, at variance with the situation found in *E. sinensis* (see Sect. 3.1), could be thus satisfactorily related to the fact that in *E. sinensis,* the posterior gills are essentially lined with an epithelium of the transporting type (with a low Na^+ permeability), while in *C. maenas,* both the respiratory (with a high Na^+ permeability) and the transporting (with low Na^+ permeability) epithelia are present. A similar situation would also occur in the blue crab *Callinectes sapidus* (Copeland and Fitzjarrel 1968).

These results lead to the interesting conclusion that the structural organization of the gills can be directly related to their functional properties, hence to the osmoregulatory capabilities of the considered species. They substantiate the idea that the model of gill structural and functional organization described for the Chinese crab *E. sinensis* can be applied quite satisfactorily to other euryhaline crabs. They moreover establish that this species is an almost unique model to study not only ion transport processes, but also the structure-function relation in a NaCl-transporting epithelium.

4 Biochemical Correlates of NaCl Transport in Gills

The biochemical approach of the gill NaCl transport which will be developed in this section deals essentially with two categories of enzymes: the membrane-bound ATPases and carbonic anhydrase.

4.1 Membrane-Bound ATPases

Several kinds of enzymes hydrolyzing ATP, i.e., ATPase, have been considered as a possible, integral part of the system(s) at work in the active transport of ions across gills. Some of them have even been reported as the enzymatic equivalent of the ion pumps. They are located within the plasma membrane itself, are activated in vitro by an ionic environment containing the ionic species transported, and exhibit a polarity of distribution which may define their function within an epithelial organization. This review will be centered mainly on the enzyme system which has been the best studied in terms of a possible enzymatic equivalent of the Na^+ pump: the $(Na^+ + K^+)$ ATPase. Attention will be paid also to another membrane-bound ATPase, the anion-dependent ATPase, while its physiological significance still remains unclear and needs further investigations.

4.1.1 ($Na^+ + K^+$) Activated ATPase

4.1.1.1 ($Na^+ + K^+$) ATPase in Crustacean Gills. Since the early works of Skou (1957) the activity of the form of ATP phosphohydrolase (E.C. 3.6.1.3), the so-called ($Na^+ + K^+$) ATPase, is thought to be involved in cation transport in tissues of many vertebrates. Its presence and activity have also been investigated in crustacean tissues by several authors and it has been shown to be widely distributed in gills (Table 5) (see also for example, Mantel and Olson 1976; Towle et al. 1976; Péqueux and Gilles 1977; Spencer et al. 1979; Neufeld et al. 1980; Péqueux and Chapelle 1982; Siebers et al. 1982; Péqueux et al. 1983; Péqueux et al. 1984; Wanson et al. 1984a, b; Holliday 1985).

The enzyme in crustacean gills is basically similar to that found in transporting tissues of vertebrates. It exhibits an absolute requirement for Mg^{2+}, Na^+, and K^+ and is inhibited by ouabain. Analysis of the molecular structure of the

Table 5. Range of ($Na^+ + K^+$) ATPase activity in crustacean gills

Species	Acclimation medium	Enzyme activity ($\mu mol\ P_i\ mg^{-1}\ h^{-1}$)	Reference
Calappa	SW	0.18	Spencer et al. (1979)
hepatica	50% SW	0.27	Spencer et al. (1979)
Callinectes	SW	4–14	Neufeld et al. (1980)
sapidus	20% SW	5–24	Neufeld et al. (1980)
Cancer	SW	1.5–3.5[a]	Péqueux et al. (unpubl. results)
pagurus	SW/2	1.8–3.8[a]	Péqueux et al. (unpubl. results)
Carcinus	SW	2.3–5.0[a]	Wanson et al. (unpubl. results)
maenas	30% SW	9.3–18.5[a]	Wanson et al. (unpubl. results)
	SW	1.5–4.1[b]	Siebers et al. (1982)
	20% SW	2.5–9.5[b]	Siebers et al. (1982)
Cyclograpsus	SW	0.85	Spencer et al. (1979)
henshawi	FW	1.40	Spencer et al. (1979)
Eriocheir	SW	0.6–1.4[a]	Péqueux et al. (1984)
sinensis	FW	2.7–4.4[a]	Péqueux et al. (1984)
Metopograpsus	SW	0.9	Spencer et al. (1979)
thukuhar	25% SW	0.8	Spencer et al. (1979)
Panopeus	SW	0.1	Spencer et al. (1979)
herbstii	25% SW	0.6	Spencer et al. (1979)
Procambarus clarkii	FW	0.9	Horiuchi (1977)
Thalamita	SW	0.5	Spencer et al. (1979)
crenata	25% SW	1.1	Spencer et al. (1979)
Uca minax	SW	1.5–4.0[a]	Wanson et al. (1984b)
	SW/8	3.0–13.0[a]	Wanson et al. (1984b)
Uca pugnax	SW	7–20	Holliday (1985)
	10% SW	10–40	Holliday (1985)

[a] 10000 x g fraction.
[b] Crude homogenates.

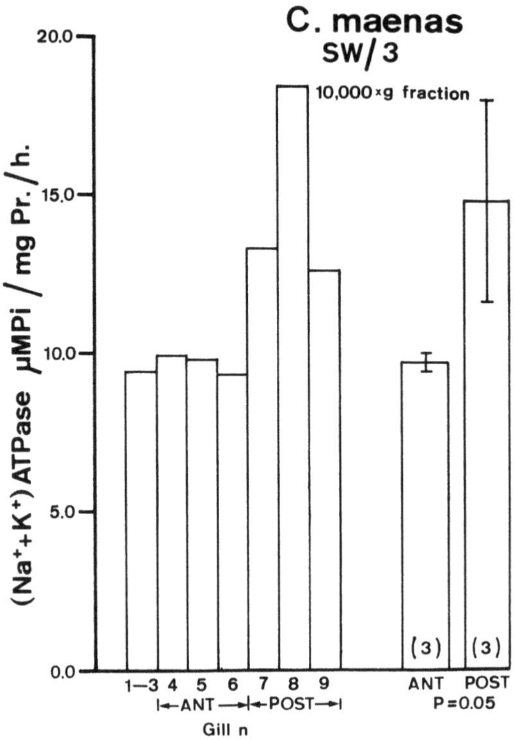

Fig. 17. ($Na^+ + K^+$) ATPase activity of individual gills (typical results) and in anterior (*ANT*) and posterior (*POST*) gills (mean data ± SD) of the shore crab *Carcinus maenas* acclimated to dilute SW (Péqueux and Gilles 1984, and unpubl.)

($Na^+ + K^+$) ATPase, as well as that of the kinetic properties of the enzyme will not be considered here as it is out of the space of this chapter. The interested reader is referred to the reviews of Towle (Towle 1984a, b).

In many animal species, the level of activity has been reported to be dependent on the type of gill considered. The results presented in Fig. 17 show that the three most posterior pairs of gills exhibit the highest activity in the shore crab *Carcinus maenas* acclimated to dilute seawater (Wanson and Péqueux, unpubl. results). Higher ($Na^+ + K^+$) ATPase activities in the posterior gills of decapods have also been reported by Mantel and Olson (1976), Mantel and Landesman (1977), Spencer et al. (1979), Neufeld et al. (1980), Siebers et al. (1982), Wanson et al. (1984a, b) and Péqueux et al. (1984b). These differences agree generally with the functional differences which have been reported as a general occurrence in these crustaceans. They nicely correlate with the fact that the mechanisms of ion uptake are located in the posterior pairs of branchiae.

4.1.1.2 ($Na^+ + K^+$) ATPase Activity and Salinity Acclimation. In hyperosmoregulating species of crustaceans, the activity of the ($Na^+ + K^+$) ATPase changes in an adaptive manner as a function of the salinity of the environment. In fact, the activity of the enzyme is greater in the gills of animals which have been acclimated to a dilute medium (*Callinectes sapidus*: Towle et al. 1976; Neufeld et al. 1980; *Carcinus maenas:* Mantel and Landesman 1977; Siebers et al. 1982; Wanson et al. 1984a; *Eriocheir sinensis:* Péqueux and Gilles 1977, 1978b;

Péqueux et al. 1984b; *Panopeus herbstii, Thalamita crenata:* Spencer et al. 1979; see also the results of Table 5 and the histograms of Fig. 18). These observations lead to the suggestion that the maintenance of osmotic and ionic balance is directly related to the level of ($Na^+ + K^+$) ATPase activity in the gills.

The relationship between enzyme activity and the salinity of the acclimation medium is, however, not always directly evident. Enzyme activity is, for example, high in semiterrestrial grapsids like *Metopograpsus thukuhar* and *Cyclograpsus henshawi,* even higher than in osmoregulators like *Thalamita crenata* and *Panopeus herbstii,* the lowest activity being found in the osmoconformer *Calappa hepatica* maintained in SW (Spencer et al. 1979). When these crabs were placed in dilute SW, a large increase in ($Na^+ + K^+$) ATPase activity occurs in *P. herbstii* and *T. crenata,* but there is no change in enzyme activity of the semiterrestrial and conforming species (Spencer et al. 1979). It may be interesting to note that the enzyme activity in osmoconforming crabs such as *C. hepatica* (Spencer et al. 1979) and *Libina emarginata* (Mantel and Landesman 1977) remains at a very low order of magnitude (about 10% of that in osmoregulating species). Another example of "discrepancy" between enzyme activity level and acclimation salinity is reported by Mantel and Landesman (1977) on the fiddler crab *Uca pugilator.* In *U. pugilator,* enzyme activity increases with acclimation to dilute medium, but maximal activity is not attained until 100 h after transfer, whereas the osmotic concentration is stabilized after 10 h.

The question which now arises, is that of whether the significant variations of enzyme activity observed in osmo-ionoregulating species facing salinity changes

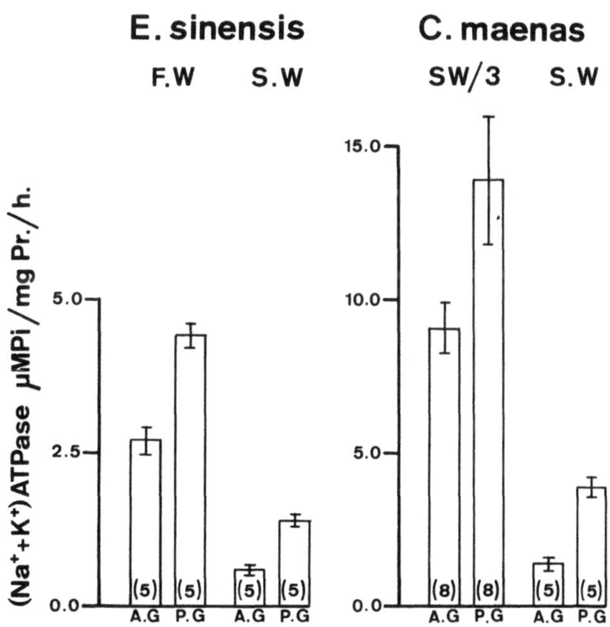

Fig. 18. ($Na^+ + K^+$) ATPase activity in the 10000 g fraction of anterior (*AG*) and posterior (*PG*) gills of the euryhaline crabs, *C. maenas* and *E. sinensis,* as a function of the salinity of their acclimation media (FW, SW, SW/3) (Péqueux and Gilles 1984, and unpubl.)

are due to the synthesis of a new enzyme, of more enzyme, or to modulation of catalytic activity of preexisting ATPase molecules. The question is far from being answered. On the Chinese crab *E. sinensis,* it has been shown that there are no significant differences between the kinetics characteristics of the enzyme from SW or FW animals (Péqueux et al. 1984). It has been considered that, in this species at least, acclimation to media of different salinity does not induce synthesis of kinetically different $(Na^{+}+K^{+})$ ATPases. The increase in activity observed upon acclimation to the dilute medium could thus be due to an increase in the amount of enzyme molecules present and/or to an activation of a preexisting enzyme. Induction of enzyme synthesis is known to occur in some species. An increase in activity due to de novo synthesis of an enzyme on cellular membranes has indeed been reported in *Artemia salina* (Conte et al. 1977).

In the Chinese crab, most of the increase in specific activity is obtained in the first 48 h of acclimation to FW (Péqueux et al. 1984). This is at variance with the results of Neufeld et al. (1980) who reported that the salinity-induced changes in activity required 1–2 weeks for completion in *Callinectes sapidus.* On the same species, Towle et al. (1976), however, reported much faster changes (3 h for 50% increase in enzyme activity after transfer from SW to 5‰ salinity medium). The results obtained on *E. sinensis* and those of Towle thus favor the idea of a rapid adjustment of the enzyme activity directly associated with the increase in the transepithelial movement of Na^{+} at the gill level. This is, however, not an argument to suggest that activation occurs by modulation of the activity of preexisting enzymes rather than by induction of synthesis of new enzyme molecules. Information on the turnover rate of the enzyme in crustacean gills is still required. In crustacean gills, there are regions in which the ultrastructure shows increased development of plasma membrane infoldings (Copeland and Fitzjarrel 1968; see also Sect. 2). This could provide an increased area supporting ion-transporting sites as well as increase in the number of the $(Na^{+}+K^{+})$ ATPase sites. Accordingly, it is possible that two levels of enzymatic control are involved in the acclimation to salinity changes: a short-term process of activation/inhibition and a longer-term one involving the synthesis of enzyme.

4.1.1.3 Localization of $(Na^{+}+K^{+})$ ATPase Activity. The bulk of evidence provided by studies conducted on several kinds of transporting epithelia supports the idea of a main and even restricted basolateral location for the $(Na^{+}+K^{+})$ ATPase (for review, see for example, Ernst et al. 1980; Hootman and Ernst 1984).

In salt-transporting tissues of osmoregulating crustaceans, information on the localization of $(Na^{+}+K^{+})$ ATPase activity is still scanty. Very recently, Barra and Péqueux have identified sites of $(Na^{+}+K^{+})$ ATPase activity on the cytoplasmic surfaces of the folds of basolateral plasma membranes of *E. sinensis* posterior gills (Fig. 19; Barra and Péqueux 1986). In their study, no electron dense precipitates were observed on the apical plasma membrane infoldings that border the apical subcuticular compartment. These observations are in full agreement with those made at the same time by Towle and Kays (1986) on thick epithelial cells from gill lamellae of two other euryhaline crabs, *Callinectes sapidus* and *Carcinus maenas.*

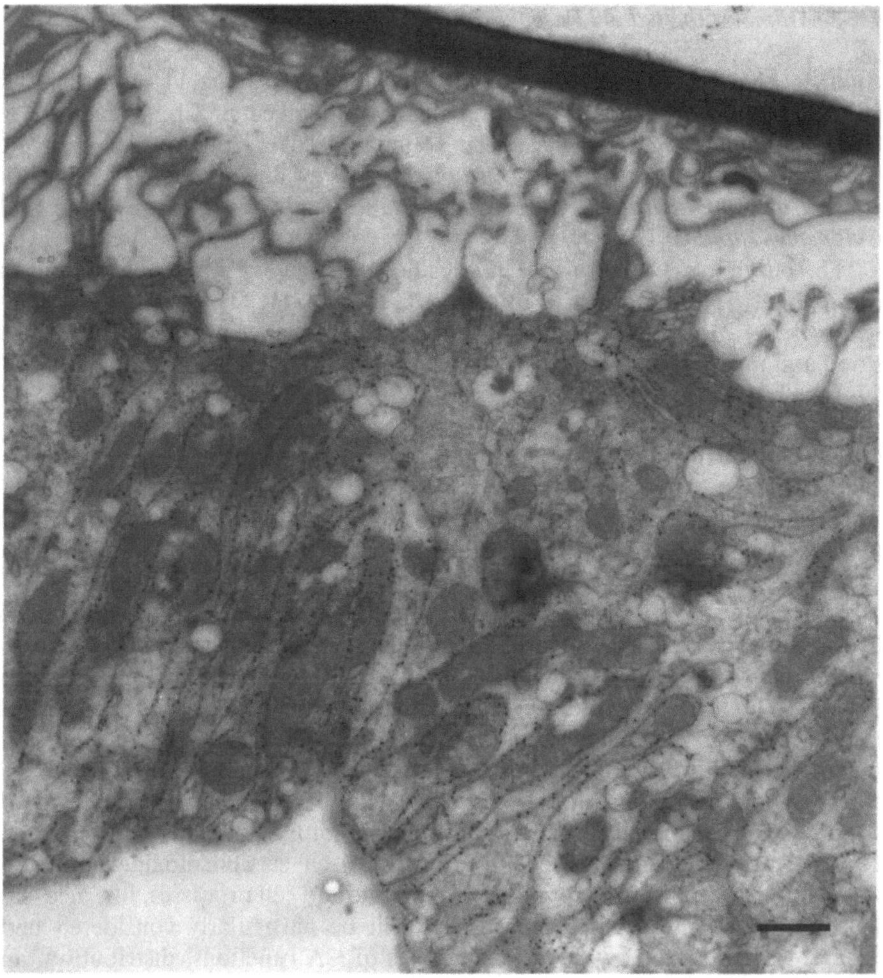

Fig. 19. Ultrastructural cytochemical localization of $(Na^+ + K^+)$ATPase in posterior gills of FW Chinese crabs, *E. sinensis* (Barra and Péqueux, unpubl.)

It thus appears from these studies that in hyperregulating crustaceans the $(Na^+ + K^+)$ ATPase exhibits also a restricted polarity of distribution within the cells of salt-transporting tissues. The enzyme would thus move Na^+ ions from the intracellular medium toward the extracellular fluid facing basolateral membranes. Thus it could provide one of the forces driving Na^+ at the apical side, even working in cooperation with other transport processes restricted to that face. This, however, does not exclude the possible existence at the level of other processes consuming ATP and favouring the uptake of Na^+ from very dilute environments toward the cytosol. Further investigations are needed to elucidate these problems.

4.1.2 Anion-Activated ATPase

Another kind of ATPase, the activity of which is stimulated by anions like HCO_3^-, has been proposed as being possibly involved in Cl^- transport. An HCO_3^- stimulated ATPase has been found indeed in fractions of crab gills containing membrane fragments (*Uca minax:* DePew and Twole 1979; terrestrial crabs, *Cardisoma carnifex, Gecarcoidea lalandii* and *Birgus latro*: Twole 1981; *Callinectes sapidus:* Lee 1982).

The enzyme has therefore been considered as playing a possible role in the transepithelial transport of Cl^- by acting on the Cl^-/HCO_3^- exchange (see previous Sect. 3.1.2). Up to now, its physiological significance is not yet fully understood and investigations in that field need further development.

4.2 Carbonic Anhydrase

This section intends to briefly review the possible implications of carbonic anhydrase (CA) in NaCl transport processes at work in gills of ionoregulating species of crustaceans.

Since its discovery in 1932 by Meldrum and Roughton, CA (carbonate deshydratase, carbonate hydro-lyase: E.C. 4.2.1.1.) has been extensively studied in vertebrates. It has also been identified in invertebrates for a long time (Ferguson et al. 1937) but has received much less attention. The primary function recognized for that enzyme, which catalyzes the reversible hydration/dehydration reaction of CO_2 and water ($CO_2+H_2O \overset{CA}{\rightleftharpoons} H^++HCO_3^-$), is a prominent role in CO_2 transport and excretion. Most of the work has therefore been concentrated on the physiological and physicochemical properties of mammalian erythrocyte CA. However, by controlling the equilibrium between carbon dioxide and bicarbonate, it has implications for many other tissue and cell processes, like acid-base balance and ion transport. This aspect will be particularly considered here. Several reviews dealing with various aspects of CA functions, distribution, and properties have already been published. The reader is also referred to them (see for example: Maren 1967a, b; Bundy 1977; Maren 1977; Wyeth and Prince 1977; Bauer et al. 1980; Henry 1984; Burnett et al. 1985).

4.2.1 Indirect Evidence of the Role of CA in Crustacean Ion Regulation

A common approach to investigate the possible implications of gill intraepithelial carbonic anhydrase in blood ion regulation has been to use one of the highly specific sulfonamide inhibitors: acetazolamide (Diamox). Other known inhibitors of CA have also been recommended, such as ethoxzolamide and methazolamide, for physiological studies; they are however, much less used (Maren 1977).

In *Pachygrapsus crassipes* acclimated to 16‰ salinity water, the lowest salinity at which they can hyperregulate their hemolymph chloride, injection of Diamox has been shown to severely impair the possibility of Cl^- regulation (Burnett et al. 1981). In contrast, Diamox had no effect on chloride regulation at high salinities at which the animal is hyporegulating. Similar results were obtained with the euryhaline blue crab *Callinectes sapidus* and the level of regulation impairment in crabs acclimated to reduced salinity was shown to be dependent upon the dose of inhibitor used (Henry and Cameron 1983). The same authors further demonstrated that SW acclimated *C. sapidus* (which are conformers in that medium) fail to survive the transfer to low salinity (where they must hyperregulate) when they are treated with Diamox (Henry and Cameron 1982a). The failure to survive transfer has been related primarily to a breakdown in the development of blood ion regulation ability.

Results from these studies thus substantiate the idea that inhibition of CA by acetazolamide severely impairs mechanisms involved in blood osmo-ionoregulation. It has been demonstrated on the crayfish and on freshwater specimens of blue crabs that acetazolamide disturbs the unidirectional ion fluxes (Ehrenfeld 1974; Cameron 1979). Injection of Diamox significantly increases influx and efflux of Cl^-. In the blue crab, stimulation of efflux is, however, larger. This results in a net outward movement of Cl^-. Interesting is also the fact that Na^+ fluxes and blood Na^+ regulation processes are also affected by Diamox treatment, although to a lower degree as Cl^- fluxes and regulation (Ehrenfeld 1974; Cameron 1979; Henry and Cameron 1983). In this case, Na^+ fluxes are inhibited, the Na^+ influx being much more decreased. Net uptake of both ions is thus reversed by Diamox treatment which explains the animal's inability to regulate its blood level of ions.

These data have been best integrated in a model of CA function in ion transport regulation, where the enzyme activity supplied H^+ and HCO_3^- which are used as counterions in the uptake of Na^+ and Cl^-.

Accordingly, Diamox-induced inhibition of CA would thus disturb indirectly Na^+/H^+ and Cl^-/HCO_3^- exchange processes. All these studies have, however, been conducted on whole animals assuming that the gill was the major tissue implicated in these exchanges. Let us now come down to the gill level.

4.2.2 CA in Crustacean Gills

In crustaceans, CA is found in a variety of tissues but it is found at its highest concentration in the gill. As for the $(Na^+ + K^+)$ ATPase, CA is distributed heterogeneously among the gill pairs in crustacean species that are capable of regulating blood ion concentration (Table 6). In the blue crab *C. sapidus* as well as in *Cardisoma guanhumi,* Henry and Cameron (1982b) reported higher enzyme activity (about 75% of the total enzyme activity) in the posterior three or four pairs of gills (Fig. 20) which are known to be predominantly responsible for salt transport. In stenohaline or conforming species which do not regulate their blood ion concentration, like *Libinia emarginata,* that difference is not found and the level of enzyme activity is even significantly lower than in regulating species (Henry and Cameron 1982b). Clearly, there is a strong correlation between the CA activity and the NaCl transport activity of the tissue considered, in this case the branchial tissue.

Table 6. Carbonic anhydrase activity in crustacean gills

Species[a]	Enzyme activity[b]	Reference
Callinectes sapidus (SW)		
A.G	0.15 mmol CO_2 min^{-1} mg $prot^{-1}$	Henry and Cameron (1982b)
P.G	0.23 mmol CO_2 min^{-1} mg $prot^{-1}$	
Callinectes sapidus (FW)		
A.G	0.25 mmol CO_2 min^{-1} mg $prot^{-1}$	Henry and Cameron (1982b)
P.G	0.84 mmol CO_2 min^{-1} mg $prot^{-1}$	
Cardisoma carnifex	0.27 E.U. mg $prot^{-1}$	Randall and Wood (1981)
Cardisoma guanhumi	0.8 mmol CO_2 min^{-1} mg $prot^{-1}$	Henry and Cameron (1982b)
Gecarcinus lateralis		
A.G	0.38 mmol CO_2 min^{-1} mg $prot^{-1}$	Henry and Cameron (1982b)
P.G	0.61 mmol CO_2 min^{-1} mg $prot^{-1}$	
Libinia emarginata		
A.G	0.05 mmol CO_2 min^{-1} mg $prot^{-1}$	Henry and Cameron (1982b)
P.G	0.06 mmol CO_2 min^{-1} mg $prot^{-1}$	
Pachygrapsus crassipes	8.1 E.U. mg $prot^{-1}$	Burnett et al. (1981)

[a] A.G and P.G, respectively, refer to anterior and posterior gills.
[b] E.U. (enzyme unit) = (Vc – Vu)/Vu where Vc: catalyzed reaction rate. Vu: uncatalyzed reaction rate.

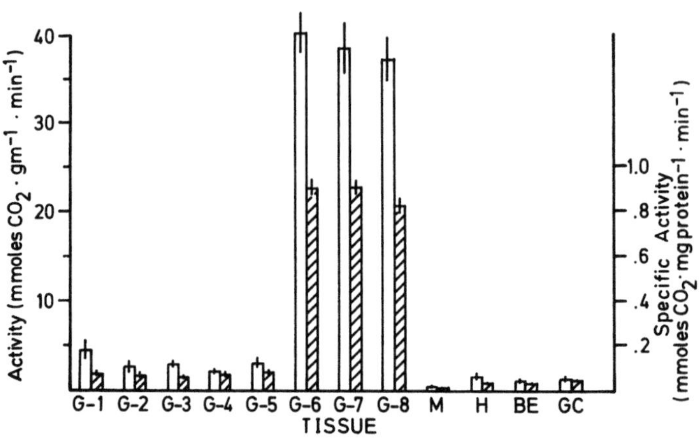

Fig. 20. Distribution of carbonic anhydrase activity in *Cardisoma guanhumi* acclimated to 950 mOsmol kg^{-1}. *Open bars* represent activity per g fresh weight of tissue and correspond to the scale on the left; *stippled bars* represent specific activity and correspond to the scale on the right. *G1 – G8*: The eight gill pairs numbered anterior to posterior. *M* muscle; *H* heart; *BE* branchial epithelium; *GG* green gland. Mean ± SE; n = 8; t: 25 °C (Henry and Cameron 1982b)

In hyperosmoregulating species, the CA activity level has also been shown to be dependent on the salinity of the acclimation environment. Henry and Cameron (1982b) indeed reported higher CA activity in the posterior gills (1.4 m mol $CO_2\,mg^{-1}\,prot^{-1}\,min^{-1}$) of blue crabs acclimated to 100 mOsmol kg^{-1} than in crabs acclimated to 850 mOsmol kg^{-1} (0.6 mmol $CO_2\,mg^{-1}\,prot^{-1}\,min^{-1}$). This observation further substantiates the idea that enzyme activity responds to the enhanced transport activity induced by environmental conditions in which blood ion concentrations are regulated at a higher level than that of the surrounding medium.

4.2.3 CA Significance in NaCl Transport Across Gills

The data reported in this section lead to a clear picture of the implications of an important fraction of the overall CA activity in NaCl transport across crustacean gill epithelium. They fit perfectly into Krogh's model, according to which Na^+ ions are transported in exchange for H^+ and/or NH_4^+, while Cl^- ions are exchanged for HCO_3^-.

That fraction of the carbonic anhydrase activity, by catalyzing the reversible hydration/dehydration reaction of CO_2 and H_2O within the transporting cells of the gill epithelium, provides H^+ and HCO_3^- ions which will be involved in these exchange processes. The other important fraction of the CA found in gill epithelium functions primarily in CO_2 excretion, but this aspect will not be considered here. An attempt is made in Fig. 21 to fit the information and hypothesis reported in the above section into a model for CA function in the NaCl-transporting cells of crustacean gill epithelium.

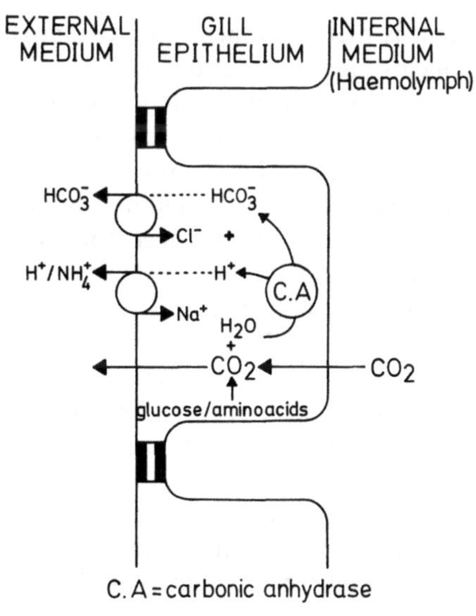

Fig. 21. A model for CA function in NaCl transporting cells of crustacean gill epithelium

5 Metabolism and NaCl Transport in Gills of Crustaceans

Up to now, only a few studies have been devoted to the metabolic cost of ion regulation in euryhaline crustaceans in terms of energy consumption for Na^+ and Cl^- uptake processes. An approach to correlate the activity of ion transport processes to cell metabolism has been to measure the influence of salinity changes on the oxygen consumption. Such experiments have been done on intact animals as well as on isolated gills and it has been clearly established that the respiration of crustaceans may be influenced by salinity.

The early work of Schlieper (1929) indicated that the respiration of the shore crab *Carcinus maenas* increases with decreasing salinity, as does the respiration of its excised gills (Poat et al. 1971). These observations have been corroborated by similar experiments conducted on excised gills from other crab species (Dehnel and McCaughran 1964; King 1965; Mantel 1967; Dehnel 1974; Engel and Eggert 1974; Engel et al. 1975). Generally, crustaceans that osmoregulate thus increase their metabolic rate as do their isolated gills.

It has also been suggested that the increase in oxygen consumption is mainly due to increased demands of energy for ionic regulation processes that must be developed by osmoregulating species upon transfer in a dilute environment. Potts and Parry (1964b), however, calculated that the increased energy available on the basis of the increased oxygen consumption exceeds largely that required for ionic regulation. The real meaning of these metabolic changes is still not well understood and remains controversial. Nevertheless, it is likely that at least part of the increased oxygen consumption is linked to ion transport. There are numerous lines of evidence supporting the existence of a tight coupling ratio between ions transported and oxygen molecules consumed (Mandel and Balaban 1981). Such a coupling necessarily implicates the $(Na^+ + K^+)$ ATPase, hence the cellular adenylate pool. From experiments carried out on excised gills of the shore crab *Carcinus maenas,* Poat et al. (1971) concluded that glycolysis does not provide much energy for Na^+/K^+ transport, but presumably this energy requirement is met prominently by oxidative phosphorylation. In agreement with that conclusion are the results of Lowy and Conte (1985) on isolated larval salt glands of the brine shrimp *Artemia salina.* The involvement of mitochondria at that level is further substantiated by morphological information on structure and concentration as we have reported in Section 2, as well as by the observations made by King (1966) on the salinity-induced changes in the mitochondria respiration rate. Accordingly, any need for a change in ion active transport activity would induce a change in the conversion rate of ATP to ADP in direct proportion to the energy requirements, and to the resulting modification of the activity of the $(Na^+ + K^+)$ ATPase system. In turn, due to a feedback action of ADP on the oxidative metabolism, there will be a concomitant increase in oxygen consumption. Engel et al. (1975) therefore suggested that rates of oxygen consumption should be inversely related to concentrations of ATP in transporting tissues. In the blue crab *C. sapidus,* these authors demonstrated that the respiration rate of anterior gills of crabs acclimated to low salinity (5‰) was lower than the respiration rate of the posterior gills only when measured at low salinity. This is consistent with the fact that, under these conditions, posterior gills are very

Table 8. Effect of salinity on the adenylate energy charge (A.E.C.) of anterior and posterior gills of *C. maenas* and *E. sinensis.*[a] (After Wanson 1983; Wanson et al. 1983; Péqueux and Gilles 1984)

	Carcinus maenas		A.E.C.	*Eriocheir sinensis*		
	Anterior gills	Posterior gills		Anterior gills	Posterior gills	
SW	0.873 ± 0.092 n = 9	0.911 ± 0.031 n = 9		0.874 ± 0.042 n = 9	0.929 ± 0.024 n = 9	SW
SW/3	0.907 ± 0.023 n = 6	0.832 ± 0.031 n = 6		0.917 ± 0.039 n = 9	0.751 ± 0.069 n = 9	FW
T-test	–	*		–	*	

[a] Mean values ± SD of n experiments. * = Significant difference at the 1‰ level. A.E.C. = ATP + 1/2 ADP/ATP + ADP + AMP.

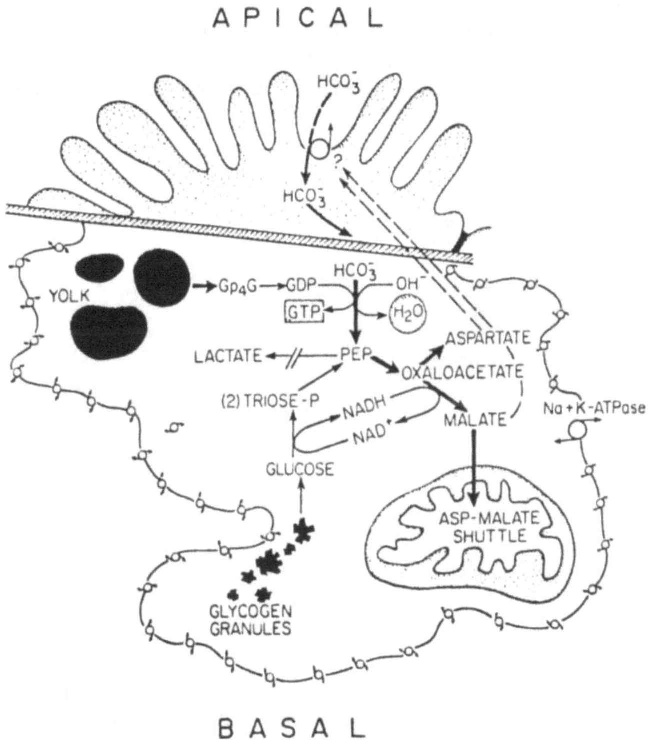

Fig. 24. The fate of exogenous bicarbonate ion and its transformation into a C-4 organic acid within the chloride cell cytoplasm of the brine shrimp *Artemia salina* (see Conte for further details) (Conte 1980)

levels at higher salinities. At the same time, the acidity of the medium increased, whereas lactate and fatty acids remained unchanged. An increased NaCl gradient thus causes a stimulation of glycolysis even though aerobic metabolism remains unaltered. Moreover, application of ouabain at concentrations known to affect the electrical and chemical gradients was shown to produce little if any change in oxidative metabolism while drastically altering the rate of glycolysis (Conte 1980). Conte thus proposed that a C-4 dicarboxylic acid pathway incorporating CO_2 from exogenous bicarbonate serves as a facultative anaerobic shunt enabling the chloride cell of the *Artemia* neck organ to switch from aerobic to anaerobic metabolism. This hypothesis is schematically represented in Fig. 24.

6 Endocrine Control of NaCl Movements Across Gills

In contrast to the extensive work done in animal groups such as worms, mollusks, insects, and fish, information on hormonal effects on osmoregulatory systems in crustaceans is still very scanty and does not lead yet to a clear visualization of the correlations binding both categories of mechanisms. There is, however, evidence in crustaceans that neuroendocrine products do affect osmotic and ionic regulation.

Evidence for neuroendocrine control of transport comes from studies conducted on intact animals where neuroendocrine centers were ligatured or removed and/or then replaced with tissues from the same animal. Further

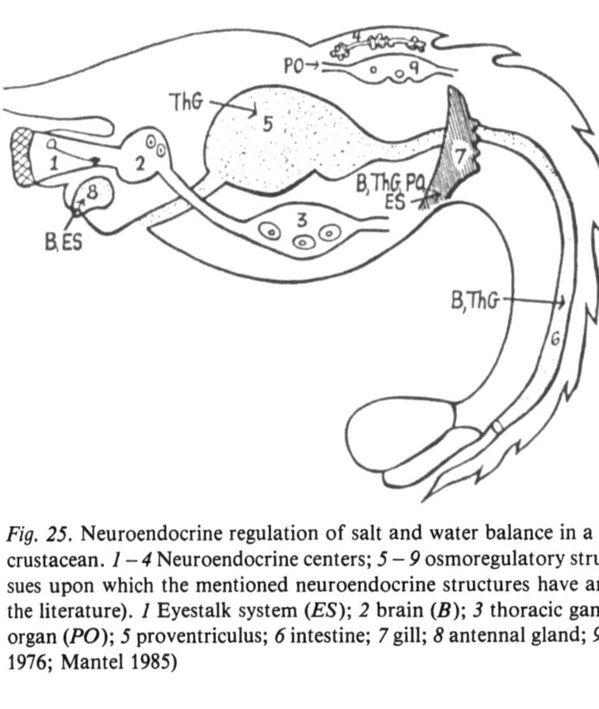

Fig. 25. Neuroendocrine regulation of salt and water balance in a diagrammatic composite decapod crustacean. *1 – 4* Neuroendocrine centers; *5 – 9* osmoregulatory structures. *Arrows* indicate target tissues upon which the mentioned neuroendocrine structures have an effect (documented examples in the literature). *1* Eyestalk system (*ES*); *2* brain (*B*); *3* thoracic ganglionic mass (*ThG*); *4* pericardial organ (*PO*); *5* proventriculus; *6* intestine; *7* gill; *8* antennal gland; *9* heart (Redrawn from Kamemoto 1976; Mantel 1985)

evidence is also to be found in the few studies carried out on isolated tissues treated with extracts of neuroendocrine structures.

The topic of neuroendocrine control of salt balance has been reviewed rather regularly in the past. It will therefore not be considered here in detail and the reader interested in an overall panorama of this literature is referred to the following works: Kamemoto (1976), Kamemoto and Oyama (1984), Mantel and Farmer (1983), and Mantel (1985). We shall only summarize (Fig. 25) the neuroendocrine structures and the organs implicated in osmotic and ionic regulation whose activity may be affected by extracts of these tissues.

As far as the gill tissue is concerned, there is evidence that it may be regulated or controlled basically by extracts of the brain, the thoracic glanglionic mass, the pericardial organ, and the eyestalk system.

In her recent and extensive account of the neurohormonal integration of osmotic and ionic regulation, Mantel (1985) concluded: "How far has the frontier of crustacean endocrinology advanced in the last few years? Not very far".

It is hoped that further experiments actually in progress on animals that change their capability for osmoregulation at different periods of their life cycle and on the effects of purified active neuroendocrine extracts on intact animals or isolated tissues, will help to advance that frontier.

Part II: Vertebrates

W. S. Marshall

1 Introduction

The present review emphasizes NaCl transport in the best-known and most intensely studied gill epithelial structure among aquatic vertebrates, the teleost gill. Recent studies are highlighted and the advantages of different experimental approaches are compared; a recent exhaustive review of gill structure and function is given by Hoar and Randall (1984). The transport of NaCl by the gill includes the function of the gill per se of seawater (SW) and freshwater (FW) stenohaline and euryhaline teleosts as well as that of accessory structures with a gill-like NaCl transport function, the skin and opercular epithelia of some species. Knowledge of the structure of the gill epithelium and its vasculature is assumed, see recent reviews by Hughes (1984) and Laurent (1984), as is a basic understanding of biological transport (e.g., see Friedman 1986).

2 The Seawater Gill

2.1 Insights from Studies in Vivo

Homer Smith (1930) demonstrated that SW eels (*Anguilla rostrata*), sculpin (*Myoxocephalus* sp.) and goosefish (*Lophius piscatorius*) secrete extrarenally a fluid hypertonic to SW; the conclusion was based on careful reconstruction of the gut and urine Mg^{2+} and SO_4^{2-} budgets and extrarenal NaCl loss by eels into distilled water. Smith suggested that because the gills are most obviously affected by rapid salinity changes, that the secretion likely occurred via the gills. The concentration gradients for Na^+, Cl^-, and K^+ favor passive entry across the gill of marine telosts. In more recent studies of gill NaCl transport, the electrical as well as the concentration gradient between plasma and SW were considered. The aim of the work was to determine which ions were actively transported and which might be passively distributed across the gill.

The trans-gill potential for many teleosts in SW is +10 to +35 mV with the environmental side grounded (*Blennius pholis,* House 1963; *Pholis gunnellus,* Evans 1969; *Gillichthys mirabilis,* Thompson 1972; *Platichthys flesus,* Potts and Eddy 1973; *Salmo gairdneri,* Greenwald et al. 1974; *A. anguilla,* House and Maetz 1974; reviewed by Potts 1984). A group of stenohaline SW teleosts and the elasmobranchs have small trans-gill potentials that range from −2 to −8 mV (Potts 1984). It is not known whether there is any relation between euryhalinity and the apparently larger trans-gill potentials in these animals. In most studies a possible junction potential between the salt bridge and the outside bath is assumed to be small and is ignored. In two studies, capillary microelectrodes were used to enter the vascular space of gill filaments to measure the gill epithelial potential locally. In *A. anguilla,* the potential measured in this manner was 17.8 ± 2.9 mV (House and Maetz 1974) and in the goby (*G. mirabilis*) the potential was 20.0 ± 1.0 mV (Thompson 1972). Others have obtained comparable results using intraperitoneal cannulae (19 ± 0.8 mV in *P. flesus* by Potts and Eddy 1973; 10.4 ± 0.9 mV in *S. gairdneri* by Greenwald et al. 1974); this assumes that the gill is the lowest resistance structure and will dominate the trans-body potential.

The electrical gradient in SW thus favors cation efflux and anion influx. Because Cl^- secretion occurs against a chemical gradient *and* electrical gradient and as marine teleosts are in steady-state Cl^- balance (i.e., have a flux ratio close to 1.0), it was clear that Cl^- was not passively distributed across the gill. The observed Cl^- flux ratio of 0.9 for SW-adapted *A. anguilla* is much smaller than the expected flux ratio of 10.6 (House and Maetz 1974) from the Ussing flux ratio equation that describes purely passive diffusion. Radiotracer studies of whole body (primarily trans-gill fluxes) revealed that Cl^- secretion by marine teleosts is inhibited by injected thiocyanate (Epstein et al. 1973, 1975) and it was argued that the effect was specific and not the result of ATPase inhibition. Because there was no significant drop in the trans-gill potential after SCN injection (House and Maetz 1974), it was thought that the potential was not a direct result of electrogenic Cl^- secretion, but rather behaved as a Goldman-type diffusion potential. It was noted that variation in Na^+ and K^+ concentration in the external medium produced the expected changes in trans-gill potential consistent with the potential being predominantly a Na^+ diffusion potential (Potts and Eddy 1973; House and Maetz 1974). Moreover, the previously reported "exchange diffusion effect" (Motais et al. 1966) of reduced Na^+ efflux after transfer of teleosts to FW was shown to be primarily the result of a large change in trans-gill potential on transfer to FW (from +10 to −40 mV in *S. gairdneri* Greenwald et al. 1974; from +17 to −36 mV in *Dormitator maculatus* Evans et al. 1974). In general, Na^+ is very near its electrochemical equilibrium in marine teleosts; the expected flux ratio is approximately 1.4, while the observed ratio is about 0.9 (House and Maetz 1974), hence it was thought that most if not all Na^+ movement was passive. Because the electrical and concentration gradients may be controlled separately in in vitro membrane chambers, the nature of Na^+ and Cl^- movements could be described definitively (vide infra). Morphological evidence pointed to a cell type that was apparently responsible for the ion transport.

2.2 Insights from Morphological Studies

2.2.1 Chloride Cells

Keys and Willmer (1932) recognized a cell type in European eel (*A. anguilla*) gill epithelium which had granular cytoplasm and was strongly acidophilic; they suggested that the cells might be secretory, hence they were termed "chloride-secreting" cells (Cl^- cells). The distribution and ultrastructure of Cl^- cells in *Fundulus* spp. by Copeland (1948) and Philpott and Copeland (1963) revealed two important features: Cl^- cells were distributed throughout the buccal cavity (not only on the gill epithelium) and there was structural variation in chloride cells between SW- and FW-adapted killifish. The latter point consolidated an association between Cl^- cells and ion transport. The variation in gill chloride cell number and/or cell size with salinity was confirmed for several euryhaline species (*A. japonica*, Jozuka 1966; Shirai and Utida 1970; *F. heteroclitus*, Karnaky et al. 1976; *Barbus filamentosus*, Zaccone 1981). The existence of extrabranchial Cl^- cells in epithelia that may be mounted as planar sheets in vitro has contributed greatly to the understanding of the mechanisms of NaCl transport by the gill.

2.2.2 Localization of Na^+, K^+-ATPase

The involvement of Na^+, K^+-ATPase in many ion transport systems is well recognized. Na^+, K^+-ATPase activity is present in the crude homogenates of gill from SW and FW teleosts (Kamiya and Utida 1968; Jampol and Epstein 1970; Pfeiler and Kirschner 1972; Zaugg and Wagner 1973; Sargent and Thomson 1974; Evans and Mallery 1975; Pfeiler 1978; Langhorne and Simpson 1986). Subcellular fractionation indicated that Na^+, K^+-ATPase activity was mostly found in a microsomal fraction, consistent with a plasma membrane location for the enzyme (Pfeiler and Kirschner 1972; Sargent and Thomson 1974). Temperature optima for the enzyme are higher than the acclimation temperature, often 40 °C or more (Pfeiler and Kirschner 1972; Pfeiler 1978) and assays are routinely run at 37 °C. Many authors suggest that Na^+, K^+-ATPase activity reflects the ability of the gill to transport ions, particularly during salinity changes (Zaugg and Wagner 1973; Evans and Mallery 1975; McCartney 1976; Thomson and Sargent 1977; Langhorne and Simpson 1986). These results must be interpreted with caution because the kinetics and substrate dependencies of the enzyme appear to vary with temperature (comparing 37 °C with physiological incubation temperatures: Pfeiler and Kirschner 1972; Pfeiler 1978) and whole gill homogenates do not distinguish Na^+, K^+-ATPase that is *actually* involved in transepithelial transport from that which is present in nonepithelial cells and in hormonally inhibited Cl^- cells. The additional fact that both uptake *and* secretory epithelial cells have substantial quantities of ^{3}H-ouabain binding sites (Ernst and Mills 1980) further clouds the issue. Hence, the lack (e.g., Langhorne and Simpson 1986) or presence (e.g., Thomson and Sargent 1977) of a correlation of SW living with high activities of Na^+, K^+-ATPase cannot be interpreted with certainty. The presence of Na^+, K^+-ATPase does, however, point to a mechanism to link metabolism directly or indirectly with gill NaCl transport.

In most epithelia, Na^+, K^+-ATPase is restricted to the basolateral membrane surface (reviewed by Ernst and Mills 1980) and can be detected by two distinct techniques. Karnaky et al. (1976) first demonstrated by light microscopic autoradiography that tritiated ouabain binding in Cl^- cells from killifish was concentrated over the cytoplasm and did not appear on the apical membrane. Using a separate kinetic method of detecting potassium-stimulated phosphatase (NPPase) activity (which would include Na^+, K^+-ATPase), Hootman and Philpott (1979) determined the ultracytochemical localization of the enzyme in

Cl^- cells in SW pinfish gills (*Lagodon rhamboides*). The enzyme appears on or near the membranes of the extensive tubular system of Cl^- cells. The tubular system is continuous with the basolateral membrane, based on thick section electron microscopy of the lead-stained tubular system of Cl^- cells in SW guppy and mullet gills (Sardet et al. 1979). The ultrastructural data are consistent with physiological evidence that Na^+,K^+-ATPase involved in Cl^- secretion is restricted to the basolateral membrane. Cl^- secretion by *G. mirabilis* skin and *F. grandis* opercular epithelium is inhibited by basally added ouabain, 10^{-5} M (Marshall 1977, 1981 a, b; Krasny 1981) and bilateral treatment in *F. heteroclitus* (Degnan et al. 1977), but not if the drug is added apically (Marshall 1977).

2.2.3 Accessory Cells

Shirai and Utida (1970) described in Japanese eel (*A. japonica*) two distinct types of chloride cells: The classical chloride cell which they called type A and a second (type B) cell that was smaller, weakly acidophilic, and contained by EM fewer tubular christae in the mitochondria. As type B cells decreased in number during SW adaptation while type A cells increased, the type B cell was thought to be a transitional stage. Their Fig. 13 clearly shows the juxtaposition of these two cell types. The characteristics of the junctional complex between adjacent chloride cells has recently attracted much interest. Hootman and Philpott (1979, 1980) demonstrated in pinfish (*L. rhomboides*) gill epithelium the presence of small mitochondria-rich cells that occurred adjacent to mature chloride cells; these were termed "accessory" cells and were distinguished as having noticeably less ultracytochemically detected Na^+, K^+-ATPase. Sardet et al. (1979) described a similar association of "adjacent" cells and mature chloride cells in mullet (*Mugil capito*) and eel gill epithelium. Importantly, they showed that one- or two-stranded apical tight junctions between accessory and Cl^- cells were permeable to lanthanum, while several stranded (more than five) tight junctions between Cl^- cells and pavement cells (or between pavement cells) were not as permeable to lanthanum. The implication was that these lanthanum "leaky" junctions may be a localized paracellular shunt for ions. Accessory cells in close apposition to Cl^- cells have also been recognized in extrabranchial locations, in the Cl^- cell-containing skin of *G. mirabilis* (Marshall and Nishioka 1980).

2.3 Isolated-Perfused Gills and Heads

There has long been a recognized "frustration because of the difficulty of obtaining clear thermodynamic data" (Maetz 1971) when one is working with an organ as complicated as the gill. It is well recognized that fish are easily stressed and respond with increased circulating catecholamines (Mazeaud et al. 1977) that, in turn, increase arterial and decrease venous blood flow through the gill (Girard and Payan 1976) and apparently decrease the rate of sodium and chloride extrusion (Pic et al. 1975). It is difficult to separate direct transport effects from changes in blood flow patterns that can also change the effective surface area for transport. Isolated-perfused gill preparations have the advantage over in vivo experiments because one can control the perfusion rate of the gills and have access to the both sides of the preparation. Originally, heart-gill preparations were used with some success; in these, the heart provided the force for perfusion of the gills, but the preparation was relatively short-lived (Keys 1931). The isolated-perfused gill arches (Shuttleworth 1972; Stagg and Shuttleworth 1982; Davis and Shuttleworth 1985a, b) and perfused head preparations (Payan et al. 1984; Perry et al. 1984) use artificially forced perfusion. In the former, paired experimental design is possible, although the method is limited to use in relatively large fish. In most work, only the branchial arterial perfusion fluid is collected (Davis and Shuttleworth 1985a, b). The latter preparation takes advantage of the twin drainage of the trout gill filaments; the efferent branchial arterial drainage is collected via cannulation of the dorsal aorta and the venous drainage, which bypasses the respiratory lamellae, is collected as overflow. Recent work has included as a refinement, venous perfusion fluid collection from a single gill arch (Olson 1984; Nekvasil and Olson 1986). While it is clear that arteriovenous anastomoses occur in some species [*Oreochromis mossambicus* (formerly *Tilapia mossambica* and *Sarotherodon mossambicus*), Vogel et al. 1974; *S. gairdneri,* Vogel et al. 1976] these structures may not be always present or necessarily operational. A nagging problem with the forcefully perfused isolated organs is the fact that damage or degeneration of the vasculature is common, indicated often by a rise in vascular resistance (Keys 1931; Perry et al. 1984). Any damage to the vasculature will create an aphysiological shunting of fluid either to the venous drainage or to the outside bathing solution. Keys (1931) noted that the heart-gill preparation showed consistently lower resistance to perfusion than the gill preparation. His speculation regarding a humoral factor from the heart as being responsible for this difference may yet prove true: the atrial natriuretic factor is vasodilatory in mammals (Winquist 1986) and could perhaps be used with perfusion preparations of fish gill.

The isolated organs, even in the event that the perfusion difficulties are overcome, still do not allow one to know or control the physiologically relevant surface areas involved in ion transport. They remain an important in vitro approach to studying gas transfer and acid-base balance, but there now exist several geometrically simple preparations that contain ion-transporting chloride cells and provide more detailed information on the mechanisms of ion transport.

2.4 Skin and Opercular Epithelia

2.4.1 Importance in NaCl Transport

Teleost skin was thought to be secretory, based on classical electrophysiological studies that demonstrated a small but reproducible current of rest (Hermann 1882; Reid 1893). Although the role of the skin in ion transport is often discounted (e.g. Maetz 1971), for some species the contribution of the general body surface to ion transport is important. In larval marine teleosts such as *F. heteroclitus,* the epithelium of the yolk sack and body skin are the sole source of ion excretion by chloride cells (Guggino 1980). In adult blennies (*Blennius pholis*) approximately half of the total body efflux of Na^+ and Cl^- occurs via the skin (Nonnotte et al. 1979), while Cl^- secretion by the skin of *G. mirabilis* (Marshall 1981a) can account for about 20–25% of the total body flux (Thompson 1972). In some air-breathing FW fish, the general body surface accounts for 75% of Na^+ uptake (Stiffler et al. 1986).

2.4.2 A Model for the Seawater Gill

The skin epithelium of *G. mirabilis* and opercular epithelia of *F. heteroclitus, F. grandis,* and *O. mossambicus* are easily dissected as planar sheets of varying sizes, from 0.07 cm^2 (*F. heteroclitus*) to 1.23 cm^2 (*G. mirabilis*) in Ussing-style membrane chambers (Karnaky et al. 1977; Marshall 1977, 1981a; Foskett et al. 1981; Krasny 1981). All of these epithelia contain mitochondria-rich chloride cells detectable by fluorescent dyes that are specific to mitochondria (diethylaminostyrylethylpyridinium iodide, DASPEI, Bereiter-Hahn 1976). The short-circuit current (I_{sc}) measured across *G. mirabilis* skin and the opercular epithelia is not significantly different from the net Cl^- flux (Degnan et al. 1977; Marshall and Bern 1980, 1981a; Foskett et al. 1981; Krasny 1981) and the I_{sc} varies linearly with the number of mitochondria-rich cells detected by DASPEI fluorescence (Marshall and Nishioka 1980), suggesting that Cl^- cells are responsible for active Cl^- secretion. Proof that only Cl^- cells are responsible for transepithelial current was provided by the vibrating probe technique (Foskett and Scheffey 1982; Scheffey et al. 1983; Foskett and Machen 1985). The probe signal is the result of a chloride current, as the signal is reduced in Cl^--free solutions (Scheffey et al. 1983). Also, the vibrating probe signal is inhibited by epinephrine and stimulated by glucagon (Foskett and Machen 1985), consistent with the effects of these substances on the I_{sc} and Cl^- active transport (Degnan et al. 1977; Marshall 1977; Marshall and Bern 1980; Foskett et al. 1983). Ultrastructurally, Cl^- cells in these epithelia are similar to the Cl^- cells of the gill epithelium (Karnaky et al. 1976; Marshall and Nishioka 1980) with some variation in cell shape because these cells typically span the entire thickness of the epithelium, which may be up to 100 μm in *G. mirabilis* skin or as little as 20 μm in *F. heteroclitus* opercular epithelium.

The positive trans-gill potential in vivo (+10 to +35 mV; +20 mV in *G. mirabilis,* vide supra) is smaller than the potential measured across *G. mirabilis* skin when a mimic of SW is placed in the outside bath ($+39 \pm 1.2$ mV; Marshall 1981a) probably because the in vivo measurement includes a low-resistance leak pathway in the gill epithelium. Evidence for this includes the potentials measured in SW across the body wall of larval *F. bermudae* (Guggino 1980) at such a stage that the gill epithelium has not yet developed; in this case, the measured potential is closer ($+40 \pm 1.3$ mV) to that measured in vitro. The potential measured with SW on the mucosal side of the *F. heteroclitus* epithelium is somewhat lower (32 ± 1.3 mV; Degnan and Zadunaisky 1980; 36.8 ± 1.3 mV, Table 1). In general, it is thought that most of the potential measured in vivo is a Na^+ diffusion potential (Potts 1984). The passive component of the transepithelial potential mea-

Table 1. $^{36}Cl^-$ fluxes in *Fundulus heteroclitus* opercular epithelium at open circuit and with "seawater" on the mucosal side (number of membrane pairs = 5)

Treatment flux	Cl^- flux ($\mu Eq\ cm^{-2}\ h^{-1}$)	G_t ($mS\ cm^{-2}$)	TEP (mV)	Jsm/Jms Observed	Jsm/Jms Predicted
Short circuit					
s→m	4.40±0.45	6.5±0.6	0		
m→s	0.47±0.07	7.3±0.2	0	10.8±2.5[c]	1.0
Net[a]	3.93±0.51				
Open circuit					
s→m	3.41±0.33	6.8±0.6	18.5±2.1		
m→s	0.74±0.06	7.8±0.2	19.7±3.4	4.8±0.5[c]	0.47
Net	2.68±0.32				
Open circuit, mucosal SW[b]					
s→m	4.73±0.47	10.9±0.9	36.8±1.3		
m→s	3.18±0.32	13.3±0.7	35.7±2.3	1.5±0.2[c]	0.08
Net	1.55±0.38[d]				

[a] Not significantly different from $I_{sc} = 4.72 \pm 0.59\ \mu Eq\ cm^{-2}\ h^{-1}$ ($P > 0.4$).
[b] NaCl added to the mucosal bath to yield $[Na^+] = 500$ mM; $[Cl^-] = 475$ mM; normal Ringer's $[Na^+] = 160$, $[Cl^-] = 137$. Other ions were the same as for brook trout Ringer's (cf. Marshall 1985).
[c] Significantly different from the predicted flux ratio ($P < 0.01$).
[d] Significantly greater than zero ($P < 0.05$).

sured in vitro with SW outside is approximately the same as the 20 mV measured in vivo (Thompson 1972). To do this, *G. mirabilis* skin, with SW on the outside was exposed to N_2 (which blocks active Cl^- secretion) and rapidly reduces the potential to 21 ± 1.7 mV (Marshall 1981a). This result also indicates that part of the potential in vitro may be associated directly or indirectly with active Cl^- transport.

If a model epithelial system is to mimic the in vivo operation of a transport system, such as the case of the skin and opercular epithelium purports to do for the gill, the bathing fluids should approximate as best as is possible the blood plasma of the animal. The use of temperatures different from the acclimation temperature, low Ringer's pH, and high CO_2 gas mixtures are common faults (cf. Perry et al. 1984) that could affect performance of the tissue in vitro. Teleostean blood is alkaline (typically pH 7.8) and blood CO_2 is much lower than in mammals, hence the use of low pH Ringer's solutions (pH 7.15–7.25; Degnan et al. 1977; Degnan and Zadunaisky 1979) and high CO_2 gas mixtures (5% CO_2/95% O_2; Degnan et al. 1977; Degnan 1985) is aphysiological. Under these conditions, the *F. heteroclitus* opercular epithelium (at open circuit and with SW on the mucosal side) had no net flux of Na^+ or Cl^- (Degnan and Zadunaisky 1979). In contrast, at physiological pH and P_{CO_2}, open circuit, and with a mimic of SW on the mucosal side, the *G. mirabilis* skin has a positive net flux for Na^+ and Cl^- (Marshall 1981a). Hence, the *G. mirabilis* preparation operates in vitro in a manner consistent with the function in vivo of epithelia containing Cl^- cells, i.e., to contribute to net salt secretion in SW. We repeated this experiment using *F. heteroclitus* opercular epithelium (at pH = 7.7, 99% O_2/1% CO_2 gas, 15 °C, membrane surface = 0.238 cm^2; other methods according to Marshall 1985) and found net secretion of Cl^- against electrochemical gradients like those in vivo (Table 1).

2.4.3 *Passive NaCl Movements Across Gill-Like Epithelia*

Na^+ distribution across the skin and opercular epithelia is almost certainly passive and diffusional, as for several preparations: (1) the net Na^+ flux in the absence of transepithelial electrochemical gradients is not significantly different from zero (Degnan et al. 1977; Krasny 1981; Marshall 1981a); (2) at open circuit, the flux ratio is not significantly different from that predicted on the basis of pure diffusion, i.e., the flux behavior of Na^+ is predicted accurately by the Ussing flux ratio equation (Degnan and Zadunaisky 1979; Marshall 1981a). (3) In the presence of Na^+ and electrical gradients combined, there is still agreement with the Ussing flux ratio equation (Marshall 1981a). (4) Voltage clamping of the epithelia to ±25 mV or ±50 mV (Degnan and Zadunaisky 1980; Krasny 1981; Fig. 1) shows that the epithelia behave as a single passive resistor to Na^+ permeation, implying that Na^+ moves by passive diffusion. Further, it would appear that Na^+ may move via a paracellular pathway, inasmuch as the epithelium behaves as a single passive resistor to Na^+ unidirectional fluxes (Fig. 1), and assuming that where transcellular transport dominates, this would cause the relation to be nonlinear (as demonstrated by the nonlinear response of unidirectional Cl^+ fluxes to voltage clamping, Fig. 1; cf. Schultz and Frizzell 1976; Marshall 1981b). In brook trout (*Salvelinus fontinalis*) opercular epithelium that lacks chloride cells, dual fluxes of ^{14}C-mannitol (which is excluded from cells) with either $^{22}Na^+$ or $^{36}Cl^-$ demonstrated that both ions move through a pathway along with mannitol in relation to their free solution mobilities (Marshall 1985).

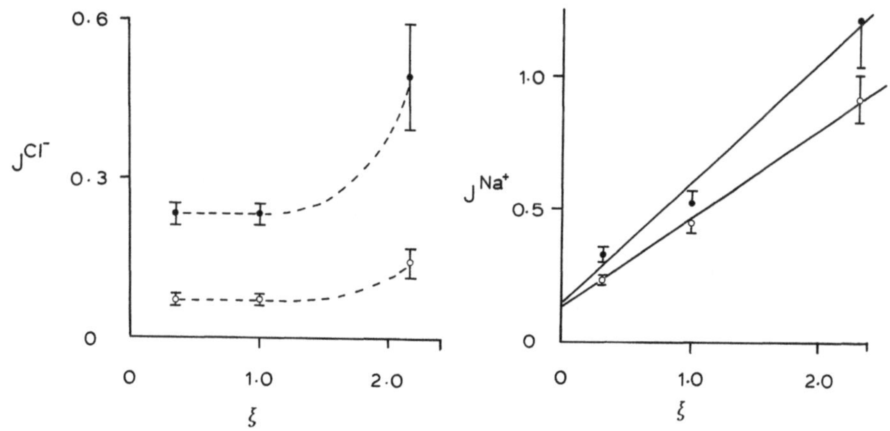

Fig. 1. Unidirectional fluxes of Na^+ and Cl^- in response to voltage clamp in the skin of *Gillichthys mirabilis*; number of membrane pairs = 5. *Abscissa* $\xi = [z_i F \Delta\psi/RT]/[\exp(z_i F \Delta\psi/RT) - 1]$, where $\Delta\psi = \psi_m - \psi_s$ for Na^+ efflux and Cl^- influx and $\Delta\psi = \psi_s - \psi_m$ for Na^+ influx and Cl^- efflux; methods according to Marshall (1981b). *Open circles* are influx, *closed circles* efflux measurements; *broken lines* fitted by eye; *solid lines* are least-squares lines-of-best-fit

This is evidence for a water-filled paracellular pathway (Dawson 1977). This shunt clearly is physiologically relevant, as the pathway is reversibly opened by acidification of the mucosal bath (Marshall 1985). Inasmuch as there is small but measurable shunt permeability in opercular epithelia that lack Cl^- cells, shunt conductance is not restricted exclusively to the single-stranded, lanthanum-permeable tight junctions between Cl^- cells and accessory cells, as previously proposed (Sardet et al. 1979). It is clear, however, that the majority of conductance and transepithelial current is associated with chloride cells, as detected by vibrating probe analysis (Foskett and Scheffey 1982; Scheffey et al. 1983; Foskett and Machen 1985).

Based on passive flux data (Marshall 1981a, b) *G. mirabilis* skin is cation-selective with $P_K > P_{Na} > P_{Cl}$ in ratio of 1.0:0.49:0.09; in *F. heteroclitus* opercular epithelium, $P_{Na} > P_{Cl}$ in the ratio of 1.0:0.4 (Degnan and Zadunaisky 1980). This is supported by results of potential changes in response to specific ion dilutions of the mucosal bath that showed 27 mV per decade change in Na^+ concentration but only 7.2 mV per decade change in Cl^- (Degnan and Zadunaisky 1980). While it is tempting to ascribe the single barrier through which Na^+ moves as a paracellular shunt (Degnan and Zadunaisky 1980; Fig. 1), this has not been definitively shown (Foskett et al. 1983). Evidence from triaminopyrimidine (TAP) inhibition of I_{sc}, conductance, Na^+ fluxes, and Cl^- efflux (Degnan and Zadunaisky 1980) could be complicated by nonspecific inhibition by TAP of NaCl-coupled transport (Frizzell et al. 1981). The lack of effects of amphotericin B and amiloride, taken as suggesting that Na^+ is not traversing a cellular path (Degnan and Zadunaisky 1980) could be related, respectively, to inaccessibility of the membrane surface because of the mucus coat and the insensitivity of teleostean conductive Na^+ pathways to amiloride (Loretz and Bern 1983). While it is clear that the passive flux of Cl^- traverses a complicated path and that both Na^+ unidirectional fluxes encounter a single diffusive barrier, it remains to be proven conclusively that the former passes through Cl^- cells and the latter passes through the apical tight junctions between Cl^- cells and accessory cells.

2.4.4 Secondary-Active Transport of Cl^-

The current model for Cl^- secretion by chloride cells involves a sodium-dependent (Marshall 1981a; Degnan 1984), ouabain-sensitive (Degnan et al. 1977; Marshall 1981a) Cl^- net flux in the absence of transepithelial electrochemical gradients where the short-circuit current is equivalent to Cl^- net flux (Degnan et al. 1977; Marshall and Bern 1980). The Cl^- secretion is inhibited by "loop" diuretics of the furosemide type (Degnan et al. 1977; Eriksson et al. 1985, 1986; Fig. 2) and by thiocyanate (Marshall 1981a). The "thiazide-type" loop diuretics are much less effective (Eriksson et al. 1985; Eriksson and Wistrand 1986). These results are consistent with "secondary active transport", proposed by Silva et al. (1977). In this model, the Cl^- transport path involves coupling of Cl^- entry across the basolateral membrane to Na^+ such that the (presumed) electrochemical gradient favoring Na^+ entry accumulates Cl^- intracellularly above its electrochemical equilibrium; Cl^- leaves the cell down its electrochemical gradient via

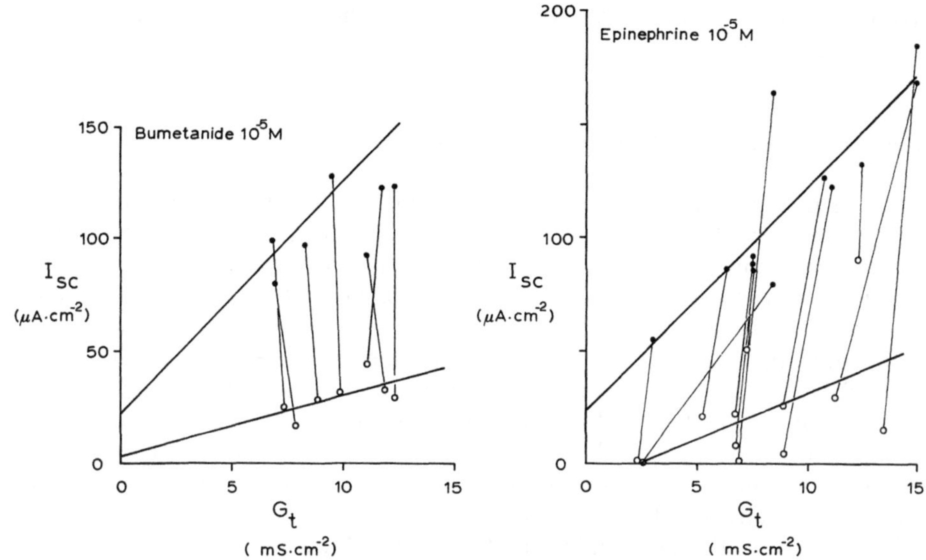

Fig. 2. Effects of epinephrine and bumetanide on the relation between short-circuit current (I_{sc}) and conductance (G_t) in *Fundulus heteroclitus* opercular epithelium (methods as in Table 1). *Solid circles* are initial and *open circles* the final measurements after addition of drug for a single epithelial preparation (initial and final points connected by *fine lines*). Lines drawn through the *solid circles* are the least-squares line-of-best-fit for the epinephrine initial values [$I_{sc} = (9.8 \pm 4.1) \cdot G_t + 22.8 \pm 41.1$; $r = 0.86$, $P < 0.001$]; this regression reflects the spontaneous variation among preparations and its y-intercept is not significantly different from zero. The same regression line is reproduced in the bumetanide figure, for reference. Note that both drugs inhibit I_{sc} with little change in G_t (i.e, steep slope), unlike the spontaneous variation in I_{sc} (and Cl^- cell number) that is accompanied by large variation in G_t (see text for details). After epinephrine treatment, the regression statistics are $I_{sc} = (4.12 \pm 4.4) \cdot G_t - 9.6 \pm 37$ ($P < 0.05$) and after bumetanide $I_{sc} = (2.62 \pm 3.8) \cdot G_t + 3.6 \pm 39$ ($P < 0.1$)

conductive pathways in the apical membrane. This pattern has been shown definitively for the corneal epithelium of rabbit (Klyce and Wong 1977), among others. The stoichiometry of the basal-coupled transport (or whether it is similar to the Na-K-2Cl coupled transporter in Cl^- uptake epithelia) is not known for Cl^- cell-containing epithelia. The electrochemical gradients for Na^+ and Cl^- across the apical and basal membranes of the Cl^- cell are also unknown.

Although Cl^- conductance is part of the model, there is little direct evidence for it, particularly as bilateral substitution of Cl^- with methylsulfate or gluconate does not reduce epithelial conductance (Degnan et al. 1977; Marshall 1981b). Exposure of the apical surface to cupric ion (0.5 mM), which in frog skin blocks Cl^- conductance, reduces opercular epithelial conductance, I_{sc} and Cl^- efflux (Degnan 1985); it was proposed that Cu^{2+} may block a Cl^- conductive pathway. It is unlikely that the copper effect is specific to Cl^- conductance, given that cupric ion in micromolar quantity strongly inhibits Na^+, K^+-ATPase and ouabain binding in SW flounder gills in vitro (Stagg and Shuttleworth 1982). Hence, it is not clear whether a Cl^- conductance is present. If Cl^- conductive pathways are present, these do not seem to be the site of hormonal regulation of Cl^- secretion (vide infra).

2.4.5 Hormonal Regulation of Cl Secretion

Because the spontaneous rate of Cl^- secretion of in vitro skin and opercular epithelia is high (Degnan et al. 1977; Marshall and Bern 1980; Foskett et al. 1982) and as the dominant adrenergic response is inhibitory and mediated by alpha-adrenergic receptors (Marshall and Bern 1980; Foskett et al. 1982; May and Degnan 1985), it is likely that the major regulation of Cl^- secretion is inhibitory. Beta-adrenergic receptors are present, as indictaed by increases in I_{sc} with addition of the beta agonist isoproterenol (Marshall and Bern 1980; May and Degnan 1985), but when the endogenous catecholamines (epinephrine and norepinephrine) are applied, the response is dominated by the alpha-receptor-mediated inhibition of Cl^- secretion. Epinephrine rapidly inhibits Cl^- secretion at physiologically realistic concentrations (Marshall and Bern 1980; Foskett et al. 1982a); there is 30–60% inhibition of I_{sc} at 10^{-7} M, a concentration below the circulating epinephrine levels in stressed salmon (Mazeaud et al. 1977). The receptor subtype is apparently alpha-2, as the opercular epithelium is sensitive to clonidine, a specific alpha-2 agonist (May and Degnan 1985) and because the receptors have an apparently higher affinity to epinephrine over norepinephrine (Marshall and Bern 1980). The mechanism of alpha-2 inhibition of Cl^- secretion is presently unknown, but does not apparently involve reductions in cAMP levels (May and Degnan 1985).

In addition to inhibition by catecholamines, there are inhibitory actions by muscarinic cholinergic receptors (May and Degnan 1985), somatostatin (Foskett and Hubbard 1981; Davis and Shuttleworth 1985a), and urotensin II (Marshall and Bern 1979, 1981), a somatostatin-like teleostean urophysial peptide (Pearson et al. 1980). It is not known whether the skin, opercular epithelium, and gill would normally be exposed to somatostatin or acetylcholine, but urotensin II has frequently been associated with teleost osmoregulation (Marshall and Bern 1981). In addition to beta-adrenergic receptors, there are stimulations of Cl^- secretion by the CRF-like peptide urotensin I (Marshall and Bern 1981) and by glucagon and VIP (Foskett et al. 1982a). The mechanism of the pharmacologically induced stimulation of Cl^- secretion by beta-adrenergic receptors has been studied in some detail: there is a significant increase in opercular epithelial cyclic-AMP content after stimulation of beta receptors using isoproterenol and after treatment with forskolin and isobutylmethylxanthine (May and Degnan 1985).

The cellular mode of the control of Cl^- transport is not known, but could feasibly include changes in the proposed Na-dependent Cl^- entry step at the basal membrane, effects on the Cl^- exit at the apical side, or possible effects on Na^+, K^+-ATPase. Whereas epinephrine-stimulated Cl^- secretion in the corneal epithelium varies with apical membrane Cl^- conductance (Klyce and Wong 1977; Marshall and Klyce 1983), in Cl^- cell-containing epithelia, large hormonally induced changes in net Cl^- flux and I_{sc} occur with little change in tissue conductance (Degnan et al. 1977; Foskett et al. 1982b; Fig. 2). This is in contrast to the large conductance changes that accompany spontaneous variation in I_{sc} (Foskett et al. 1982b; Fig. 27) and the reduction in I_{sc} by prolactin (Foskett et al. 1982b). The spontaneous variation and prolactin effect are attributed to the close association of active transport and leak pathways with individual Cl^- cells (Foskett and Machen 1985) and variation in the density of Cl^- cells (Marshall and Nishioka 1980).

It has been shown (Frizzell and Schultz 1972) that a re-expression of the statement of equivalent conductance at zero membrane potential relates conductance (G_i) to unidirectional ion flux from side 1 to side 2 ($J_{i_{12}}$):

$$G_i = (J_{i_{12}} z_i^2 F^2)/(RT) \ ,$$

where z_i, F, R, and T have their thermodynamic definitions. Because Cl^- influx is unaffected by epinephrine (Marshall and Bern 1980), the change in efflux rate is equivalent to the change in net flux and I_{sc} ($I_{sc} = z_i F J_{net}$), $\Delta J_{net} = \Delta J_{i_{12}}$. Substituting, the relation becomes:

$$G_t = I_{sc} (z_i F/RT) \ ,$$

and the slope of the I/G plot should be 25 mV, *if only conductive paths that contribute to I_{sc} were changing.* Such a low slope is seen with spontaneous variation in I_{sc} (25.8 mV Foskett et al. 1982b; 9.8 mV in Fig. 2). But much steeper slopes are associated with epinephrine inhibition (75.3 ± 9.2 mV, Fig. 2) and there is effectively no conductance change after bumetanide (Fig. 2). The suggestion is that epinephrine (and isobutylmethylxanthine, Foskett et al. 1982a) act primarily on a nonconductive step, possibly via the bumetanide-inhibitable step.

3 The Freshwater Gill

3.1 Ion Exchanges and Active Transport

Freshwater teleosts maintain blood ion concentrations much higher than the environment; most species can do so for varying times without dietary salt intake, implying that the animals are in steady-state ion balance. The four most commonly studied species are the goldfish *Carassius auratus,* eel, *A. rostrata* and *A. anguilla,* and rainbow trout, *S. gairdneri.* The prevailing model (e.g., see Kirschner 1983) for Na^+ and Cl^- uptake is by relatively independent, electrically neutral ion exchanges (Na^+/H^+ and/or Na^+/NH_4^+ in parallel with Cl^-/OH^- or Cl^-/HCO_3^-) that may be driven indirectly by a basolateral Na^+, K^+-ATPase. Cl^- uptake shows stimulation by HCO_3^- addition on the *trans* side (Kerstetter and Kirschner 1972) and inhibition by addition to the *cis* side (Perry et al. 1981) as well as inhibition by SITS (Perry et al. 1981), indicating Cl^-/HCO_3^- exchange. Cl^- uptake (Epstein et al. 1973; de Renzis 1975) but not Na^+ uptake (Epstein et al. 1973) is inhibited by thiocyanate ion, implying that the two ion uptake mechanisms are independent. In *S. gairdneri,* the K_m for the saturable uptake of Na^+ is 0.46 mM (Kerstetter et al. 1970) and for Cl^- the K_m is 0.25 mM (Kerstetter and Kirschner 1972).

Na^+ uptake is linked to NH_4^+ excretion (Kerstetter and Keeler 1976; Payan and Matty 1975; Payan et al. 1975) as long as external Na^+ is kept at low (~1 mM) concentrations. Na^+ uptake is inhibited by amiloride (Kerstetter and Keeler 1976; Payan 1978; Perry et al. 1981) associated with pronounced acidosis (Perry et al. 1981) and reduction in NH_4^+ excretion (Payan 1978), implying an amiloride-sensitive $Na^+/(H^+$ or $NH_4^+)$ exchange. Inhibition of Na^+ uptake by acetazolamide (Payan et al. 1975; Kerstetter and Keeler 1976) suggests a link between the two exchangers, but this drug is ineffective on Cl^- influx (Kerstetter and Kirschner 1972). There exists a tight linkage of ion exchanges in FW to acid-base balance (Heisler 1982; Perry et al. 1981) and in the models of acid-base balance in the gill, the principal driving forces for Na^+ and Cl^- uptake appear to be the (respective) gradients favoring H^+ (NH_4^+) and HCO_3^- excretion, rather than an ATPase. Results with ouabain are equivocal: in some cases there is apparent inhibition of Na^+ uptake (Payan et al. 1975), in other cases, no effect (Kerstetter and Keeler 1976).

3.1.1 Anion ATPase

There is some disagreement as to whether a plasmalemmal anion-stimulated ATPase could also be involved. This enzyme appears to be restricted to mitochondrial membranes in some preparations of *S. gairdneri* gill, as determined by differential centrifugation on sucrose density gradients and the use of marker enzymes, cytochrome oxidase, and 5′-nucleotidase (Van Amelsvoort et al. 1977). The anion-stimulated ATPase distribution followed closely that of cytochrome oxidase in all fractions, hence the appearance of these two enzymes in small amounts in the microsomal fraction was attributed to mitochondrial membrane contamination. Others suggest that the coincidence of 5′-nucleotidase with anion ATPase in the microsomal fraction is evidence of the plasmalemmal location for some of the enzyme (Bornancin et al. 1980). The K_m for Cl^- of the ATPase, in the presence of HCO_3^-, is 1 – 14 mM, depending on the fraction being assayed; the lower K_m (1 mM in *C. auratus*, 3.3 mM in *S. gairdneri*) appears in the microsomal fraction (de Renzis and Bornancin 1977; Bornancin et al. 1980). The pH optimum for the enzyme is rather broad, around 8.2 (de Renzis and Bornancin 1977; Van Amelsvoort et al. 1977). It is interesting that HCO_3^- alone (but not Cl^- alone) can stimulate the enzyme (Kerstetter and Kirschner 1974; de Renzis and Bornancin 1977; Bornancin et al. 1980). Until the question of the subcellular distribution of the enzyme can be resolved definitively, the role of the anion ATPase in ion uptake by FW teleosts will remain in doubt.

3.1.2 Freshwater Cl^- Cells

It is well known that Cl^- cells with the typical ultrastructural features of numerous mitochondria and an extensive tubular system, are present in the gill epithelia of FW-adapted euryhaline and stenohaline FW fish (Shirai and Utida 1970; Kikuchi 1977). These Cl^- cells show significant binding of tritiated ouabain

(Karnaky et al. 1976), as do several epithelia involved in ion uptake (Ernst and Mills 1980). FW Cl^- cells have shallow apical pits (Philpott and Copeland 1963; Hossler et al. 1985) and have well-developed apical tight junctions between adjacent Cl^- cells (Karnaky 1986), in contrast to the deep apical pits and single-stranded tight junctions of their SW counterparts. The functional significance of the former characteristic is unknown, but the latter feature could account in part for the reduced permeability of the gill in FW (Karnaky 1986). In *O. mossambicus* opercular epithelium, FW chloride cells are small, contain a poorly developed tubular system and often have no contact at the apical surface (Foskett et al. 1981); this is correlated with high tissue resistance and very low I_{sc}. Whereas the relation between the SW Cl^- cell and Cl^- secretion is now strong (vide supra), it is not clear if or how Cl^- cells contribute to FW active ion uptake.

3.2 Opercular Epithelium and Skin

Attempts to establish a model epithelial structure that mimics the operation of the FW gill in teleosts have thus far been unsuccessful. The skin and opercular epithelia of euryhaline marine teleosts such as *F. heteroclitus* and *G. mirabilis* (Degnan et al. 1977; Marshall 1977) still retain (albeit reduced) Cl^- secretory capacity even after the animals have been adapted to dilute 1–5% SW for extended periods. Presumably, the cells are not actively involved in Cl^- secretion in FW, although the mechanism or hormone that prevents this is not known. Substitution of Cl^- for an impermeant anion on the mucosal side decreases Cl^- efflux in both epithelia (Degnan 1984; Marshall 1981a); this effect is not one involving Cl^-/Cl^- exchange diffusion, since the removal of serosal Cl^- has no effect on Cl^- influx (Marshall 1981a; Degnan and Zadunaisky 1980), hence Cl^- self-exchange is minimal. If FW is placed on the mucosal side of epithelia from 5% SW-adapted fish, there still is at open circuit a net Na^+ and Cl^- flux in the secretory direction (Marshall unpublished). The possible existence of an inhibitory hormone or hormones that would inhibit ion secretion and stimulate ion uptake in FW-adapted animals has not been examined in detail. While prolactin is well known to have growth and osmoregulatory functions in FW fish (reviewed by Bern 1975), this hormone apparently does not have rapid-acting effects on several osmoregulatory epithelia (Bern et al. 1981). The slower adaptational response of SW animals to prolactin clearly is a reduction in the number and size of Cl^- cells with a concomitant reduction in Cl^- secretion (Foskett et al. 1982b). In any case, net ion uptake apparently does not occur spontaneously in skin and opercular epithelia of FW-adapted euryhaline *marine* fish.

The opercular epithelia of two euryhaline *freshwater* teleosts, *O. mossambicus* and *S. fontinalis,* have been examined. The former preparation contains few small Cl^- cells (Foskett et al. 1981) and has no net flux of Cl^- and very low I_{sc}. The latter preparation has no Cl^- cells detectable by fluorescence microscopy, no I_{sc}, and no net transport of Cl^- (Marshall 1985). However, these experiments, with both sides of the tissue bathed in Ringer's may not detect a small capacity NaCl uptake (cf. Kirschner 1983); studies with dilute media bathing the mucosal side would be more definitive. Whereas the gills of most FW fish appear to be the source of ion uptake, recently Stiffler et al. (1986) found that a large component (75%) of Na^+ uptake of the FW air-breathing teleost *Synbranchus marmoratus* occurs via the general body surface. They also showed that the Na^+ flux ratio was greater than that predicted by the Ussing flux ratio equation, suggesting that Na^+ was accumulated against its electrochemical gradient. The skin of this species, which contains some mitochondria-rich cells, may prove to be an important model for Na^+ and Cl^- transport in the FW gill.

4 Conclusions and Future Directions

Chloride cells in gills, opercular and skin epithelia of SW teleosts actively secrete Cl^- via a basally-located bumetanide-sensitive, Na^+-coupled mechanism. Transmembrane Na^+ electrochemical gradients likely provide the driving force and these gradients are maintained by basolaterally-located Na^+,K^+-ATPase. Cl^- extrusion at the apical membrane is less-well understood, but could involve Cl^- conductive pathways. Intracellular ion activity measurements and patch clamp studies of the apical membrane are needed to resolve these mechanisms unequivocally. Na^+ secretion is passive, paracellular, and driven by a serosa-positive transepithelial potential; the most feasible location for the Na^+ shunt is locally via leaky junctions between accessory cells and chloride cells.

Recent studies have expanded our knowledge of the multihormonal control of Cl^- secretion by chloride cells. The rapid stimulation of ion transport by β-adrenergic agonists, glucagon and PGE_1 is via activation of adenylate cyclase (Guibbolino and Lahlou 1987), however the side and mode of action for inhibition of Cl^- transport remains poorly understood. Also, arachidonic acid and locally-produced eicosanoids (leucotrienes LTC_4, LTD_4 and LTE_4) stimulate Cl^- secretion, while PGE_2 is inhibitory (Van Praag et al. 1987). Among the slower-acting hormones, cortisol stimulates chloride cell development and after transfer to SW, cortisol titres in plasma are often elevated and are associated with concomitant changes in high-affinity nuclear and cytosolic cortisol receptors in the gills (Weisbart et al. 1987). Thyroid hormone function is often depressed after SW transfer (Omeljaniuk and Eales 1986; Weisbart et al. 1987), suggestive of a longterm inhibitory control (in FW by the thyroid) of Cl^- cells. As we have seen earlier, prolactin inhibits growth and development of SW-type Cl^- cells. Clearly, the endocrine control of Cl^- cell function is as complicated as it is important to the osmoregulation of the animal. Work in this area is of great interest, as the chloride cell is unique among vertebrate ion transporting systems in the variety and complexity of hormonal responses.

References to Parts I and II

André M (1960) Les écrevisses françaises. Savoir en histoire naturelle, vol 25. Lechevalier, Paris, p 293

Babula A (1977) Ultrastructure of gill epithelium in *Mesidothea entomon* (Crustacea, Isopoda). Acta Med Pol 18:281–282

Balaban RS, Mandel LJ, Benos DJ (1979) On the cross-reactivity of amiloride and 2,4,6 triaminopyrimidine (TAP) for the cellular entry and tight junctional cation permeation pathways in epithelia. J Memb Biol 49:363–390

Baldwin GF, Kirschner LB (1976) Sodium and chloride regulation in *Uca* adapted to 10% sea water. Physiol Zool 49:172–180

Barra JA, Péqueux AJR (1986) Ultrastructural and biochemical localization of Na^+ K^+ ATPase in salt-transporting gills of the Chinese crab. In: Comparative physiology of environmental adaptations. Abstr 8th Conf ESCPB, Strasbourg, France. ESCPB Publ, p 21

Barra JA, Péqueux A, Humbert W (1983) A morphological study on gills of a crab acclimated to fresh water. Tissue Cell 15(4):583–596

Bauer C, Gros G, Bartels H (1980) Biophysics and physiology of carbon dioxide. Springer, Berlin Heidelberg New York, 453 pp

Bereiter-Hahn J (1976) Dimethylaminostyrylmethylpyridiniumiodine (DASPMI) as a fluorescent probe for mitochondria in situ. Biochim Biophys Acta 423:1–14

Berridge MJ, Oschman JL (1972) Transporting epithelia. Academic Press, London New York, p 91

Bern HA (1975) Prolactin and osmoregulation. Am Zool 15:937–948

Bern HA, Bisbee CA, Collie NL, Foskett JK, Hughes B, Loretz CA, Marshall WS (1981) Failure of ovine prolactin to elicit rapid responses by osmoregulatory surfaces. Gen Comp Endocrinol 44:128–130

Bielawski J (1964) Chloride transport and water intake into isolated gills of crayfish. Comp Biochem Physiol 13:423–432

Bielawski J (1971) Ultrastructure and ion transport in gill epithelium of the crayfish, *Astacus leptodactylus* Esch. Protoplasma 73:177–190

Bock F (1925) Die Respirationsorgane von *Potamobius astacus* Leach (*Astacus fluviatilis* Fabr.) Z Wiss Zool 124:51–117

Bornancin M, Renzis G de, Naon R (1980) Cl^--HCO_3^--ATPase in gills of the rainbow trout: evidence for its microsomal localization. Am J Physiol 238:R251–R259

Bryan GW (1960) Sodium regulation in the crayfish *Astacus fluviatilis*. J Exp Biol 34:71–84

Bubel A (1976) Histological and electron microscopical observations on the effects of different salinities and heavy metal ions on the gills of *Jaera nordmanni* (Rathke) (Crustacea, Isopoda). Cell Tissue Res 167:65–95

Bubel A, Jones MB (1974) Fine structure of the gills of *Jaera nordmanni* (Rathke) (Crustacea, Isopoda). J Mar Biol Assoc UK 54:737–743

Bundy HF (1977) Review: carbonic anhydrase. Comp Biochem Physiol 57B:1–7

Burggren WW, McMahon BR, Costerton JW (1974) Branchial water- and blood-flow patterns and the structure of the gill of the crayfish *Procambarus clarkii*. Can J Zool 52:1511–1518

Burnett LE, Woodson PBJ, Rietow MG, Vilicich VC (1981) Crab gill intra-epithelial carbonic anhydrase plays a major role in hemolymph CO_2 and chloride ion regulation. J Exp Biol 92: 243–254

Burnett LE, Dunn TN, Infantino RL (1985) The function of carbonic anhydrase in crustacean gills. In: Gilles R, Gilles-Baillien M (eds) Transport processes, iono- and osmoregulation. Springer, Berlin Heidelberg New York, pp 159–168

Cameron JN (1978a) NaCl balance in blue crabs, *Callinectes sapidus*, in fresh water. J Comp Physiol 123:127–135

Cameron JN (1978b) Effects of hypercapnia on blood acid-base status, NaCl fluxes and trans-gill potential in freshwater blue crabs, *Callinectes sapidus*. J Comp Physiol 123:137–141

Cameron JN (1979) Effects of inhibitors on ion fluxes, trans-gill potential and pH regulation in freshwater blue crabs, *Callinectes sapidus* (Rathbun). J Comp Physiol 133:219–225

Cioffi M (1984) Comparative ultrastructure of arthropod transporting epithelia. Am Zool 24:139–156

Compère Ph, Péqueux, A, Goffinet G, Gilles R (1985) Ultrastructure of the branchial epithelium of the green crab *Carcinus maenas* as a function of the salinity of the environment. Abstr 3rd Colloq SBCF, Toulouse, Sept 1985. Biology of the Cell 54:7a
Conte FP (1980) Role of C-4 pathway in crustacean chloride cell function. Am J Physiol 238:R269 – R276
Conte FP, Droukas P, Ewing R (1977) Development of sodium regulation and de novo synthesis of Na + K-activated ATPase in larval brine shrimp, *Artemia salina*. J Exp Zool 202:339 – 361
Conte FP, Lowy J, Carpenter J, Edwards A, Smith R, Ewing RD (1980) Aerobic and anaerobic metabolism of nauplii *Artemia salina* as a function of salinity. In: Persoone G, Sorgeloos P, Roels O, Jaspers R (eds) The brine shrimp *Artemia*, vol 2. Universa Press, Wetteren, pp 125 – 136
Copeland DE (1948) The cytological basis of chloride transfer in the gills of *Fundulus heteroclitus*. J Morphol 82:201 – 227
Copeland DE (1967) A study of salt secreting cells in the brine shrimp (*Artemia salina*). Protoplasma 63:363 – 384
Copeland DE (1968) Fine structure of salt and water uptake in the land crab, *Gecarcinus lateralis*. Am Zool 8:417 – 432
Copeland DE, Fitzjarrell AT (1968) The salt absorbing cells in the gills of the blue crab (*Callinectes sapidus* Rathbun) with notes on modified mitochondria. Z Zellforsch 92:1 – 22
Croghan PC (1958) The mechanism of osmotic regulation in *Artemia salina* (L): the physiology of the branchiae. J Exp Biol 35:234 – 242
Croghan PC, Curra RA, Lockwood APM (1965) The electrical potential difference across the epithelium of isolated gills of the crayfich *Austropotamobius pallipes* (Lereboullet). J Exp Biol 42:463 – 474
Curra RA (1965) Notes on the morphology of the branchiae of the crayfish *Austropotamobius pallipes* (Lereboullet). Zool Anz 174:313 – 323
Davis MS, Shuttleworth TJ (1985 a) Pancreatic peptides regulate Cl^- secretion in the marine teleost gill. Peptides 6:379 – 382
Davis MS, Shuttleworth TJ (1985b) Peptidergic and adrenergic regulation of electrogenic ion transport in isolated gills of the flounder *Platichthys flesus* L. J Comp Physiol B155:471 – 478
Dawson DC (1977) Na and Cl transport across the isolated turtle colon: parallel pathways for transmural ion movement. J Membr Biol 37:213 – 233
Degnan KJ (1984) Sodium and chloride dependence of chloride secretion by the opercular epithelium. J Exp Zool 231:11 – 17
Degnan KJ (1985) The role of K^+ and Cl^- conductances in chloride secretion by opercular epithelium. J Exp Zool 236:119 – 125
Degnan KJ, Zadunaisky JA (1979) Open-circuit sodium and chloride fluxes across isolated opercular epithelia from the teleost *Fundulus heteroclitus*. J Physiol (London) 294:483 – 495
Degnan KJ, Zadunaisky JA (1980) Passive sodium movements across the opercular epithelium: the paracellular shunt pathway and ionic conductance. J Membr Biol 55:175 – 185
Degnan KJ, Karnaky KJ Jr, Zadunaisky JA (1977) Active chloride transport in the in vitro opercular skin of a teleost *Fundulus heteroclitus*, a gill-like epithelium rich in chloride cells. J Physiol (London) 271:155 – 191
Dehnel PA (1974) Gill tissue respiration in the crab *Eriocheir sinensis*. Can J Zool 52:923 – 937
Dehnel PA, McCaughran DA (1964) Gill tissue respiration in two species of estuarine crabs. Comp Biochem Physiol 44B:869 – 879
DePew EF, Towle DW (1979) Bicarbonate-stimulated ATPase in plasma membrane fractions of fiddler crab (*Uca minax*) gill. Mar Biol Lett 1:59 – 67
Dunel-Erb S, Massabuau JCh, Laurent P (1982) Organisation fonctionnelle de la branchie d'écrevisse. C R Soc Biol 176(3):248 – 258
Ehrenfeld J (1974) Aspects of ionic transport mechanisms in crayfish *Astacus leptodactylus*. J Exp Biol 61:57 – 70
Engel DW, Eggert LD (1974) The effect of salinity and sex on the respiration rates of excised gills of the blue crab *Callinectes sapidus*. Comp Biochem Physiol 47A:1005 – 1011
Engel DW, Ferguson RL, Egger LD (1975) Respiration rates and ATP concentrations in the excised gills of the blue crab as a function of salinity. Comp Biochem Physiol 52A:669 – 673
Epstein FH, Maetz J, Renzis G de (1973) Active transport of chloride by teleost gill: inhibition by thiocyanate. Am J Physiol 224:1295 – 1299

Epstein FH, Silva P, Forrest JN, Solomon RJ (1975) Chloride transport and its inhibition by thiocyanate in gills of seawater teleosts. Comp Biochem Physiol 52A:681–683

Eriksson O, Wistrand PJ (1986) Chloride transport inhibition by various types of loop diuretics in fish opercular epithelium. Acta Physiol Scand 126:93–101

Eriksson O, Mayer-Gostan N, Wistrand PJ (1985) The use of isolated fish opercular epithelium as a model tissue for studying intrinsic activities of loop diuretics. Acta Physiol Scand 125:55–66

Ernst SA, Mills JW (1980) Autoradiographic localization of tritiated ouabain-sensitive sodium pump sites in ion transporting epithelia. J Histochem Cytochem 28:72–77

Ernst SA, Riddle CV, Karnaky KJ Jr (1980) Relationship between localization of Na^+-K^+ ATPase cellular fine structure, and reabsorptive and secretory electrolyte transport. In: Bronner F, Kleinzeller A (eds) Current topics in membranes and transport, vol 13. Academic Press, London New York, pp 355–385

Evans DH (1969) Sodium, chloride and water balance of the intertidal teleost, *Pholis gunnellus*. J Exp Biol 50:179–190

Evans DH (1977) Further evidence for Na^+/NH_4^+ exchange in marine teleost fish. J Exp Biol 70:213–220

Evans DH (1980) Salt transport mechanisms in branchial epithelia. In: Gilles R (ed) Animals and environmental fitness, vol 1. Pergamon Press, Oxford, pp 61–78

Evans DH, Cameron JN (1986) Gill ammonia transport. J Exp Zool 239:17–23

Evans DH, Mallery CH (1975) Time course of sea water acclimation by the euryhaline teleost, *Dormitator maculatus:* correlation between potassium stimulation of sodium efflux and Na/K activated ATPase activity. J Comp Physiol 96:117–122

Evans DH, Carrier JC, Bogan MB (1974) The effect of external potassium ions on the electrical potential measured across the gills of the teleost *Domitator maculatus.* J Exp Biol 61:277–283

Ferguson JKW, Lewis L, Smith J (1937) The distribution of carbonic anhydrase in certain marine invertebrates. J Cell Comp Physiol 10:395–400

Finol HJ, Croghan PC (1983) Ultrastructure of the branchial epithelium of an amphibious brackish-water crab. Tissue Cell 15:63–75

Fisher JM (1972) Fine structural observations on the gill filaments of the fresh-water crayfish *Astacus pallipes* (Lereboullet) Tissue Cell 4:287–299

Foskett JK, Hubbard GM (1981) Hormonal control of chloride secretion by teleost opercular membrane. Ann NY Acad Sci 372:643

Foskett JK, Machen TE (1985) Vibrating probe analysis of teleost opercular epithelium: correlation between active transport and leak pathways of individual chloride cells. J Membr Biol 85:23–35

Foskett JK, Scheffey C (1982) The chloride cell: definitive identification as the salt-secretory cell in teleosts. Science 215:164–166

Foskett JK, Logsdon D, Turner T, Machen TE, Bern HA (1981) Differentiation of the chloride extrusion mechanism during seawater adaptation of a teleost fish, the cichlid *Sarotherodon mossambicus.* J Exp Biol 93:209–224

Foskett JK, Hubbard GM, Machen TE, Bern HA (1982a) Effects of epinephrine, glucagon and vasoactive intestinal polypeptide on chloride secretion by teleost opercular membrane. J Comp Physiol 146:27–34

Foskett JK, Machen TE, Bern HA (1982b) Chloride secretion and conductance of teleost opercular membrane: effects of prolactin. Am J Physiol 242:R380–R389

Foskett JK, Bern HA, Machen TE, Conner M (1983) Chloride cells and the hormonal control of teleost fish osmoregulation. J Exp Biol 106:255–281

Foster CA, Howse HD (1978) A morphological study on gills of the brown shrimp, *Penaeus aztecus.* Tissue Cell 10:77–92

Friedman MH (1986) Principles and models of biological transport. Springer, Berlin Heidelberg New York Tokyo, 260 pp

Frizzell RA, Schultz SG (1972) Ionic conductances of extracellular shunt pathway in rabbit ileum. J Gen Physiol 59:318–346

Frizzell RA, Smith PL, Field M (1981) Sodium chloride absorption by flounder intestine: a model for the renal thick ascending limb. In: Dinno MA, Callahan AB (eds) Membrane biophysics: structure and function in epithelia. AR Liss, New York

Gilles R (1980a) Cell volume control as studied in crustacean tissues. In: Gilles R (ed) Animals and environmental fitness, vol 2. Pergamon Press, Oxford New York, pp 3–24

Gilles R (1980b) Mechanisms of volume regulation and control of the level of intracellular osmotic effectors. In: Engell HC (ed) Metabolic response to surgery and injury. Am J Part Nutr
Gilles R, Péqueux AJR (1981) Cell volume regulation in crustaceans: relationship between mechanism for controlling the osmolality of extracellular and intracellular fluids. J Exp Zool 215:351 – 362
Gilles R, Péqueux AJR (1983) Interactions of chemical and osmotic regulation with the environment. In: Vernberg FJ, Vernberg WB (eds) The Biology of Crustacea (Dorothy Bliss, Editor-in-chief), vol. 8. Academic Press, London New York, pp 109 – 177
Gilles R, Péqueux AJR (1985) Ion transport in crustacean gills: Physiological and ultrastructural approaches. In: Gilles R, Gilles-Baillien M (eds) Transport processes, iono- and osmoregulation. Springer Verlag, Berlin Heidelberg, p 136 – 158
Gilles R, Péqueux AJR (1986) Physiological and ultrastructural studies of NaCl transport in crustaceans gills. Boll Zool 53:173 – 182
Girard JP, Payan P (1976) Effect of epinephrine on vascular space of gills and head of rainbow trout. Am J Physiol 230:1555 – 1560
Gocha N, Péqueux A, Wanson S, Gilles R (1987) Cl^- fluxes across isolated, perfused gills of the chinese crab *Eriocheir sinensis* (M. Edw) acclimated to fresh water. Comp Biochem Physiol 88A:581 – 584
Greenaway P (1972) Calcium regulation in the freshwater crayfish *Austropotamobius pallipes* (Lereboullet). I. Calcium balance in the intermoult animal. J Exp Biol 57:471 – 487
Greenaway P (1976) The regulation of haemolymph calcium concentration of the crab *Carcinus maenas* (L). J Exp Biol 64:149 – 157
Greenwald L, Kirschner LB, Sanders M (1974) Sodium efflux and potential differences across the irrigated gill of sea water adapted rainbow trout (*Salmo gairdneri*). J Gen Physiol 64:135 – 147
Guggino WB (1980) Salt balance in embryos of *Fundulus heteroclitus* and *F. bermudae* adapted to seawater. Am J Physiol 238:R42 – R49
Guibbolini M, Lahlou B (1987) Basal and stimulated adenylate cyclase activity in the gill epithelium of the rainbow trout. Gen Comp Endocrinol 67:178 – 188
Heisler N (1982) Transepithelial ion transfer processes as mechanisms for fish acid-base regulation in hypercapnia and lactacidosis. Can J Zool 60:1108 – 1122
Henry RP (1984) The role of carbonic anhydrase in blood ion and acid-base regulation. Am Zool 24:241 – 251
Henry RP, Cameron JN (1982a) Acid-base balance in the euryhaline blue crab, *Callinectes sapidus*, during acclimation from high to low salinity. J Exp Biol 101:255 – 264
Henry RR, Cameron JN (1982b) The distribution and partial characterization of carbonic anhydrase in selected aquatic and terrestrial decapod crustaceans. J Exp Zool 221:309 – 321
Henry RP, Cameron JN (1983) The role of carbonic anhydrase in respiration, ion regulation and acid-base balance in the aquatic crab *Callinectes sapidus* and the terrestrial crab *Gecarcinus lateralis*. J Exp Biol 103:205 – 223
Hermann L (1882) Neue Untersuchungen über Hautströme. Pfluegers Arch 27:280 – 304
Hoar WS, Randall DJ (eds) (1984) Fish physiology, vol 10. Gills, part B. Ion and water transfer. Academic Press, London New York
Holliday Ch (1985) Salinity-induced changes in gill Na, K-ATPase activity in the mud fiddler crab, *Uca pugnax*. J Exp Zool 233:199 – 208
Horiuchi S (1977) Characterization of gill sodium-potassium ATPase in the freshwater crayfish *Procambarus clarkhii*. Comp Biochem Physiol 56:135 – 138
Hootman SR, Ernst SA (1984) Ultrastructural localization of ($Na^+ + K^+$) ATPase in specialized membranes of salt transporting cells in marine vertebrates. In: Péqueux A, Gilles R, Bolis L (eds) Osmoregulation in estuarine and marine animals. Lecture notes on coastal and estuarine studies, vol 9. Springer, Berlin Heidelberg New York, pp 171 – 189
Hootman SR, Philpott CW (1979) Ultracytochemical localization of Na^+, K^+-activated ATPase in chloride cells from the gills of a euryhaline teleost. Anat Rec 193:99 – 130
Hootman SR, Philpott CW (1980) Accessory cells in teleost branchial epithelium. Am J Physiol 238:R199 – R206
Horton SL (1984) The effect of salinity acclimation on sodium ion regulation in the euryhaline amphipod *Crangonyx pseudogracilis* (Bousfield). Abstr 1st Congr Comp Physiol Biochem (Sect IUBS), Liège (Belgium), Aug 27 – 31, 1984, pp B146 – B147

Horton SL (1985) Aspects of sodium ion and water regulation and gill ultrastructure in *Crangonyx pseudogracilis* (Bousfield) (Crustacea: Amphipoda). Ph D Thesis, Univ Leicester, p 294

Hossler FE, Musil G, Karnaky KJ Jr, Epstein FH (1985) Surface ultrastructure of the gill arch of the killifish, *Fundulus heteroclitus,* from seawater and freshwater, with special reference to the morphology of apical crypts of chloride cells. J Morphol 185:377–386

House CR (1963) Osmotic regulation in the brackish water teleost, *Blennius pholis.* J Exp Biol 40:87–104

House CR, Maetz J (1974) On the electrical gradient across the gill of the sea water-adapted eel. Comp Biochem Physiol 47A:917–924

Hughes GM (1982) An introduction to the study of gills. In: Houlihan DF, Rankin JC, Shuttleworth TJ (eds) Gills. Cambridge Univ Press, Cambridge, pp 1–24

Hughes GM (1984) General anatomy of the gills. In: Hoar WS, Randall DJ (ed) Fish physiology, vol 10A. Academic Press, London New York, pp 1–72

Jampol LM, Epstein FH (1970) Sodium-potassium-activated adenosine triphosphatase and osmotic regulation by fishes. Am J Physiol 218:607–611

Jozuka K (1966) Chloride-excreting cells and mucus-secreting cells in the gills of the Japanese common eel, *Anguilla japonica.* Annot Zool Jpn 39:202–210

Kamemoto FI (1976) Neuroendocrinology of osmoregulation in decapod crustacea. Am Zool 16:141–150

Kamemoto FI, Oyama SN (1984) Neuroendocrine influence on effector tissues of hydromineral balance in crustaceans. In: Lofts B (ed) 9th Int Symp Comp Endocrinol Proc. Hong Kong Univ Press

Kamiya M, Utida S (1968) Changes in activity of sodium-potassium activated adenosine triphosphatase in gills during adaptation of the Japanese eel to sea water. Comp Biochem Physiol 26: 675–685

Karnaky KJ Jr (1986) Structure and function of the chloride cell of *Fundulus heteroclitus* and other teleosts. Am Zool 26:209–224

Karnaky KJ Jr, Kinter LB, Kinter WB, Stirling CE (1976) Teleost chloride cell. II. Autoradiographic localization of gill Na, K-ATPase in killifish *Fundulus heteroclitus* adapted to low and high salinity environments. J Cell Biol 70:157–177

Karnaky KJ Jr, Degnan KJ, Zadunaisky JA (1977) Chloride transport across isolated opercular epithelium of killifish: a membrane rich in chloride cells. Science 195:203–205

Kerstetter TH, Keeler M (1976) On the interaction of NH_4^+ and Na^+ fluxes in the isolated trout gill. J Exp Biol 64:517–525

Kerstetter TH, Kirschner LB (1972) Active chloride transport by the gills of rainbow trout *Salmo gairdneri.* J Exp Biol 56:263–272

Kerstetter TH, Kirschner LB (1974) HCO_3^- dependent ATPase activity in the gills of rainbow trout *Salmo gairdneri.* Comp Biochem Physiol 48B:581–589

Kerstetter TH, Kirschner LB, Rafuse DD (1970) On the mechanisms of sodium ion transport by the irrigated gills of rainbow trout *Salmo gairdneri.* J Gen Physiol 56:342–359

Keys A (1931) The heart-gill preparation of the eel and its perfusion for the study of a natural membrane in situ. Z Vergl Physiol 15:352–363

Keys A, Willmer EN (1932) Chloride secreting cells in the gills of fishes, with special reference to the common eel. J Physiol (London) 76:368–378

Kikuchi S (1977) Mitochondria-rich (chloride) cells in the gill epithelia from four species of stenohaline fresh water teleosts. Cell Tissue Res 180:87–98

Kikuchi S (1982) A unique cell membrane with a lining of repeating subunits on the cytoplasmic side of presumably ion-transporting cells in the gill epithelium of *Daphnia magna* (Crustacea: cladocera). J Submicrosc Cytol 14:711–715

Kikuchi S (1983) The fine structure of the gill epithelium of a fresh-water flea, *Daphnia magna* (Crustacea: Phyllopoda) and changes associated with acclimation to various salinities. Cell Tissue Res 229:253–268

King EN (1965) The oxygen consumption of intact crab and excised gills as a function of decreased salinity. Comp Biochem Physiol 15:93–102

King EN (1966) Oxidative activity of crab gill mitochondria as a function of osmotic concentration. Comp Biochem Physiol 17:245–258

King EN, Schoffeniels E (1969) In vitro preparation of crab gill for use in ion transport studies. Arch Int Physiol Biochim 77:105–111

Kirschner LB (1979) Control mechanisms in crustaceans and fishes. In: Gilles R (ed) Mechanisms of osmoregulation in animals. Wiley Interscience, Chichester, pp 157–222
Kirschner LB (1983) Sodium chloride absorption across the body surface: frog skins and other epithelia. Am J Physiol 244:R429–R443
Kirschner LB, Greenwald L, Kerstetter TH (1973) Effect of amiloride on sodium transport across body surfaces of freshwater animals. Am J Physiol 224:832–837
Kirschner LB, Greenwald L, Sanders M (1974) On the mechanism of sodium extrusion across the irrigated gill of sea water-adapted rainbow trout *Salmo gairdneri*. J Gen Physiol 64: 148–165
Klyce SD, Wong RKS (1977) Site and mode of adrenaline action on chloride transport across the rabbit corneal epithelium. J Physiol (London) 266:777–799
Koch HJ (1934) Essai d'interprétation de la soi-disant réduction vitale des sels d'argent par certains organes d'Arthropodes. Ann Soc Sci Med Nat Brux Ser B 54:346–361
Koch HJ (1954) Cholinesterase and active transport of sodium chloride through the isolated gills of the crab *Eriocheir sinensis* (M. Edw). In: Kitching JA (ed) Recent developments in cell physiology. Butterworths, London, pp 15–27
Koch HJ, Evans J, Schicks E (1954) The active absorption of ions by the isolated gills of the crab *Eriocheir sinensis* (M. Edw). Meded K Vlaam Acad Wet Kl Wet 16:1–16
Kormanik GA, Cameron JN (1981) Ammonia excretion in the seawater blue crab (*Callinectes sapidus*) occurs by diffusion, and not Na^+/NH_4^+ exchange. J Comp Physiol 141:457–462
Krasny JE Jr (1981) Ion transport properties of the opercular epithelium of *Fundulus grandis*. Ph D Diss, Univ Miami USA, 164 pp
Krogh A (1938) The active absorption of ions in some fresh water animals. Z Vergl Physiol 25:335–350
Krogh A (1939) Osmotic regulation in animals. Cambridge Univ Press, Cambridge
Langhorne P, Simpson TH (1986) The interrelationship of cortisol gill (Na + K) ATPase, and homeostasis during parr-smolt transformation of Atlantic salmon *Salmo salar* L. Gen Comp Endocrinol 61:203–213
Laurent P (1984) Gill internal morphology. In: Hoar WS, Randall DJ (ed) Fish physiology, vol 10A. Academic Press, London New York, pp 73–183
Lee SH (1982) Salinity adaptation of HCO_3^- dependent ATPase activity in the gills of the blue crab (*Callinectes sapidus*). Biochim Biophys Acta 689:143–154
Leray C (1984) Cellular energy metabolism and its regulation in gills of fish and crustacea. In: Péqueux A, Gilles R, Bolis L (eds) Osmoregulation in estuarine and marine animals. Lecture notes on coastal and estuarine studies vol 9. Springer, Berlin Heidelberg New York, pp 145–156
Lignon JM (1987) Structure and permeability of decapod crustacea cuticle. In: Kirsch R, Lahlou B (eds) Comparative physiology of environmental adaptations, vol 1. Part I: Adaptations to salinity and dehydration. Karger, Basel, pp 178–187
Lockwood APM (1962) The osmoregulation of crustacea. Biol Rev 37:257–305
Lockwood AMP (1977) Transport and osmoregulation in crustacea. In: Gupta BL, Oschman JL, Moreton RB, Wall BJ (eds) Transport of ions and water in animals. Academic, London New York, pp 673–707
Lockwood AMP, Inman CBE (1973) Water uptake and loss in relation to the salinity of the medium in the amphipod crustacean *Gammarus duebeni*. J Exp Biol 58:149–163
Loretz CA, Bern HA (1983) Control of ion transport by *Gillichthys mirabilis* urinary bladder. Am J Physiol 245:R45–R52
Lowy RJ, Conte FP (1985) Isolation and functional characterization of crustacean larval salt gland. Am J Physiol 248:R702–R708
Lucu C, Siebers D (1986) Amiloride-sensitive sodium flux and potentials in perfused *Carcinus* gill preparations. J Exp Biol 122:25–35
Maetz J (1971) Fish gills: mechanisms of salt transfer in fresh water and sea water. Philos Trans R Soc London Ser B 262:209–249
Maetz J (1973) Na^+/NH_4^+ Na^+/H^+ exchange and NH_3 movement across the gill of *Carassius auratus*. J Exp Biol 58:255–275
Maetz J (1974) Aspects of adaptation to hypo-osmotic and hyper-osmotic environments. In: Malins DC, Sargent JR (eds) Biochemical and biophysical perspectives in marine biology. Academic Press, London New York, pp 1–167

Maetz J, Garcia-Romeu F (1964) The mechanisms of sodium and chloride uptake by the gills of a fresh water fish, *Carassius auratus*. II. Evidence for NH_4^+/Na^+ and HCO_3^-/Cl^- exchanges. J Gen Physiol 47:1209–1227

Mandel LJ, Balaban RS (1981) Stoichiometry and coupling of active transport to oxidative metabolism in epithelial tissues. Am J Physiol 240:F357–F371

Mangum CP (1982) The functions of gills in several groups of invertebrate animals. In: Houlihan DF, Rankin JC, Shuttleworth TJ (eds) Gills. Cambridge Univ Press, Cambridge, pp 77–97

Mantel LH (1967) Asymmetry potentials, metabolism and sodium fluxes in gills of the blue crab, *Callinectes sapidus*. Comp Biochem Physiol 20:743–753

Mantel LH (1985) Neurohormonal integration of osmotic and ionic regulation. Am Zool 25:253–263

Mantel LH, Farmer LL (1983) Osmotic and ionic regulation. In: Mantel LH (ed) The biology of crustacea (Dorothy Bliss, editor-in-chief), vol 5. Academic Press, London New York, pp 53–161

Mantel LH, Landesman J (1977) Osmotic regulation and Na-K-activated ATPase in the green crab, *Carcinus maenas* and the spider crab *Libinia emarginata*. Biol Bull (Woods Hole Mass) 153:437–438

Mantel LH, Olson JR (1976) Studies on the $Na^+ + K^+$-activated ATPase of crab gills. Am Zool 16:269 (Abstr, p 223)

Maren TH (1967a) Carbonic anhydrase: chemistry, physiology and inhibition. Physiol Rev 47(4):595–781

Maren TH (1967b) Carbonic anhydrase in the animal kingdom. Fed Proc 26(4):1097–1103

Maren TH (1977) Use of inhibitors in physiological studies of carbonic anhydrase. Am J Physiol 232:F291–F297

Marshall WS (1977) Transepithelial potential and short-circuit current across the isolated skin of *Gillichthys mirabilis* (Teleostei: Gobiidae), acclimated to 5% and 100% seawater. J Comp Physiol 114:157–165

Marshall WS (1981a) Sodium dependency of active chloride transport across isolated fish skin (*Gillichthys mirabilis*). J Physiol (London) 319:165–178

Marshall WS (1981b) Active transport of Rb^+ across skin of the teleost *Gillichthys mirabilis*. Am J Physiol 241:F482–F486

Marshall WS (1985) Paracellular ion transport in trout opercular epithelium models osmoregulatory effects of acid precipitation. Can J Zool 63:1816–1822

Marshall WS, Bern HA (1979) Teleostean urophysis: urotensin II and ion transport across the isolated skin of a marine teleost. Science 204:519–521

Marshall WS, Bern HA (1980) Ion transport across the isolated skin of the teleost, *Gillichthys mirabilis*. In: Lahlou B (ed) Epithelial transport in the lower vertebrates. Cambridge Univ Press, Cambridge, pp 337–350

Marshall WS, Bern HA (1981) Active chloride transport by the skin of a marine teleost is stimulated by urotensin I and inhibited by urotensin II. Gen Comp Endocrinol 43:484–491

Marshall WS, Klyce SD (1983) Cellular and paracellular pathway resistances in the "tight" Cl^--secreting epithelium of the rabbit cornea. J Membr Biol 73:275–282

Marshall WS, Nishioka R (1980) Relation of mitochondria-rich chloride cells to active chloride transport in the skin of a marine teleost. J Exp Zool 214:147–156

Martelo MJ, Zanders IP (1986) Modifications of gill ultrastructure and ionic composition in the crab *Goniopsis cruentata* acclimated to various salinities. Comp Biochem Physiol 84A(2): 383–389

May SA, Degnan KJ (1985) Converging adrenergic and cholinergic mechanisms in the inhibition of Cl secretion in fish opercular epithelium. J Comp Physiol B 156:183–189

Mazeaud MM, Mazeaud F, Donaldson EM (1977) Primary and secondary effects of stress in fish: some new data with a general review. Trans Am Fish Soc 106:201–212

McCartney TH (1976) Sodium-potassium dependent adenosine triphosphatase activity in gills and kidneys of Atlantic salmon (*Salmo salar*). Comp Biochem Physiol 53A:351–353

Meldrum NV, Roughton FJW (1932) The CO_2 catalyst present in blood. J Physiol (London) 75:15

Milne DJ, Ellis RA (1973) The effect of salinity acclimation on the ultrastructure of the gills of *Gammarus oceanicus* (Segerstråle, 1947) (Crustacea: Amphipoda). Z Zellforsch 139:311–318

Morse HC, Harris PJ, Dornfeld EJ (1970) *Pacifastacus leniusculus:* fine structure of Arthrobranch with reference to active ion uptake. Trans Am Microsc Soc 89(1):12–27

Motais R, Garcia-Romeu F, Maetz J (1966) Exchange diffusion effect and euryhalinity in teleosts. J Gen Physiol 50:391 – 422
Nekvasill NP, Olson KR (1986) Extraction and metabolism of circulating catecholamines by the trout gill. Am J Physiol 250:R526 – R531
Neufeld GJ, Holliday ChW, Pritchard JB (1980) Salinity adaptation of gill Na, K-ATPase in the blue crab *Callinectes sapidus*. J Exp Zool 211:215 – 224
Nichols DJ, Weisbart M (1985) Cortisol dynamics during seawater adaptation of Atlantic salmon *Salmo salar*. Am J Physiol 248:R651 – R659
Nonnotte G, Nonnotte L, Kirsch R (1979) Chloride cells and chloride exchange in the skin of a seawater teleost, the shanny (*Blennius pholis* L.). Cell Tissue Res 199:387 – 396
Olson KR (1984) Distribution of flow and plasma skimming in isolated perfused gills of three teleosts. J Exp Biol 109:97 – 108
Omeljaniuk RJ, Eales JG (1986) Effect of 3,5,3'-triiodo-L-thyronine on gill Na^+/K^+-ATPase of rainbow trout *Salmo gairdneri*, in fresh water. Comp Biochem Physiol 84A:427 – 429
Parry G, Potts WTW (1965) Sodium balance in the fresh water prawn *Palaemonetes antennarius*. J Exp Biol 42:415 – 421
Payan P (1978) A study of Na^+/NH_4^+ exchange across the gill of the perfused head of the trout (*Salmo gairdneri*). J Comp Physiol 124:181 – 188
Payan P, Matty AJ (1975) The characteristics of ammonia excretion by a perfused isolated head of trout (*Salmo gairdneri*): effect of temperature and CO_2-free ringer. J Comp Physiol 96:167 – 184
Payan P, Matty AJ, Maetz J (1975) A study of sodium pump in the perfused head preparation of the trout *Salmo gairdneri* in freshwater. J Comp Physiol 104:33 – 48
Payan P, Girard JP, Mayer-Gostan N (1984) Branchial ion movements in teleosts: the roles of respiratory and chloride cells. In: Hoar WS, Randall DJ (eds) Fish physiology, vol. 10B. Academic Press, London New York, pp 39 – 63
Pearson D, Shiveley JE, Clarke BC, Geschwind II, Barkley M, Nishioka RS, Bern HA (1980) Urotensin II: a somatostatin-like peptide in the caudal neurosecretory system of fishes. Proc Natl Acad Sci USA 77:5021 – 5024
Péqueux A, Barra JA (1980) Fine structure of the gill epithelium of the euryhaline crab *Eriocheir sinensis* as a function of the salinity of the acclimation medium. Eur J Cell Biol 22(1):470
Péqueux A, Chapelle S (1982) $(Na^+ + K^+)$ ATPase activity and phospholipids in two euryhaline crabs as related to changes in the environmental salinity. Mar Biol Lett 3:43 – 52
Péqueux A, Gilles R (1977) Osmoregulation of the Chinese crab *Eriocheir sinensis* as related to the activity of the $(Na^+ + K^+)$ ATPase. Arch Int Physiol Biochim 85:426 – 427
Péqueux A, Gilles R (1978a) Na^+/NH_4^+ co-transport in isolated perfused gills of the Chinese crab *Eriocheir sinensis* acclimated to fresh water. Experientia 34:1593 – 1594
Péqueux A, Gilles R (1978b) Osmoregulation of the euryhaline Chinese crab *Eriocheir sinensis*. Ionic transports across isolated perfused gills as related to the salinity of the environment. In: McLusky DS, Berry AJ (eds) Physiology and behaviour of marine organisms. Pergamon Press, Oxford, pp 105 – 111
Péqueux A, Gilles R (1981) Na^+ fluxes across isolated perfused gills of the Chinese crab *Eriocheir sinensis*. J Exp Biol 92:173 – 186
Péqueux A, Gilles R (1984) Control of extracellular fluid osmolality in crustaceans. In Péqueux A, Gilles R, Bolis L (eds) Osmoregulation in estuarine and marine animals. Springer, Berlin Heidelberg New York, pp 18 – 34
Péqueux A, Gilles R (1988) The transepithelial potential difference of isolated perfused gills of the Chinese crab *Eriocheir sinensis* acclimated to fresh water. Comp Biochem Physiol 89A:163 – 172
Péqueux A, Barra JA, Nonnotte L, Kirsch R, Gilles R (1980) Fine structure of salt-absorbing epithelium in gills of the crab *Eriocheir sinensis* acclimated to fresh water. Comp Biochem Physiol Suppl 2. In: Gilles R (ed) Animals and environmental fitness, Abstr vol. Pergamon Press, Oxford, pp 13 – 14
Péqueux A, Chapelle S, Wanson S, Goffinet G, François C (1983) $(Na^+ + K^+)$ ATPase activity and phospholipid content of various fractions of the posterior gills of *Carcinus maenas* and *Eriocheir sinensis*. Mar Biol Lett 4:267 – 279
Péqueux A, Barra JA, Goessens G, Wanson S, Gilles R (1984a) Fine structure of gill epithelium in three crabs, as related to their ionoregulation performances. Abstr 1st Congr Comp Physiol Biochem (Sect IUBS), Liège (Belgium), Aug 27 – 31, 1984, p B-135

Péqueux A, Marchal A, Wanson S, Gilles R (1984b) Kinetic charactistics and specific activity of gill ($Na^+ + K^+$) ATPase in the euryhaline Chinese crab *Eriocheir sinensis* during salinity acclimation. Mar Biol Lett 5:35–45

Perry SF, Haswell MS, Randall DJ, Farrell AP (1981) Branchial ionic uptake and acid-base regulation in the rainbow trout, *Salmo gairdneri*. J Exp Biol 92:289–303

Perry SF, Davie PS, Daxboeck C, Ellis AG, Smith DG (1984) Perfusion methods for the study of gill physiology. In: Hoar WS, Randall DJ (eds) Fish physiology, vol. 10B. Academic Press, London New York, pp 325–388

Pfeiler E (1978) Na-ATPase and NaK-ATPase activities in gills of rainbow trout (*Salmo gairdneri*). J Comp Physiol 124:97–104

Pfeiler E, Kirschner LB (1972) Studies on gill ATPase of rainbow trout (*Salmo gairdneri*). Biochim Biophys Acta 282:301–310

Philpott CW, Copeland DE (1963) Fine structure of chloride cells from three species of *Fundulus*. J Cell Biol 18:389–404

Pic P, Mayer-Gostan N, Maetz J (1975) Branchial effects of epinephrine in the seawater-adapted mullet. II. Na and Cl extrusion. Am J Physiol 228:441–447

Poat PC, Thabrew MI, Munday KA (1971) Cation transport and metabolism as a function of salinity in the excised gill of *Carcinus maenas*. Comp Biochem Physiol 39A:699–707

Potts WTW (1984) Transepithelial potentials in fish gills. In: Hoar WS, Randall DJ (eds) Fish physiology, vol. 10B. Academic Press, London New York, pp 105–128

Potts WTW, Eddy F (1973) Gill potentials and sodium fluxes in the flounder *Platichthys flesus*. J Comp Physiol 87:29–48

Potts WTW, Parry G (1964a) Sodium and chloride balance in the prawn *Palaemonetes varians*. J Exp Biol 41:591–601

Potts WTW, Parry G (1964b) Respiration and electrolyte regulation. In: Osmotic and ionic regulation in animals. Macmillan, New York, pp 330–341

Pressley TA, Graves JS, Krall AR (1981) Amiloride-sensitive ammonium and sodium ion transport in the blue crab. Am J Physiol 241 (10):R370–R378

Randall DJ, Wood CM (1981) Carbon dioxide excretion in the land crab *Cardisoma carnifex*. J Exp Zool 218:37–44

Reid EW (1893) The electromotive properties of the skin of the common eel. Philos Trans R Soc London Ser B184:335–365

Renzis G de (1975) The branchial chloride pump in the goldfish *Carassius auratus*: relationship between Cl^-/HCO_3^- and Cl^-/Cl^- exchanges and the effect of thiocyanate. J Exp Biol 63: 587–602

Renzis G de, Bornancin M (1977) A Cl^-/HCO_3^--ATPase in the gills of *Carassius auratus*, its inhibition by thiocyanate. Biochim Biophys Acta 467:192–207

Robinson GD, Potts WTW (1979) Ion fluxes and diffusion potentials in the Dungeness crab *Cancer magister*. J Comp Physiol 131:285–292

Sardet C, Pisam M, Maetz J (1979) The surface epithelium of teleostean fish gills: cellular and junctional adaptations of the chloride cell in relation to salt adaptation. J Cell Biol 80: 96–117

Sargent J, Thomson A (1974) The nature and properties of the inducible sodium-plus-potassium ion-dependent adenosine triphosphatase in the gills of eels (*Anguilla anguilla*) adapted to fresh water and sea water. Biochem J 144:69–75

Scheffey C, Foskett JK, Machen TE (1983) Localization of ionic pathways in the teleost opercular membrane by extracellular recording with a vibrating probe. J Membr Biol 75:193–203

Schlieper C (1929) Über die Einwirkung niederer Salzkonzentrationen auf marine Organismen. Z Vergl Physiol 9:478–514

Schultz SG, Frizzell RA (1976) Ionic permeability of epithelial tissues. Biochim Biophys Acta 443:181–189

Shaw J (1959a) The absorption of sodium ions by the crayfish *Astacus pallipes* Lereboullet. I. The effect of external and internal sodium concentrations. J Exp Biol 36:136–144

Shaw J (1959b) Salt and water balance in the East African freshwater crab, *Potamon niloticus* (M. Edw). J Exp Biol 36:157–176

Shaw J (1960a) The absorption of ions by the crayfish. II. The effect of external anion. J Exp Biol 37:534–547

Shaw J (1960b) The absorption of sodium ions by the crayfish *Astacus pallipes* Lereboullet. III. The effect of other cations in the external solution. J Exp Biol 37:548–556

Shaw J (1960c) The absorption of chloride ions by the crayfish *Astacus pallipes* Lereboullet. J Exp Biol 37:557–572

Shaw J (1961) Studies on ionic regulation in *Carcinus maenas*. J Exp Biol 38:135–152

Shaw J, Sutcliffe DW (1961) Studies on sodium balance in *Gammarus duebeni* Lilljeborg and *G. pulex pulex* (L.). J Exp Biol 38:1–15

Shirai N, Utida S (1970) Development and degeneration of the chloride cell during seawater and freshwater adaptation of the Japanese eel, *Anguilla japonica*. Z Zellforsch 103:247–264

Shuttleworth TJ (1972) A new isolated perfused gill preparation for the study of the mechanisms of ionic regulation in teleosts. Comp Biochem Physiol 43A:59–64

Siebers D, Leweck K, Markus H, Winkler A (1982) Sodium regulation in the shore crab *Carcinus maenas* as related to ambient salinity. Mar Biol 69:37–43

Siebers D, Winkler A, Lucu C, Thedens G, Weichart D (1985) Na-K-ATPase generates an active transport potential in the gills of the hyperregulating shore crab *Carcinus maenas*. Mar Biol 87:185–192

Siebers D, Lucu C, Winkler A, Dalla Venezia L, Wille H (1986) Active uptake of sodium in the gills of the hyperregulating shore crab *Carcinus maenas*. Helgol Meeresunters 40:151–160

Silva P, Solomon R, Spokes K, Epstein F (1977) Ouabain inhibition of gill Na-K-ATPase: relationship to active chloride transport. J Exp Zool 199:419–426

Skou JC (1957) The influence of some cations on an adenosine triphosphatase from peripheral nerves. Biochim Biophys Acta 23:394–402

Smith DS, Linton JR (1971) Potentiometric evidence for the active transport of sodium and chloride across excised gills of *Callinectes sapidus*. Comp Biochem Physiol 39A:367–378

Smith HW (1930) The absorption and excretion of water and salts by marine teleosts. Am J Physiol 93:480–505

Smith PG (1969) The ionic regulations of *Artemia salina* (L.) I. Measurements of electrical potential difference and resistance. J Exp Biol 51:727–738

Soltoff SP, Mandel LJ (1983) Amiloride directly inhibits the Na, K-ATPase activity of rabbit kidney proximal tubules. Science 220:957–959

Spaargaren DH (1977) On the anion-cation mobility ratio in the body wall of the shore crab *Carcinus maenas*. Neth J Sea Res 11 (2):200–207

Spencer AM, Fielding AH, Kamemoto FI (1979) The relationship between gill Na, K-ATPase activity and osmoregulatory activity in various crabs. Physiol Zool 52:1–10

Stagg RM, Shuttleworth TJ (1982) The effects of copper on ionic regulation by the gills of the seawater-adapted flounder (*Platichthys flesus* L.). J Comp Physiol 149:83–90

Stiffler DF, Graham JB, Dickson KA, Stockmann W (1986) Cutaneous ion transport in the freshwater teleost *Synbranchus marmoratus*. Physiol Zool 59:406–418

Talbot P, Clark WH, Lawrence AL (1972) Light and electron microscopic studies on osmoregulatory tissue in the developing brown shrimp *Penaeus aztecus*. Tissue Cell 4:271–286

Taylor HH, Greenaway P (1979) The structure of the gills and lungs of the arid-zone crab *Holthuisana* (*Austrothelphusa*) *transversa* (Brachyura: Sundathelphusidae) including observations on arterial vessels within the gills. J Zool 189:359–384

Thompson R (1972) Mechanisms of osmoregulation in a euryhaline goby, *Gillichthys mirabilis*: the role of active and passive transport of sodium and chloride across the gills. Ph D Diss, Univ Calif, San Diego, 105 pp

Thomson AJ, Sargent JR (1977) Changes in the levels of chloride cells and (Na^+, K^+)-dependent ATPase in the gills of yellow and silver eels adapting to seawater. J Exp Zool 200:33–40

Towle DW (1981) Transport-related ATPases as probes of tissue function in three terrestrial crabs of Palau. J Exp Zool 218:89–95

Towle DW (1984a) Regulatory functions of ($Na^+ + K^+$) ATPase in marine and estuarine animals. In: Péqueux A, Gilles R, Bolis L (eds) Osmoregulation in estuarine and marine animals. Lecture notes on coastal and estuarine studies, vol 9. Springer, Berlin Heidelberg New York, pp 157–170

Towle DW (1984b) Membrane-bound ATPases in arthropod ion transporting tissues. Am Zool 24:177–185

Towle DW, Kays WT (1986) Basolateral localization of $Na^+ + K^+$-ATPase in gill epithelium of two osmoregulating crabs, *Callinectes sapidus* and *Carcinus maenas*. J Exp Biol 239:311–318

Towle DW, Palmer GE, Harris JL (1976) Role of gill $Na^+ - K^+$-dependent ATPase in acclimation of blue crabs (*Callinectes sapidus*) to low salinity. J Exp Zool 196:315–322

Van Amelsvoort JMM, Pont JJHHM de, Bonting SL (1977) Is there a plasma membrane-located anion-sensitive ATPase? Biochim Biophys Acta 466:283–301

Van Praag D, Farber SJ, Minkin E, Primor N (1987) Production of eicosanoids by the killifish gills and opercular epithelia and their effect on active transport of ions. Gen Comp Endocrinol 67:50–57

Vogel WOP (1978) Arteriovenous anastomoses in the afferent region of trout gill filaments (*Salmo gairdneri* Richardson, Teleostei). Zoomorphologie 90:205–212

Vogel W, Vogel V, Schlote W (1974) Ultrastructural study of arteriovenous anastomoses in gill filaments of *Tilapia mossambica*. Cell Tiss Res 155:491–512

Vogel W, Vogel V, Pfautsch M (1976) Arteriovenous anastomoses in rainbow trout gill filaments. A scanning and transmission electron microscopy study. Cell Tiss Res 167:373–385

Wanson S (1983) Purine nucleotides patterns and energy charge in *Carcinus maenas* and *Eriocheir sinensis* gills during osmoregulation. Abstr. 5th ESCPB Conf ESCPP, Liège (Belgium), pp 154–155

Wanson S, Péqueux A (1981) Na^+ transport by isolated perfused gills of the shore-crab *Carcinus maenas* acclimated to dilute and concentrated media. In: Exogenous and endogenous influences on metabolic and neural controls, Abstr 3rd Congr ESCPB, Pergamon Press, Oxford, pp 183–184

Wanson S, Péqueux A, Leray C (1983) Effect of salinity changes on adenylate energy charge in gills of two euryhaline crabs. Arch Int Physiol Biochim 91:81–82

Wanson S, Péqueux A, Gilles R (1984a) Kinetic study of gill $(Na^+ + K^+)$ ATPase in green crabs *Carcinus maenas* facing salinity changes. Abstr 1st Congr Comp Physiol Biochem (Sect IUBS), Liège (Belgium), Aug 27–31, 1984, pp B-136

Wanson S, Péqueux A, Roer R (1984b) Na^+ regulation and $(Na^+ + K^+)$ ATPase activity in the euryhaline fiddler crab *Uca minax* (Leconte). Comp Biochem Physiol 79A (4):673–678

Weisbart M, Chakraborti PK, Gallivan G, Eales JG (1987) Dynamics of cortisol receptor activity in the gills of brook trout, *Salvelinus fontinalis*, during seawater adaptation. Gen Comp Endocrinol 68:440–448

Winkler A (1986) The role of the transbranchial potential difference in hyperosmotic regulation of the shore crab *Carcinus maenas*. Helgol Meeresunters 40:161–175

Winquist RJ (1986) Possible mechanisms underlying the vasorelaxant response to atrial natriuretic factor. Fed Proc 45:2371–2375

Wyeth P, Prince R (1977) Carbonic anhydrase. Inorg Perspect Biol Med 1:37–71

Zaccone G (1981) Effect of osmotic stress on the chloride and mucous cells in the gill epithelium of the fresh-water teleost *Barbus filamentosus* (Cypriniformes, Pisces). A structural and histochemical study. Acta Histochem 68:147–159

Zanders IP (1980) Regulation of blood ions in *Carcinus maenas* (L.). Comp Biochem Physiol 65A:97–108

Zaugg WS, Wagner HH (1973) Gill ATPase activity related to parr-smolt transformation and migration in steelhead trout (*Salmo gairdneri*): Influence of photoperiod and temperature. Comp Biochem Physiol 45B:955–965

Chapter 2

Sodium and Chloride Transport Across the Invertebrate Intestine

G. A. Gerencser

Contents

1 Introduction

The genial experiments of Ussing and his collaborators (Ussing and Zerahn 1951) on frog skin in the mid-1950s ushered in the modern era of the study of ion transport across epithelia. These studies defined the nature of the bioelectric potential across isolated epithelia by demonstrating the interrelationship between the bioelectric potential and active ion transport. Another Dane, Skou (1965), molecularly defined the Na^+ transport process with his brilliant experiments delineating the transport nature of the $(Na^+ + K^+)$-stimulated ATPase enzyme. For several years thereafter, Na^+ transport across epithelia occupied the collective focus of epithelial transport physiologists with Cl^- assuming a secondary role of importance. However, within the last decade there has been a resurrection of interest in Cl^- transport, primarily because Cl^- has been found to be actively transported in a very wide range of species and cells (Frizzell et al. 1979; Gerencser 1983 a).

Both mechanisms of transepithelial Cl^- transport that have been reasonably well established are secondary active transport processes. The first broadly documented epithelial Cl^- transport process involves Cl^-/HCO_3^- antiport and is found, for example, in anal papillae of mosquito larvae (Stobbart 1967), fish gills (Maetz and Garcia-Romeu 1964; deRenzis and Maetz 1973; Kerstetter and Kirschner 1972), frog skin (Watlington et al. 1977), urodele intestine (Gunter-Smith and White 1979), turtle bladder (Leslie et al. 1973), rat intestine (Hubel

Department of Physiology, College of Medicine, University of Florida, Gainesville, FL, USA

Advances in Comparative and Environmental Physiology
Vol. 1. Edited by R. Greger

1968), rabbit colon (Frizzell et al. 1976), and human small intestine (Turnberg et al. 1970). The energy source for this process is unknown, but it has been suggested that uphill Cl^- transport is energized by a favorable downhill electrochemical potential gradient for HCO_3^- (Frizzell et al. 1979). The second well-documented Cl^- transport process is active by nature and is thought to be effected through an electrically neutral Na^+ symport which drives Cl^- uphill into epithelial cells via the inward flow of Na^+ down a favorable electrochemical gradient. This NaCl symport is located within the mucosal membrane if Cl^- is actively absorbed by the epithelium, or is located within the basolateral membrane if Cl^- is actively secreted. Epithelia which exemplify NaCl symport absorption include prawn intestine (Ahearn et al. 1977), flounder intestine (Field et al. 1978; Duffey et al. 1979), sculpin intestine (House and Green 1965), marine eel intestine (Skadhauge 1974), flounder urinary bladder (Renfro 1977), trout urinary bladder (Lahlou and Fossat 1971), *Necturus* gallbladder (Graf and Giebisch 1979; Ericson and Spring 1982), *Necturus* proximal tubule (Spring and Kimura 1978), bullfrog small intestine (Quay and Armstrong 1969a), frog skin (Watlington et al. 1977; Nagel et al. 1981), bovine rumen (Chien and Stevens 1972), rat colon (Binder and Rawlins 1973), rabbit gallbladder (Frizzell et al. 1975), rabbit ileum (Nellans et al. 1973), and human intestine (Turnberg et al. 1970). Those epithelia in which Na^+-coupled Cl^- secretion has been demonstrated include killifish operculum (Degnan et al. 1977), pinfish gills (Farmer and Evans 1981), shark rectal gland (Silva et al. 1977), frog stomach (Sachs et al. 1978), frog cornea (Candia 1972), rabbit ileum (Nellans et al. 1974), and dog trachea (Al-Bazzaz and Al-Awqati 1979). In these systems Na^+ is thought to be actively recycled at the basolateral membrane by the Na^+ pump [i.e., $(Na^+ + K^+)$-ATPase] while Cl^- moves energetically downhill from cytosol to the mucosa via a cAMP-enhanced Cl^- conductance (Klyce and Wong 1977).

However, a considerable amount of Cl^- transport data has accumulated in the literature that does not conform to any of the two well-established models described above. For example, White (1980) described an electrogenic Cl^- uptake mechanism located in the mucosal membrane of *Amphiuma* intestine which is independent of mucosal Na^+ or HCO_3^-, and similarly, Hanrahan and Phillips (1983) have provided evidence for an electrogenic Cl^- accumulative mechanism located in the mucosal membrane of locust rectal epithelium. Gerencser (1983b) has presented evidence that is consistent with an active Cl^- extrusion process which exists in the basolateral membrane of *Aplysia* enterocytes. This mechanism is electrogenic and is independent of Na^+.

Even though Frizzell et al. (1979) and DePont and Bonting (1981) have stated that there is no compelling evidence for primary active Cl^- transport in animal epithelia, the possibility exists because of transport data as exemplified, in part, by *Amphiuma* intestine (White 1980), locust rectum (Hanrahan and Phillips 1983), *Aplysia* gut (Gerencser 1983b), bullfrog intestine (Armstrong et al. 1972), and, also, because of additional reports of anion-stimulated ATPase sensitive to Cl^- in epithelial systems known to transport Cl^- (DePont and Bonting 1981; Gerencser 1983a; Gerencser and Lee 1983).

The present discussion focuses on some recent and continuing studies in the author's laboratory rather than being a complete, comprehensive exposition on intestinal Cl^- transport processes. It is primarily concerned with the impact of recent electrophysiological and biochemical studies on the development of current ideas concerning the nature of net Cl^- transport by molluscan intestinal tissue. It is hoped that the results and conclusions presented herein, though restricted in large measure to studies with *Aplysia* intestine, will be of interest within the broader context of comparative epithelial transport phenomena.

2 Transport Phenomena in *Aplysia* Intestine

2.1 Transcellular Transport Characteristics

2.1.1 Basal Electrical Characteristics

Generally speaking, intestinal preparations of vertebrates bathed in a substrate-free Na^+-containing Ringer solution generate spontaneous transepithelial potential differences (ψ_{ms}) of the magnitude 1 – 5 mV, the serosal surface being positive relative to the mucosal surface (Barry et al. 1965; Clarkson et al. 1961; Quay and Armstrong 1969a). The short-circuit current (SCC) in these vertebrate intestines was accounted for by means of active Na^+ absorption in the absence of, or in excess of, active Cl^- absorption (Barry et al. 1965; Quay and Armstrong 1969a).

Several observations indicate that some aspects of ion transport by invertebrate intestine differ from those of ion transport across vertebrate intestine. For example, the ψ_{ms} across isolated intestine of the moth, *Hyalophora cecropia*, was observed to be serosa-negative (Harvey et al. 1968). This electrical orientation was accounted for by an active electrogenic K^+ pump oriented in a secretory direction (Harvey and Nedergaard 1964). Additionally, the ψ_{ms} across the intestine of the locust rectum was observed to be serosa-negative, while the SCC was largely accounted for by an absorptive flux of Cl^- when Na^+ was removed from the bathing medium (Williams et al. 1978).

Ion transport across *Aplysia californica* intestine also appears to differ from ion transport across vertebrate intestine. For example, *Aplysia californica* intestine bathed in a Na^+-containing seawater medium elicits a spontaneous ψ_{ms} (0.5 – 3.0 mV) such that the serosal surface is negative relative to the mucosal surface (Gerencser 1978b, 1981a, 1982a, 1983b). The SCC across *Aplysia californica* intestine was accounted for by active absorptive mechanisms for both Na^+ and Cl^-, the net absorptive Cl^- transport exceeding the net absorptive Na^+ transport. These results were qualitatively the same as those observed with *Aplysia juliana* intestine (Gerencser and Hong 1977; Gerencser et al. 1977a) and account for the negative serosal ψ_{ms} observed in this intestinal preparation. Both Na^+ and Cl^- transport mechanisms were partially dependent upon the other co-ion being present in the bathing medium for substituting either Na^+ or Cl^- with another cation or anion, respectively, and this experimental condition reduced the temporal life or ψ_{ms} and SCC (Gerencser 1978b, 1983b). The replacement of Cl^- by SO_4^{-2} in the seawater bathing medium significantly depolarized but did not abolish ψ_{ms} and SCC across *Aplysia californica* intestine (Gerencser 1979a, b, 1981b). Similarly, the replacement of Na^+ with $Tris^+$ in the seawater medium significantly hyperpolarized the ψ_{ms} across *Aplysia* intestine (Gerencser 1983b). These results suggested electrogenic components of net, active Cl^- and Na^+ absorption across the isolated *Aplysia* intestine. These results also suggested the possibility of independent parallel transport processes for Na^+ and Cl^- across this preparation unlike that demonstrated for Na^+-Cl^- transport coupling in rabbit gallbladder (Frizzell et al. 1975) and other vertebrate epithelia (Schultz 1977; Frizzell et al. 1979).

By definition, primary active transport mechanisms are directly coupled to metabolism. The observations that the metabolic inhibitors 2,4-dinitrophenol, potassium cyanide, anoxia and iodoacetate inhibited ψ_{ms} and SCC further supported the hypothesis of active ion transport mechanisms (namely, Na^+ and Cl^-) maintaining the transmural electrical characteristics (Gerencser 1978b). As with several vertebrate intestinal preparations (Armstrong et al. 1972; Newey et al. 1965), the *Aplysia californica* intestine maintained ψ_{ms} and SCC with active ion transport that was predominantly energized through aerobic metabolism.

These findings coupled with the observations that serosally applied ouabain, a specific inhibitor of Na^+-K^+-dependent ATPase activity (Skou 1965), and mucosally applied furosemide, a Cl^- transport inhibitor (Burg and Green 1973), inhibited ψ_{ms} and SCC (Gerencser 1978b, 1981b, c), suggested that either intracellular Na^+ stimulates Na^+-K^+-dependent ATPase activity for the production of energy for Cl^- transport processes similar to that suggested for rabbit kidney (Gerencser and Hong 1975; Gerencser et al. 1977b), or that active Cl^- absorption is coupled to active Na^+ absorption. Therefore, ouabain inhibition of a Na^+-K^+-dependent ATPase would necessarily inhibit active Cl^- absorption.

However, recent observations suggest, for some time at least, that Cl^- absorption is independent of Na^+ absorption (Gerencser 1984a, e). Therefore, Cl^- absorption would be independent of Na^+-K^+-dependent ATPase activity. The ψ_{ms} measured in a Na^+-free seawater medium (Tris Cl) was stable for 3 – 5 h and the electrical orientation of ψ_{ms} was serosa negative relative to the mucosal solution. The magnitude of ψ_{ms} was significantly higher in the Na^+-free medium as opposed to the Na^+-containing medium (Gerencser 1983b). In the absence of Na^+ in the bathing medium, the SCC was identical to the net mucosal to serosal (M → S) Cl^- flux (Gerencser 1984a). Additionally, mucosally applied thiocyanate or acetazolamide inhibited ψ_{ms}, SCC, and net M → S Cl^- flux (Table 1).

Table 1. Chloride fluxes in Tris Cl seawater media[a]

	J_{ms}	J_{sm}	J_{ms}^{NET}	SCC
Before thiocyanate addition	216.7 ± 14.2 (6)	180.1 ± 13.2 (6)	36.6 ± 6.8	30.3 ± 7.9 (6)
After thiocyanate addition	175.7 ± 8.1 (6)	173.9 ± 8.6 (6)	1.8 ± 1.3	0.8 ± 0.7 (6)

[a] Values are means ± SEM in $nEq \cdot cm^{-2} \cdot min^{-1}$. Number of experiments shown in parentheses (Gerencser 1984a)

These results suggested that active Cl^- absorption is electrogenic and can occur in the absence of Na^+; and, that in the presence of Na^+ the ψ_{ms} is attenuated by the simultaneous electrogenic absorption of Na^+.

2.1.2 Sugar Effects

Actively transported sugars instantaneously enhance the serosa-positive ψ_{ms} when added to the mucosal aspect of vertebrate intestine (Quay and Armstrong 1969b; Schultz 1977; Schultz and Curran 1970). In fact, the change in ψ_{ms} induced by the actively absorbed sugar is a saturable function of the concentration of the added sugar (Quay and Armstrong 1969b). This change in ψ_{ms} is solely due to enhanced absorption of Na^+ in the vertebrate intestine (Quay and Armstrong 1969b; Schultz 1977; Schultz and Curran 1970).

Little effort has been directed toward the absorption and electrical characteristics of invertebrate intestine. In fact, Bidder (1950) was the first to suggest absorption of nutrients from the intestine rather than the digestive gland of molluscan cephalopods. Active absorption of several sugars across the gut wall of the mollusk, *Cryptochiton stelleri*, was demonstrated by Lawrence and Lawrence (1967). Furthermore, Lawrence and Mailman (1967) observed a serosal-negative ψ_{ms} across the anterior intestine of *Cryptochiton*; while Lawrence and his colleagues (Lawrence et al. 1972) demonstrated that mucosally applied monosaccharides hyperpolarized the *Cryptochiton*'s serosal-negative ψ_{ms}. These observations are strongly suggestive of an enhanced anion absorption caused by mucosal sugars.

Studies involving the effects of various hexoses on Na^+ and Cl^- transport across the isolated intestine of *Aplysia californica* revealed that the mucosally applied, actively absorbed sugars, D-glucose, D-galactose, or D-3-O-methyl-glucose, instantaneously hyperpolarized the serosal-negative ψ_{ms} and concomitantly increased SCC (Gerencser 1978a, 1981c, 1982a). These electrical effects were accounted for by means of an increase in the unidirectional $M \rightarrow S$ flux for both Na^+ and Cl^-, the enhanced absorptive Cl^- flux exceeding the enhanced absorptive Na^+ flux (Gerencser 1978a). The addition of these same sugars to the serosal aspect of the gut, at the same concentration, had virtually no effect on the transmural electrical characteristics, ψ_{ms} and SCC.

Furthermore, the addition of high concentrations (80 – 100 mM) of either L-glucose or D-sorbose (a nonactively transported, virtually nonmetabolizable hexose) to either the mucosal or serosal compartments of *Aplysia californica* intestine caused no change in ψ_{ms} or SCC (Gerencser 1978a, 1981c, 1982a). These results suggest a stereospecific membrane effect elicited by actively transported sugars on the electrical characteristics rather than a nonspecific electrokinetic effect.

Mucosal phlorizin or serosal ouabain partially inhibited the hexose-driven ψ_{ms} and SCC of *Aplysia californica* intestine (Gerencser 1978a, 1981c, 1982a), suggesting both an interaction of these sugars with the mucosal membrane (Alvarado and Crane 1962) and involvement of a Na^+-K^+-dependent ATPase (Skou 1965) in the apparent simultaneous transport of Na^+ and Cl^- from

M → S. D-galactose accumulation in *Aplysia* intestinal epithelia was shown to be absolutely dependent upon the presence of Na^+ in the seawater bathing medium, suggesting a reciprocal relationship between Na^+ and sugar transport (Gerencser 1978a, 1982a), an observation that is no different from that observed in vertebrate intestine (Alvarado and Crane 1962; Schultz 1977; Schultz and Curran 1970).

These results of the effects of luminally applied sugars on Cl^- absorption in the *Aplysia californica* intestine are unique to the animal kingdom relative to intestinal absorption, except for similar observations made with *Aplysia juliana* (Gerencser and Hong 1977). The sugar-induced electrogenicity of the active Cl^- absorptive process in *Aplysia californica* intestine underlines a basic difference in *Aplysia* versus the response of vertebrate intestinal preparations to the mucosal application of actively transported hexoses (Barry et al. 1965; Quay and Armstrong 1969b; Schultz 1977; Schultz and Curran 1970).

2.1.3 Amino Acid Effects

In all vertebrate intestinal preparations studied to date mucosally applied, actively transported amino acids hyperpolarize the serosal-positive ψ_{ms} (Lerner 1978; Quay and Armstrong 1969b; Schultz 1977; Schultz and Curran 1970). This solely reflects an increased Na^+ absorption (Rose and Schultz 1970; Schultz and Curran 1968).

When glycine was added to the mucosal bathing medium of isolated *Aplysia californica* intestine, the spontaneous serosal-negative ψ_{ms}, SCC, and unidirectional M → S Cl^- flux were all enhanced (Gerencser 1981a, c, 1982a). Serosally applied glycine had no effect on ψ_{ms}, SCC, or on either the unidirectional M → S or S → M Cl^- flux. These observations of amino acid stimulating active electrogenic Cl^- transport in *Aplysia californica* intestine have not been reported in any other intestinal preparation to date except where Quay and Armstrong (1969a) reported that mucosally applied L-alanine stimulated the unidirectional S → M flux of Cl^- across bullfrog gut.

The *Aplysia californica* intestine was shown to possess a low affinity glycine transport mechanism, for kinetic studies revealed an apparent K_m for glycine stimulation of SCC to be 37 mM (Gerencser 1981a). Amino acid concentrations in the *Aplysia* intestine are relatively high (52 mM) in the natural feeding state, which would satisfy the low affinity, high capacity amino acid uptake mechanism (Gerencser 1981a), rather than a high affinity, low capacity uptake mechanism found in the vertebrates and other animals (Lerner 1978). D-alanine and D-valine had virtually equivalent maximal stimulating effects on ψ_{ms} and SCC as did their L-counterparts when added to the mucosal medium; however, the D-amino acids had much lower affinities than the L-amino acids for the mucosal transport site, strongly suggesting that the amino acid transport sites were not stereospecific as were the sugar transport sites (Gerencser 1978a). Presently, these observations may mean a crossover point, in terms of specificity in the evolution of the transport systems for organic solutes in the intestine of *Aplysia californica*. I am not aware of any report that shows a stereospecificity for one class of organic compounds and not another in intestinal epithelia.

In short-circuited *Aplysia* intestine actively transported sugars and amino acids appear to utilize different mucosally located mechanisms in enhancing Cl^- absorption. Upon virtual maximal stimulation of ψ_{ms} and SCC by mucosal D-glucose, a further increase in the electrical characteristics could be obtained by addition of glycine to the mucosal medium, or vice versa (Gerencser 1981 a). Additive effects of amino acids and sugars delineating separate transport mechanisms for both types of organic solutes in both vertebrate and invertebrate preparations have also been reported (Lerner 1978; Quay and Armstrong 1969 b; Schultz 1977; Schultz and Curran 1970).

Ouabain was shown to reduce the unidirectional $M \rightarrow S$ Cl^- flux in short-circuited *Aplysia* intestine such that the net $M \rightarrow S$ flux of Cl^- was no different than zero (Gerencser 1981 a). Upon glycine addition to the mucosal medium, the unidirectional $M \rightarrow S$ Cl^- flux and ψ_{ms} were enhanced. The net Cl^- flux was identical to the SCC under these conditions. These results suggest both an independent, electrogenic Cl^- transport pathway from that of Na^+ and an energy source that is utilized by the Cl^- transport mechanism that is not coupled to the Na^+-K^+-stimulated ATPase. This observation is in contrast with that observed in vertebrate intestine where net Cl^- absorption ceases when ouabain is present on the serosal aspect of the intestine (Frizell et al. 1979; Schultz 1977). It is thought by some investigators that all active Cl^- transport in vertebrate intestine is ultimately dependent on Na^+-K^+-dependent ATPase activity (Armstrong 1975, 1979 a; Frizzell et al. 1979; Schultz 1977).

Progressive reduction of the extracellular to intracellular Na^+ gradient across the mucosal membrane of *Aplysia* intestinal epithelial cells by serosal ouabain application progressively lessens the mucosal glycine stimulatory effect on Cl^- absorption across this preparation (Gerencser 1981 a). This result suggests the involvement of the Na^+ gradient across the mucosal membrane as an integral part of the mechanism by which amino acids enhance active, electrogenic Cl^- absorption across *Aplysia* intestine (Gerencser 1981 a, c, 1982 a).

The finding that aminoisobutyric acid (AIB), a nonmetabolizable amino acid, is actively accumulated and stimulates both the unidirectional $M \rightarrow S$ flux of Cl^- and SCC across *Aplysia* intestine strongly suggests that AIB transport, and not its metabolism, is the cause of enhanced Cl^- absorption (Gerencser 1981 a). Assuming a 1:1 coupling for AIB and Na^+ at the mucosal membrane, then the increase in the unidirectional $M \rightarrow S$ flux of Cl^-, by mucosal AIB application, should be greater than the increase in the unidirectional $M \rightarrow S$ flux of Na^+ as predicted by the increase in serosal negativity of ψ_{ms}. This phenomenon is observed by Gerencser (1981 a), where the increase in the unidirectional $M \rightarrow S$ flux of Cl^- was greater than the unidirectional $M \rightarrow S$ flux of AIB. Continuing this argument, if AIB is coupled one-to-one to a net positive charge flow from the extra- to the intracellular space across the mucosal membrane, then the algebraic sum of the positive charge flow and that of the increase in $M \rightarrow S$ flux of Cl^- should approximate the change in SCC, which it does (Gerencser 1981 a). These data suggest an electrical coupling between the sodium-dependent, actively transported AIB zwitterion (primary event) and chloride (secondary event), which move in the same direction, mucosa to serosa.

As with sugar stimulation of Cl^- absorption in *Aplysia* intestine, amino acid stimulation of salt transport is unique relative to other intestinal preparations (Lerner 1978; Quay and Armstrong 1969b; Schultz 1977; Schultz and Curran 1970). Mucosal amino acid effects on ψ_{ms} in vertebrate intestine always result in the serosa becoming more positive (Lerner 1978; Schultz 1977; Schultz and Curran 1970). This observation is entirely due to an enhanced Na^+ absorption (Quay and Armstrong 1969b; Schultz 1977). However, in *Aplysia* intestine actively absorbed amino acids stimulate both Na^+ and Cl^- absorptive fluxes, the enhanced Cl^- flux exceeding the enhanced Na^+ flux. This is reflected in the hyperpolarization of the serosal-negative ψ_{ms} in the mucosal presence of these compounds. These observations underscore the fundamental differences that exist between *Aplysia californica* intestine and other intestinal preparations studied relative to mucosal amino acid effects on electrogenic salt transport.

2.2 Intracellular Transport Characteristics

2.2.1 Membrane Potential

The cytosol of *Aplysia californica* enterocytes is negative (-68.2 ± 2.2 mV) relative to the mucosal solution as determined by measurements of mucosal membrane potential difference, ψ_m (Gerencser 1981b, 1982a, 1983b; Gerencser and White 1980). Since ψ_{ms} was simultaneously measured with ψ_m and found to be serosa negative (-0.4 ± 0.1 mV), the calculated serosal membrane potential difference (ψ_s) was less negative than ψ_m according to the equation: $\psi_{ms} = \psi_m + \psi_s$. The intracellular negativity of *Aplysia californica* enterocytes demonstrates that these cells belong to the "well"-type potential profile as opposed to the "staircase" model where the inside of the cell is positive relative to the mucosal solution (Armstrong 1975). The well-type potential profile has been well documented amongst vertebrate enterocytes (Rose and Schultz 1970; White 1977; White and Armstrong 1971). However, the finding in *Aplysia* that ψ_m was more negative than ψ_s was different in terms of electrical orientation from that found for bullfrog (White and Armstrong 1971) or rabbit (Rose and Schultz 1970) enterocytes where ψ_s was more negative than ψ_m.

Simultaneous measurements of ψ_m and ψ_{ms} showed ψ_m to depolarize from -68.2 ± 2.2 mV to -62.1 ± 2.5 mV and ψ_{ms} to hyperpolarize from -0.4 ± 0.1 mV to -1.5 ± 0.1 mV after mucosal D-glucose addition (Table 2). The change in $\psi_m(\Delta\psi_m)$ was significantly greater than the change in $\psi_{ms}(\Delta\psi_{ms})$. The calculated change in $\psi_s(\Delta\psi_s)$ was significantly greater than the observed $\Delta\psi_m$. Measurement of ψ_m and ψ_s and their respective responses to actively transported sugars in bullfrog (White and Armstrong 1971) and rabbit small intestine (Rose and Schultz 1970) demonstrated that the primary electrical event elicited by these solutes is a depolarization of ψ_m. This depolarization was not observed in the absence of Na^+ or in the presence of solutes whose mucosal influx was Na^+-independent (Rose and Schultz 1970). Simultaneous measurement of $\Delta\psi_{ms}$ induced by these solutes demonstrated that the ratio $\Delta\psi_m/\Delta\psi_{ms}$ was consistently greater than one, indicating that, besides there being a decrease in ψ_m, there was

Table 2. ψ_m, ψ_{ms}, and calculated ψ_s in NaCl seawater medium[a]

	ψ_m (mV)	ψ_s (mV)	ψ_{ms} (mV)	n
Before glucose addition	-68.2 ± 2.2 (61)	$+67.8 \pm 2.0$ (61)	-0.4 ± 0.1 (61)	6
After glucose addition	-62.1 ± 2.5 (61)	$+60.6 \pm 2.1$ (61)	-1.5 ± 0.1 (61)	6
	$P<0.01$	$P<0.01$	$P<0.01$	

[a] Values are means ± SE, ψ_m, mucosal membrane potentials; ψ_{ms}, transmural potentials; ψ_s, serosal membrane potentials. Numbers in parentheses are numbers of observations; n is number of animals. Polarity of ψ_m, ψ_{ms} are relative to mucosal solution. Polarity of calculated ψ_s is relative to cytoplasm (Gerencser and White 1980)

a smaller but significant decrease in ψ_s (Rose and Schultz 1970; White and Armstrong 1971). These results were explained as follows: the decrease in ψ_m elicited by actively transported organic solutes (sugars, amino acids) was consistent with the notion that linked entry of Na^+ and organic solutes across the mucosal membrane of the intestinal epithelial cell was rheogenic (Rose and Schultz 1970). In the absence of coupling between ψ_m and ψ_s, one would predict an equivalent change in both ψ_m and ψ_{ms}. However, if ψ_m and ψ_s are coupled by a relatively low resistance extracellular pathway, a change in ψ_m will induce an associated, though not necessarily equal, change in ψ_s (Rose and Schultz 1970; White and Armstrong 1971).

If, as in vertebrate intestine, there is a tight coupling between Na^+ and Cl^- transport at the mucosal membrane of *Aplysia* intestinal cells, then it would be expected that replacement of Cl^- with nonabsorbed anions would have relatively little effect on ψ_m or ψ_{ms}, or replacement of Na^+ with nonabsorbed cations would likewise have little initial effect on ψ_{ms} and, if anything, lead to a depolarization of ψ_m (Frizzell et al. 1979). These predictions are based upon the notion that active Cl^- absorption is absolutely dependent on the mucosal presence and transport of Na^+ (Armstrong et al. 1979a; Frizzell et al. 1979; Schultz 1977). Nellans et al. (1973) elegantly demonstrated these precepts in rabbit ileum, thereby precluding either electrical coupling between Na^+ and Cl^- across the mucosal membrane or any independent transport processes for these ions across the same membrane.

However, in *Aplysia* intestine ψ_m and ψ_{ms} were significantly reduced from -67.8 ± 1.3 and -0.6 ± 0.1 mV, respectively, to -7.2 ± 3.3 and -0.2 ± 0.1 mV, respectively, when Cl^- was replaced in the seawater bathing medium with SO_4^{-2} (Gerencser 1981b). Conversely, ψ_m and ψ_{ms} were significantly elevated when Na^+ was replaced with Tris-hydroxyaminomethane ($Tris^+$) in the seawater bathing medium (Table 3). These results exclude the notion of a coupled, mediated entry for Na^+ and Cl^- across the mucosal membrane as described for various vertebrate epithelia, including intestine (Frizzell et al. 1979). These results also suggest that the mucosal membrane of the *Aplysia californica* enterocyte is highly permeable to Cl^- (almost a Cl^- electrode) and that there is a finite permeability of this membrane to Na^+.

Table 3. Mucosal membrane potentials (ψ_m), transmural potentials (ψ_{ms}), and calculated serosal membrane potentials (ψ_s) in NaCl and Tris Cl seawater media[a]

	ψ_m (mV)	ψ_s (mV)	ψ_{ms} (mV)	n
NaCl	-66.3 ± 1.6 (83)	$+65.5 \pm 1.3$ (83)	-0.8 ± 0.1 (83)	11
Tris Cl	-70.1 ± 1.9 (79)	$+68.5 \pm 1.3$ (79)	-1.6 ± 0.2 (79)	11
NaCl	-64.8 ± 2.0 (76)	$+64.3 \pm 1.6$ (76)	-0.5 ± 0.1 (76)	11
	$P<0.05$	$P<0.05$	$P<0.01$	

[a] Values are means ± SE. Numbers in parentheses are numbers of observations; n, number of animals. Polarities of ψ_m and ψ_{ms} are relative to mucosal solution. Polarity of calculated ψ_s is relative to cytoplasm (Gerencser 1983 b)

The mean voltage divider ratio obtained in a substrate-free, Na^+ Cl^- seawater medium was 1.62 ± 0.11 (Gerencser 1981 b, c). This value was significantly different from 1.00, indicating that the rate-limiting step for ion entry into the *Aplysia californica* enterocyte appears to be the mucosal membrane. Approximately 62% of the transcellular resistance to ion movement arises from this membrane barrier. The finding that the voltage divider ratio, defined as the ratio of mucosal membrane resistance to that of serosal membrane resistance, is not equal to one suggests that the two membranes in series are asymmetrical with regard to ionic conductance properties.

Mucosal D-glucose significantly reduced input resistance of the *Aplysia* intestinal epithelial cells from 15.2 ± 1.6 to 10.1 ± 1.5 MΩ, suggesting rheogenic entry of ion or ions (probably Na^+) across the mucosal membrane which would tend to increase the conductivity of that membrane (Gerencser 1981 b). That the basolateral membrane is electrically unresponsive to D-glucose (Gerencser 1978 a, 1982 a) lends further credence to this hypothesis.

2.2.2 *Intracellular Activity Measurements*

Reports of intracellular Cl^- activity (a^i_{Cl}) in vertebrate enterocytes showed that Cl^- is accumulated across the mucosal membrane such that the a^i_{Cl} was two to three times that predicted for electrochemical equilibrium across that membrane (Armstrong et al. 1977, 1979 a; Duffey et al. 1979). These studies concluded that Cl^- uphill movement across the mucosal membrane was coupled to the simultaneous downhill movement of Na^+ and it was this extracellular to intracellular Na^+ electrochemical gradient across the mucosal membrane that was the driving force responsible for intracellular Cl^- accumulation.

The mean a^i_{Cl} determined in *Aplysia californica* intestinal epithelial cells bathed in a NaCl seawater medium devoid of substrate was 10.1 ± 0.5 mM. The mean a^i_{Cl} significantly increased after mucosal glucose addition to 14.2 ± 0.6 mM (Table 4, lower panel). Both a^i_{Cl} values, before and after D-glucose addition, were significantly less than those predicted by the electrochemical equilibrium for Cl^- across the mucosal membrane. These results were interpreted as if there was

Table 4. Intracellular Cl activities and membrane potentials in the presence and absence of glucose in NaCl seawater medium[a]

	a_{Cl} (mM)	ψ_m (mV)	n
Glucose-free medium	12.1 ± 0.3 (29)	−64.4 ± 1.4 (35)	5
Before glucose addition	10.1 ± 0.5 (10)	−70.7 ± 0.7 (11)	2
After glucose addition	14.2 ± 0.6 (6)	−62.6 ± 0.5 (19)	2
	$P<0.01$	$P<0.01$	

[a] Values are means ± SE. Numbers in parentheses are number of observations; n is number of animals. The observations reported in glucose-free media in the top half of the table incorporate the observations in the lower half of the table (before glucose addition) made under the same conditions. Polarity of ψ_{ms} is relative to the mucosal solution (Gerencser and White 1980)

Table 5. Potential profile, intracellular K^+ and Na^+ activities, and transmucosal K^+ and Na^+ electrochemical potential differences in *Aplysia* intestine bathed in NaCl seawater medium[a]

ψ_{ms} (mV)	ψ_m (mV)	a^i_K (mM)	a^i_{Na} (mM)	$\Delta\bar{\mu}_K/F$ (mV)	$\Delta\bar{\mu}_{Na}/F$ (mV)
−1.3 ± 0.4	−55.3 ± 2.2 (89)	383 ± 15 (42)	17.2 ± 2.5 (16)	43	−129

[a] Values are means ± SEM. Numbers in parentheses are number of observations. Numbers of animals used were: ψ_{ms}, $\psi_m = 11$; $a^i_K = 7$; $a^i_{Na} = 4$. Polarities of ψ_{ms} and ψ_m are relative to the mucosal solution (Gerencser 1983c)

Table 6. Intracellular chloride activities and mucosal membrane potentials in NaCl and Tris Cl seawater media[a]

	a^i_{Cl} (mM)	a^{eq}_{Cl} (mM)	ψ_m (mV)	n
NaCl	13.9 ± 0.5 (30)	28.6	−64.6 ± 1.3 (48)	7
Tris Cl	9.1 ± 0.3 (32)	21.4	−72.2 ± 1.4 (51)	7
	$P<0.01$		$P<0.01$	

[a] Values are means ± SE. Numbers in parentheses are number of observations; n is number of animals. Polarity of ψ_m is relative to the mucosal solution (Gerencser 1983b)

an active Cl^- transport mechanism located in the basolateral membrane (Gerencser and White 1980).

Chloride accumulation requires an adequate amount of free energy implicit within the Na^+ electrochemical gradient directed from the mucosal solution into the cytosol of intestinal epithelial cells (Armstrong et al. 1979a, b; Frizzell et al. 1979; Gerencser 1984c, d; Schultz 1977) if indeed, there is a mechanistic coupling between these two ions at the mucosal membrane. Table 5 shows that the intracellular Na^+ activity (a^i_{Na}) in *Aplysia* intestinal cells is 17 mM; this a^i_{Na} translates into an electrochemical driving force of 129 mV directed from the mucosal solution into the cytoplasm of these cells (Gerencser 1983c). This amount of energy is sufficiently adequate for the work involved in accumulating Cl^- inside the *Aplysia* enterocyte (Gerencser 1983b) as calculated for vertebrate intestine (Armstrong et al. 1979a, b). In fact, assuming a coupling ratio of 1:1 between Na^+ and Cl^-, an accumulation ratio of 2.3 could be achieved if the coupling efficiency for Na^+ and Cl^- were 100% at the mucosal membrane. However, Cl^- was not accumulated in these cells (Gerencser 1983b; Gerencser and White 1980).

If active Cl^- transport across *Aplysia* intestine involves the application of the Na^+ electrochemical gradient across the mucosal membrane, then the abolition of the Na^+ electrochemical gradient should abolish active Cl^- transport (Armstrong et al. 1979a, b; Schultz 1977; Schultz and Curran 1968). The a^i_{Cl}, under these experimental conditions, should approximate an electrochemical equilibrium value (a^{eq}_{Cl}). Table 6 shows that this is not the case! With Na^+ in the seawater medium, a^i_{Cl} equaled 13.9 mM; whereas in the absence of Na^+, a^i_{Cl} was 9.1 mM. Both of these a^i_{Cl} values (in the presence and absence of Na^+) are significantly less than those intracellular electrochemical equilibrium values for Cl^- (Table 6). These results suggest that the Na^+-gradient hypothesis is untenable as applied to Cl^- entry across the mucosal membrane of *Aplysia* enterocytes and that Cl^- entry across this barrier involves, most likely, a Na^+-independent, passive (downhill) transport pathway.

If Cl^- entry is driven solely by its electrochemical potential difference across the mucosal membrane (i.e., the flow of Cl^- is not coupled to any nonconjugate driving force) the electromotive force for Cl^- across that membrane (E^m_{Cl}) must be described by:

$$E^m_{Cl} = \frac{RT}{Z_{Cl}F} \cdot \ln \frac{a^0_{Cl}}{a^i_{Cl}} ,$$

(R, T, Z, and F having their usual physical meanings), regardless of the underlying transport mechanism (e.g., simple diffusion, single-file diffusion, or carrier-mediated facilitated diffusion). Utilizing data from Gerencser and White (1980) it can be shown that in the absence of musosal D-glucose, the calculated E^m_{Cl} equaled -88.6 mV, while in the presence of mucosal D-glucose E^m_{Cl} equaled -80.3 mV for *Aplysia californica* villus enterocytes, which is in good agreement with the E^m_{Na} values calculated for both *Amphiuma* urinary bladder (Fromter and Gebler 1977) and toad urinary bladder (Sudou and Hoshi 1977). As can be determined, the driving force ($E^m_{Cl} - \psi_m$) for Cl^- movement is directed across the mucosal membrane from the extracellular to the intracellular aspect. This is direct evidence for the existence of a Cl^- electrochemical "well" in the cytosol of the *Aplysia californica* villus enterocyte.

The Cl^- permeability of the *Aplysia californica* enterocyte mucosal membrane (P^m_{Cl}) in both the absence and presence of D-glucose, may be calculated using the Goldman-Hodgkin-Katz equation, i.e.:

$$P^m_{Cl} = -\frac{RT(M \rightarrow S)^{NET}_{Cl}}{F\psi_m} \cdot \frac{1 - \exp(-F\psi_m/RT)}{a^i_{Cl} - a^0_{Cl}\exp(-F\psi_m/RT)} .$$

These calculations, utilizing data from Gerencser (1978a) and Gerencser and White (1980), yielded P^m_{Cl} of 0.010 and 0.027 cm h^{-1}, in the absence and presence of mucosal D-glucose, respectively. These values are in reasonable agreement with the value of 0.03 cm h^{-1} estimated by Fuchs et al. (1977) for the Na^+ permeability of the outer membrane of frog skin. These results strongly suggest that Na^+-glucose co-transport initiates significant mucosal membrane permeability changes to Cl^-. Utilizing these calculated P^m_{Cl} values and data from Gerencser and White (1980) the resistance of the mucosal membrane to Cl^- (R^m_{Cl}) can be calculated using the equation:

$$R^m_{Cl} = RT/P^m_{Cl} \cdot Z^2_{Cl} \cdot F^2 \cdot a^i_{Cl} \ .$$

The R^m_{Cl} values in the absence and presence of mucosal D-glucose were 5951 and 1568 $\Omega \cdot cm^2$, respectively.

The intriguing features of these calculations further strengthen the ideas of an active, electrogenic Cl^- transport across the basolateral membrane and an electrical coupling between D-glucose-induced ψ_m depolarization and associated mucosal membrane resistance changes to Cl^-, and therefore enhanced Cl^- entry across the mucosal membrane of *Aplysia californica* enterocytes.

2.3 Paracellular Transport Characteristics

The very low ψ_m of *Aplysia californica* intestine (Gerencser 1978a, b, 1981c, 1982a, 1983b) and the low tissue resistance of approximately 25 $\Omega \cdot cm^2$ strongly suggests that a low resistance extracellular shunt pathway exists in *Aplysia californica* intestine and that it plays a major role in ionic conductance across the tissue. Quantitatively and qualitatively these transmural electrical characteristics were similar to those found in *Aplysia juliana* intestine (Gerencser and Hong 1977; Gerencser et al. 1977a). Strengthening this idea was the previous finding (Gerencser 1978b) which demonstrated that the S→M fluxes for both Na^+ and Cl^- were much greater than the net active M→S fluxes of these ions, which was similar to that found in *Aplysia juliana* intestine (Gerencser and Hong 1977). Recently, Gerencser and Loughlin (1983) reported that for both Na^+ and Cl^- the extracellular shunt component was greater than 90% of the total S→M movement of these ions across the *Aplysia californica* intestinal tissue. This buttresses the notion that the extracellular pathway, and not the cellular pathway, is the predominant route for ionic conductance across the tissue. Similarly, Gerencser (1982b) reported that more than 90% of the total ionic conductance across *Aplysia juliana* intestine was via the extracellular route. Therefore, a portion of the electrical dialogue between the mucosal and basolateral membranes, in the presence of mucosal D-glucose, is probably permitted via the low resistance extracellular shunt. As with vertebrate intestinal absorptive cells, the primary event involving electrical communication between ψ_m and ψ_s in *Aplysia californica* intestinal epithelial cells is probably the rheogenic and electrogenic entry of Na^+ coupled to the entry of D-glucose at the mucosal membrane which depolarizes ψ_m (Gerencser and White 1980). Since the $\Delta\psi_s$ is greater than the concomitant $\Delta\psi_m$, it is possible that stimulation of an electrogenic Cl^- pump also contributes to the electrical dialogue between ψ_m and ψ_s, for it has been demonstrated in this preparation that active Cl^- transport exceeds active Na^+

transport in both the presence and absence of mucosally applied D-glucose (Gerencser 1978a, 1981a,c, 1982a).

2.4 Enterocyte Plasma Membrane Vesicles

2.4.1 Existence and Location of Anion-ATPase

Presently, without question, the greatest argument regarding Cl^--stimulated ATPase activity is its localization within the subarchitecture of cells. It seems that Cl^--stimulated ATPase activity resides in both microsomal and mitochondrial fractions (DePont and Bonting 1981) of cell homogenates. However, DePont and Bonting (1981) and Schuurmans Stekhoven and Bonting (1981) have pontificated that microsomal or plasma membrane localization of this enzyme is entirely due to mitochondrial contamination, hence the dispute. If Cl^--stimulated ATPase activity is exclusively of mitochondrial origin, it is very difficult to conceive a mechanism which is the ATPase that can drive net Cl^- transport across plasma membranes. On the other hand, if the Cl^--stimulated ATPase is located in the plasma membrane, it would not be difficult to envision primary active Cl^- transport by this enzyme analogous to the $(Na^+ + K^+)$-stimulated ATPase and its role in the net transport of Na^+ and K^+ across the plasma membranes (Skou 1965).

Without argument, the primary site of anion (specifically Cl^-) stimulated ATPase activity within animal cells appears to be in the mitochondria. Obviously, the key question is: what is the origin of the Cl^--stimulated ATPase activity of nonmitochondrial organelles? Is it as van Amelsvoort et al. (1977) have dogmatically stated, that all nonmitochondrial organelles which exhibit Cl^--stimulated ATPase activity have been contaminated with the mitochondrial-based enzyme, or is there a true, separate Cl^--stimulated ATPase that is localized within the cellular plasma membranes and which, therefore, can possibly act as the prime mover of net Cl^- transport between the intracellular and extracellular space?

Gerencser and Lee (1983, 1985a) presented results which indicated that the basolateral membranes of *Aplysia* enterocytes contain $Cl^- - HCO_3^-$-ATPase activity. Their finding that the subcellular membrane fraction has a high specific activity in $Na^+ - K^+$-ATPase, but has no perceptible cytochrome c oxidase activity and a significantly reduced succinic dehydrogenase activity, supports this conclusion (Table 7). The observation that there was very little NADPH-cytochrome c reductase activity in the membrane fraction (Table 7) suggests that the basolateral membranes in this fraction were also relatively free from endoplasmic reticulum and Golgi body membrane contamination (Mircheff et al. 1979). The failure of oligomycin to inhibit $Cl^- - HCO_3^-$-ATPase activity in the basolateral membrane fraction is also consistent with the nonmitochondrial origin of the anion ATPase (Table 8). Supporting this contention is the corollary finding that oligomycin inhibited mitochondrial anion-stimulated ATPase activity (Table 8). The finding that efrapeptin, a direct inhibitor of mitochondrial F_1-ATPase activity (Bullough et al. 1982), significantly inhibited Mg^{2+}-ATPase activity in

Table 7. Distribution of marker enzymes and anion-stimulated ATPase during preparation of plasma membranes from *Aplysia* enterocytes[a]

Enzyme	H (Homogenate)	P_2 (Mitochondria)	S-III (Basolateral membranes)
Total protein (mg)	208.59 ± 24.95	15.13 ± 1.95	5.47 ± 1.09
Na^+/K^+-ATPase	0.85 ± 0.11	0.25 ± 0.23	5.26 ± 1.96
Alkaline phosphatase	0.91 ± 0.29	1.02 ± 0.30	1.96 ± 0.36
5′-nucleotidase	0.41 ± 0.17	0.37 ± 0.11	0.89 ± 0.33
Cytochrome c oxidase	0.50 ± 0.12	0.97 ± 0.23	n.d.
Succinic dehydrogenase	22.30 ± 4.90	83.40 ± 27.60	4.17 ± 2.40
NADPH-cytochrome c reductase	3.73 ± 0.28	7.44 ± 1.25	1.32 ± 0.18
Mg^{2+}-ATPase	2.69 ± 0.57	5.10 ± 0.84	9.16 ± 1.82
HCO_3^--ATPase	3.85 ± 0.83	7.26 ± 0.91	14.12 ± 2.02
Cl^--ATPase	2.51 ± 0.59	2.50 ± 0.79	5.77 ± 2.26

[a] Values are means ± SE from 9 to 11 different preparations. Enzyme activity is expressed as μmol h^{-1} mg^{-1} protein for Na^+/K^+-ATPase, alkaline phosphatase, and 5′-nucleotidase; Δ log (ferrocytochrome c) min^{-1} mg^{-1} protein for cytochrome c oxidase; μmol min^{-1} mg^{-1} protein for succinic dehydrogenase; nmol min^{-1} mg^{-1} protein for NADPH-cytochrome c reductase; μmol 15 min^{-1} mg^{-1} protein for Mg^{2+}-, HCO_3^--, and Cl^--ATPase. H, homogenate; P_2, pellets from 9,500 × g centrifugation; S-III, 40 to 50% sucrose interface; n.d., not detectable. Starting intestinal mucosal was approximately 1.5 g (Gerencser and Lee 1985 a)

the mitochondrial and not in the basolateral membrane fraction (Table 9) also supports the notion that the plasma membrane fraction is of extramitochondrial origin.

Taken together, all of these observations support the hypothesis that anion ($Cl^- - HCO_3^-$)-stimulated ATPase activity probably exists in other subcellular loci than mitochondria. It appears that in numerous epithelia, which transport anions, anion ($Cl^- - HCO_3^-$)-stimulated ATPase activity forms an integral part of the plasma membrane (Gerencser and Lee 1983; Gerencser 1984b, 1985, 1986b; Table 7).

2.4.2 *Functions of Anion-ATPase*

There is, as of yet, no direct proof of primary active anion transport; that is, an ATPase which translocates anions up their respective electrochemical gradients powered by the simultaneous hydrolysis of ATP. The main argument for the anion-stimulated ATPase moving anions up their respective energy gradients is based upon parallel observations of anion-stimulated ATPase activity and anion transport phenomena in the same tissue. This was initially exemplified by the following extant observations: (1) anion-stimulated ATPase activity and H^+ secretion are inhibited in parallel by inhibitors such as thiocyanate and hydrazine in gastric mucosa (Sachs et al. 1972); (2) anion-stimulated ATPase activity and

Table 8. Effect of oligomycin on Cl^-/HCO_3^--ATPase activity[a]

	Oligomycin 10^{-9} M	Mg^{2+}-ATPase		HCO_3^--ATPase		Cl^--ATPase	
		Sp. Act.	Inh. (%)	Sp. Act.	Inh. (%)	Sp. Act.	Inh. (%)
H	0	4.38±0.16	0	4.84±2.09	0	1.46±0.10	0
(Homogenate)	6.8	2.80±0.49	36	3.36±1.81	31	1.28±0.47	12
	13.6	2.10±0.07	52	2.77±0.86	43	0.71±0.04	51
P_2	0	4.24±1.41	0	4.00±1.68	0	1.06±0.26	0
(Mitochondria)	6.8	2.90±0.41	32	4.14±2.05	0	1.13±0.27	0
	13.6	1.77±1.33	58	2.45±0.17	39	0.68±0.02	36
S-III	0	14.65±3.21	0	14.63±3.41	0	2.97±0.43	0
(Basolateral	6.8	12.65±2.85	14	14.79±4.71	0	2.37±0.47	20
membranes)	13.6	9.47±2.26	35	14.04±3.31	4	3.16±0.91	0

[a] Oligomycin was preincubated with enzymes in the reaction mixture at the indicated concentrations for 10 min at 30 °C. Values are means ± SE from three different experiments. Sp. Act. = specific activity, μmol 15 min^{-1} mg^{-1} protein; Inh. (%) = % of inhibition (Gerencser and Lee 1985 a)

Table 9. Effect of efrapetin on Cl^-/HCO_3^--ATPase activity[a]

	Efrapeptin ($g\,ml^{-1}$)	Mg^{2+}-ATPase		HCO_3^--ATPase		Cl^--ATPase	
		Sp. Act.	Inh. (%)	Sp. Act.	Inh. (%)	Sp. Act.	Inh. (%)
H	0	2.44±0.68	0	4.13±0.23	0	1.21±0.54	0
(Homogenate)	0.05	1.64±0.56	33	3.23±0.95	22	0.82±0.09	31
	0.10	1.12±0.66	58	2.30±0.75	44	0.77±0.27	37
P_2	0	5.85±1.47	0	8.12±1.14	0	0.62±0.36	0
(Mitochondria)	0.05	4.66±1.42	28	6.46±1.70	21	0.56±0.24	12
	0.10	1.73±0.32	66	6.32±0.45	26	0.39±0.23	38
S-III	0	5.02±1.08	0	7.81±1.50	0	1.91±1.41	0
(Basolateral	0.05	4.76±1.08	4	6.63±1.93	15	1.73±0.87	12
membranes)	0.10	4.84±1.34	3	4.74±1.27	42	1.37±0.14	30

[a] Efrapeptin was preincubated with enzymes in the reaction mixture at the indicated concentrations for 10 min at 30 °C. Values are means ± SE from three different preparations. Sp. Act. = specific activity, μmol 15 min^{-1} mg^{-1} protein; Inh. (%) = % of inhibition (Gerencser and Lee 1985 a)

HCO_3^- transport are inhibited in parallel by thiocyanate and parachloromercurobenzoate in pancreatic epithelium (Simon 1972); and (3) HCO_3^- secretion and anion-stimulated ATPase activity were depressed in parallel during metabolic acidosis in salivary duct epithelium, whereas HCO_3^- transport and anion-stimulated ATPase activity increased in parallel during metabolic alkalosis of this tissue (Wais and Knauf 1975).

It was not until the following observations that HCO_3^--stimulated ATPase activity was linked with Cl^- pumping because no Cl^- activation of this enzyme had been observed. De Renzis and Bornancin (1977) were the first to demonstrate the membrane presence of a (Cl^- + HCO_3^-)-stimulated ATPase in goldfish gill epithelium and suggested that the enzyme could participate in the branchial Cl^-/HCO_3^- exchange mechanism. Bornancin et al. (1977) confirmed these results in freshwater eel gill epithelium as did Bornancin et al. (1980) in freshwater trout gill epithelium. Kinetic studies in these three gill epithelial systems strongly suggested that a (Cl^- + HCO_3^-)-stimulated ATPase is involved in the Cl^-/HCO_3^- exchange mechanism and therefore in the acid-base regulation of freshwater fish. In addition, the finding that Cl^- activation of anion-stimulated ATPase activity was inhibited by thiocyanate (de Renzis and Bornancin 1977) was consistent with transport studies which showed that Cl^- influxes were inhibited by thiocyanate (de Renzis 1975). These studies on gill epithelium strongly support the hypothesis that the Cl^--stimulated ATPase is involved in gill anion exchanges that are related to mineral and acid-base homeostasis in freshwater fish.

The fiddler crab gill has been shown to actively absorb Cl^- from low salinities (Baldwin and Kirschner 1976a) and actively extrude Cl^- in high salinity media (Baldwin and Kirschner 1976b). In concert with these findings, DePew and Towle (1979) demonstrated the existence of an anion-stimulated ATPase in the gill cell plasma membrane of fiddler crab and suggested that this enzyme is so situated with its environment that it is highly accessible to Cl^- and HCO_3^-, and thus may play a direct role in active Cl^-/HCO_3^- exchange.

Lee (1982) used an additional approach to the question concerning correspondence between transport and anion-stimulated ATPase activity. After it was established that anion-stimulated ATPase activity existed in the plasma membrane of blue crab gill epithelium, the animals were adapted to low salinities. This thinking presumed that Cl^-/HCO_3^- exchange should increase under these osmotically stressful conditions, therefore this transport activity could be reflected in an increase in the activity of anion-stimulated ATPase activity. This was indeed the case and Lee (1982) suggested that anion-stimulated ATPase activity appears likely to play an important role in anion transport for osmoregulation and/or acid-base homeostasis in marine organisms.

Komnick et al. (1980) reported the presence of (Cl^- + HCO_3^-)-stimulated ATPase activity in plasma membranes of larval dragonfly rectum. The Cl^--stimulated ATPase activity was inhibited by thiocyanate as was the Cl^- influx into the rectal epithelia. These results suggested the possible existence of an ATPase-mediated, active Cl^- transport mechanism located in the plasma membrane of larval dragonfly rectal epithelial cells.

In the eel (*Anguilla japonica*) intestine, relatively recent electrophysiological experiments have shown that active transport of Cl^- coupled with water transport markedly increases during seawater adaptation (Ando 1975; Ando et al. 1975). The observed increase in Cl^- absorption raised the question of an associated increase in activity of an enzyme contributing to the transport process. It was demonstrated by Morisawa and Utida (1976) that anion-stimulated ATPase activity existed in an oligomycin-insensitive, thiocyanate-sensitive membrane fraction of eel intestinal enterocytes that was also relatively deficient of cytochrome c oxidase activity. Seawater adapation increased the enzyme activity commensurate with changes in Cl^- and water transport. From these considera-

Table 10. Effects of inhibitors on Cl^-/HCO_3^--ATPase activity[a]

Inhibitor	Concentration (mM)	Mg^{2+}-ATPase (Sp. Act.)	Mg^{2+} + HCO_3^--ATPase (Sp. Act.)	HCO_3^--ATPase (Sp. Act.)	Inh. (%)	Mg^{2+} + Cl^--ATPase (Sp. Act.)	Cl^--ATPase (Sp. Act.)	Inh. (%)
Control	–	10.1 ± 1.6	25.4 ± 0.9	15.3	0	16.8 ± 0.3	6.7	0
Thiocyanate	10.1	3.2 ± 1.2	11.2 ± 1.0	8.0	48	7.7 ± 0.6	4.5	33
Acetazolamide	1.0	9.7 ± 1.6	16.6 ± 2.2	6.9	55	11.1 ± 1.0	1.4	79
Acetazolamide	2.0	7.3 ± 1.2	8.7 ± 3.3	1.4	91	7.7 ± 2.0	0.4	94
Ouabain	1.0	9.5 ± 1.6	24.8 ± 0.8	15.3	0	16.5 ± 0.5	7.0	0
Vanadate	0.5	9.6 ± 1.6	21.9 ± 2.2	12.3	20	14.0 ± 1.4	4.4	34
Vanadate	1.0	8.0 ± 1.4	15.6 ± 2.3	7.6	50	10.5 ± 1.1	2.5	63

[a] The inhibitors was preincubated with enzymes in the reaction mixture at the indicated concentrations for 10 min at 30 °C. Values are means ± SE from three to five different experiments. Specific activity (Sp. Act.) is expressed as μmol 15 min^{-1} mg^{-1} protein. Inh. (%) is percentage of inhibition (Gerencser and Lee 1985 a)

tions, these authors concluded that the anion-stimulated ATPase played a direct role in Cl^- transport in the eel intestine.

Gerencser and Lee (1983, 1985 a) and Gerencser (1984 b, 1985) demonstrated the existence of a Cl^--stimulated ATPase activity in *Aplysia* enterocyte basolateral plasma membranes, suggesting a cause-and-effect relationship between ATPase activity and Cl^- transport. The ATPase activity stimulated by Cl^- was strongly inhibited by thiocyanate, vanadate, and acetazolamide (Table 10).

These results with inhibitors (Table 10) strongly suggest a possible participation by the $Cl^- - HCO_3^-$-stimulated ATPase in net chloride absorption by the *Aplysia* intestine (Gerencser and Lee 1985 a). The findings that anion-stimulated ATPase is inhibited by thiocyanate but not by ouabain have also been demonstrated in many tissues known to perform active anion transport and to contain anion-stimulated ATPase activity (Katz and Epstein 1971; Sachs et al. 1965). Additionally, thiocyanate inhibits anion transport across a variety of epithelial systems (Epstein et al. 1973; Katz and Epstein 1971). The parallel between ouabain insensitivity and thiocyanate sensitivity to Cl^--stimulated ATPase activity and net active Cl^- absorption in the *Aplysia* intestine warrants speculation that the active Cl^- absorptive mechanism could be driven by a $Cl^- - HCO_3^-$-stimulated ATPase found in the enterocyte basolateral plasma membrane. Additional support for this contention rests with the present finding that $Cl^- - HCO_3^-$-stimulated ATPase activity of the basolateral plasma membranes is inhibited by vanadate. Vanadate has been shown to inhibit ion-translocating ATPases while having no effect on the mitochondrial anion-sensitive ATPase (Akera et al. 1983). These results suggest that the $Cl^- - HCO_3^-$-stimulated ATPase is an ion-translocating ATPase of the $E_1 - E_2$ variety rather than the $F_0 - F_1$ type.

Acetazolamide inhibited $Cl^- - HCO_3^-$-stimulated ATPase activity in the *Aplysia* intestine (Table 10). This finding has also been demonstrated in blue crab gill HCO_3^--ATPase (Lee 1982). Although acetazolamide has been shown to be a specific inhibitor of carbonic anhydrase (Maren 1977), it has also been demonstrated to be a good Cl^- transport inhibitor (White 1980). Thus, our data further strengthen the notion that the $Cl^- - HCO_3^-$-stimulated ATPase, which is inhibited by acetazolamide, may be involved in Cl^- transport across the molluscan gut.

In addition, Gerencser (1984b, 1985) and Grencser and Lee (1985b) have demonstrated an ATP-dependent Cl^- uptake in *Aplysia* inside-out enterocyte basolateral plasma membrane vesicles which was inhibitable by thiocyanate, vanadate, and also, by acetazolamide (Gerencser 1986a). The ATP-driven Cl^- uptake was obtained in the absence of Na^+, K^+, HCO_3^-, or a pH gradient between the intra- and extravesicular space, which is strong suggestive evidence that the $(Na^+ + K^+)$-ATPase enzyme (Schuurmans Stekhoven and Bonting 1981), Na^+/Cl^- symport (Frizzell et al. 1979; Gerencser and Lee 1983), K^+/Cl^- symport (Greger and Schlatter 1984), $Na^+/K^+/Cl^-$ symport (Greger and Schlatter 1984), Cl^-/HCO_3^- or Cl^-/OH^- antiport (Gerencser and Lee 1983; Gerencser 1986b), or K^+/H^+ antiport (Schuurmans Stekhoven and Bonting 1981) are not mechanisms that are involved in the accumulation of Cl^- within the vesicles in the presence of ATP. It was also demonstrated that acetazolamide could inhibit the ATP plus Cl^--dependent intravesicular negative potential (measured by lipophilic cation distribution at steady state) in *Aplysia* enterocyte basolateral membrane vesicles (Gerencser 1986a).

In an attempt to further elucidate the electrogenicity of the ATP-dependent Cl^- transport process, several experimental maneuvers were performed by Gerencser (1986b) as follows: first, an inwardly directed valinomycin-induced K^+ diffusion potential, making the vesicle interior electrically positive, enhanced ATP-driven Cl^- uptake compared with vesicles lacking the ionophore (Fig. 1); second, an inwardly directed FCCP-induced H^+ diffusion potential, making the vesicle interior less negative, increased ATP-dependent Cl^- uptake compared to control; third, ATP plus Cl^- was required for an increase in intravesicular negativity measured by lipophilic $TPMP^+$ distribution across the vesicular membrane. Vanadate, thiocyanate, and valinomycin-K^+ inhibited the ATP plus Cl^--dependent intravesicular negativity (Gerencser 1986b). These results are strongly consistent with the hypothesis that the active electrogenic Cl^- transport mechanism in *Aplysia* intestine is an electrogenic, Cl^--transporting, Cl^--stimulated ATPase found in the enterocyte-basolateral membrane.

In summary, it is quite apparent that in the past few years there has been an increasing number of convincing studies in a variety of animal tissues that have provided indirect, correlative evidence that active Cl^- transport is primary by nature. The active translocation of Cl^- by an enzyme that directly utilizes the energy from ATP hydrolysis is not unlike that observed in plants (Hill and Hanke 1979; Auffret and Hanke 1981). Indeed, the evidence for primary active Cl^- transport in these simple living things is almost as convincing as that presented for $(Na^+ + K^+)$-stimulated ATPase and $(Ca^{2+} + Mg^{2+})$-stimulated ATPase in their respective roles for actively transferring Na^+, K^+, and Ca^{2+} across animal plasma membranes. As alluded to by DePont and Bonting (1981) and Gerencser (1986b) future experimental steps in assuring that an animal Cl^--

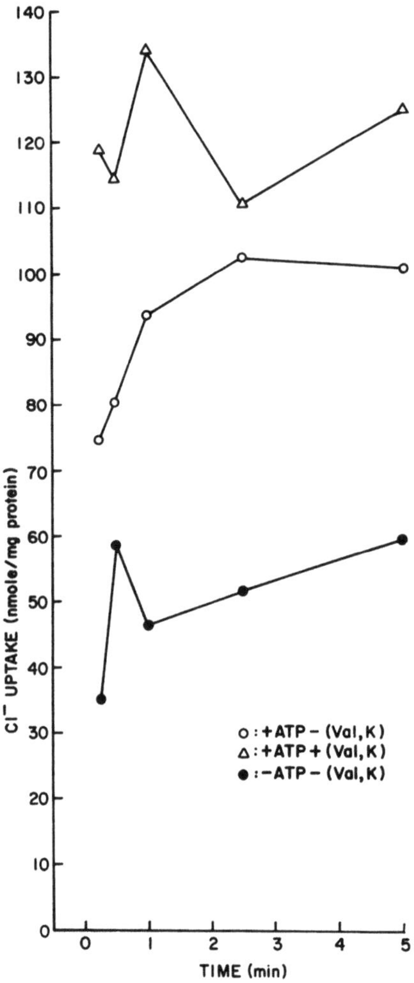

Fig. 1. Effect of valinomycin-induced transmembrane K^+ diffusion potential on time course of ATP-dependent Cl^- uptake by *Aplysia* EPMV. *Triangles,* vesicles whose reaction mixture contained 40 mM K^+ and 0.18 mM valinomycin, otherwise they were under isoosmotic equilibrium tracer exchange conditions. *Open circles*: same as above except that no valinomycin nor K^+ was present on either side of the vesicular membrane. *Solid circles*: same as above, except that no valinomycin, K^+, nor ATP was present on either side of the vesicular membrane. pH 7.8, on both sides of the vesicular membrane, was used throughout these experiments. The extra- and intravesicular medium composition was as follows: 10 mM Tris/HEPES (pH 7.8), 250 mM sucrose, 3 mM $MgSO_4$, and 25 mM choline chloride. Values reported are means of nine different experiments (40–45 animals)

stimulated ATPase is involved in primary Cl^- transcellular movement should approximate the following: (1) a specific inhibitor for the enzyme should be found or synthesized (e.g., an antibody) and this inhibitor should inhibit the transport process; and (2) the Cl^--stimulated ATPase should be biochemically isolated and after its incorporation in liposomes should then be shown to support active Cl^- transport.

3 Model for Chloride Absorption

Mucosal D-glucose caused a depolarization of ψ_m, a hyperpolarization of ψ_{ms}, and an increase in a^i_{Cl}, which were predictions of the electrical coupling hypothesis that the author previously formulated (Gerencser 1978a; Gerencser and

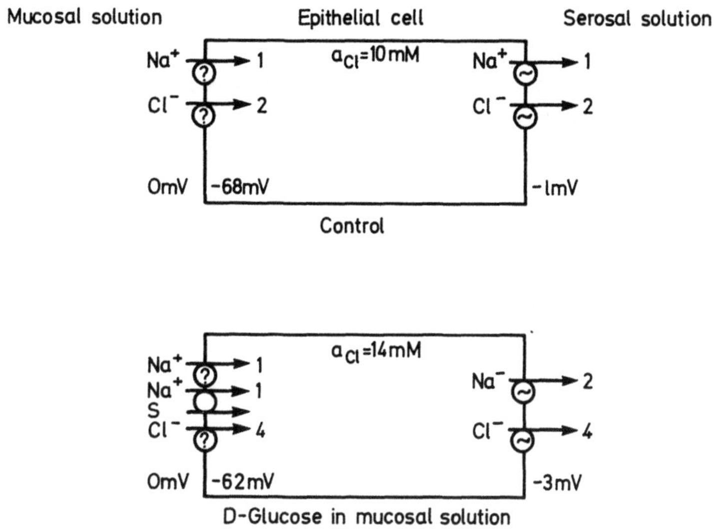

Fig. 2. Models depicting the relationship between transport of sodium and chloride, in the absence and presence of D-glucose, across the brush border of in vitro *Aplysia* intestine and active transmural transport. The *arrows* at the mucosal and serosal membranes designate net movements. The mechanisms at the serosal membrane are active transport processes (Gerencser 1981a)

Hong 1977). Figure 2 (lower panel) illustrates diagrammatically the proposed electrical coupling model which describes how mucosally applied, actively transported, organic solutes (e.g., sugars, amino acids) enhance Cl^- absorption across the *Aplysia* intestine. If a Na^+, sugar cellular entry depolarized the mucosal membrane as observed in Table 3, this would lower the electrochemical gradient opposing intracellular Cl^- entry. This would then allow Cl^- to move across the mucosal membrane into the intracellular space with greater ease as has been demonstrated with input resistance measurements and via membrane conductivity calculations (vide supra). The finding that a^i_{Cl} is significantly higher in the presence of mucosal D-glucose than in its absence lends strong support for this argument (Gerencser and White 1980). The Cl^- pump as described by Gerencser (1978a, b) would then accommodate the increased intracellular thermodynamic activity of Cl^- by increasing its rate of work much as a variable output device. This was previously demonstrated both by an increase in SCC and unidirectional $M \rightarrow S$ Cl^- flux after mucosal D-glucose addition (Gerencser 1978a). Since it has been shown that the major portion of the SCC before and after D-glucose addition to the mucosal solution is a Cl^- current (Gerencser 1978a), there would be a decrease in the negativity of the ψ_s (Gerencser and White 1980) by the increased Cl^- pump activity; which, linked with ψ_m and ψ_s coupling through a low resistance extracellular shunt (Gerencser and Loughlin 1983), would lead to a greater serosal-negative ψ_{ms} (Schultz 1977). The active Na^+ transport system would likewise accommodate the increased intracellular Na^+ transport pool by increasing its activity. This was also previously demonstrated by an increase in the unidirectional $M \rightarrow S$ Na^+ flux after mucosal glucose addition (Gerencser 1978a).

The finding that removal of Na^+ from the bathing medium hyperpolarized both ψ_m and ψ_{ms} suggests that: (1) the mucosal membrane has a finite conductance to Na^+, and (2) Na^+ extrusion across the basolateral membrane of the *Aplysia californica* enterocyte is electrogenic. The observation that Cl^- removal from the bathing medium depolarized both ψ_m and ψ_{ms} suggests that the mucosal membrane has a large conductance to Cl^-, and (2) the Cl^- extrusion mechanism located within the basolateral membrane is electrogenic. Collectively, these results indicate that there are separate and independent transport mechanisms for both Na^+ and Cl^- located within the mucosal membrane (Fig. 2, upper panel). Since the electrochemical potential for both of these ions is less in the cytosol than in the mucosal bathing solution, no active transport system for either ion need be invoked at the mucosal membrane level (Gerencser 1982a, 1983b). The active extrusion mechanism for Cl^- located within the basolateral membrane could be the Cl^--stimulated ATPase that was observed in the basolateral membrane fraction of *Aplysia californica* enterocytes (Gerencser and Lee 1983, 1985a).

4 Conclusions

The results and conclusions presented in this chapter emphasize the electrical and biochemical transport characteristics of *Aplysia* intestine relative to solute transport in other epithelial preparations.

With respect to Na^+ transport, the results presented herein strongly suggest that the underlying mechanisms involved in the net $M \rightarrow S$ transfer of this ion by *Aplysia* intestine do not differ fundamentally from the corresponding processes in vertebrate intestine (Quay and Armstrong 1969a; Rose and Schultz 1970; Schultz and Curran 1970). Among other similarities observed is the enhanced capacity for net $M \rightarrow S$ Na^+ transport, under suitable conditions, to actively transported sugars. Other similarities also observed include the size and orientation of the electrical potential differences across the mucosal and serosal boundaries of the epithelial cells (Gerencser 1981b; Gerencser and White 1980) and that Na^+ exists in an electrochemical "well" in the *Aplysia* enterocyte (Gerencser 1983c). Also, Na^+/K^+-ATPase activity exists in the basolateral membrane fraction of *Aplysia* enterocytes (Table 7), a finding no different for numerous epithelia (Schuurmans Stekhoven and Bonting 1981).

However, there seem to be several fundamental differences between the Cl^- transfer processes of *Aplysia* intestine and the corresponding processes in vertebrate intestine. Cl^- absorption across vertebrate intestine has been shown to be active and linked to the movement of Na^+ in the same direction (Frizzell et al. 1979; Quay and Armstrong 1969a). This observation has been strengthened and confirmed by recent intracellular Cl^- activity measurements which show an accumulation of intracellular Cl^- across the mucosal membrane (Armstrong et al. 1977; Duffey et al. 1979; White 1980). In contrast, Cl^- accumulation was not observed in *Aplysia* intestine (Gerencser and White 1980; Gerencser 1983b). In fact, intracellular Cl^- was found to be at a lower electrochemical potential than in the mucosal medium (Gerencser and White 1980; Table 4). Also, the intestine of *Aplysia* responds to the mucosal presence of actively transported solutes such as sugars or amino acids in a way that is unique when compared with observations from other intestinal preparations. In intestine from this animal, sugars and amino acids stimulate Cl^- absorption as well as Na^+ absorption, and the increase in Cl^- absorption exceeds that of Na^+ (Gerencser 1978a, 1981a). This is reflected in ψ_{ms}. In all vertebrate preparations studied to date, actively transported sugars and amino acids elevate ψ_{ms}, i.e., cause the serosa to become more positive (Quay and Armstrong 1969b; Schultz 1977; Schultz and Curran 1970). This reflects enhanced Na^+ absorption (Schultz 1977; Schultz and Curran 1970). In contrast, these sugars and amino acids reduce ψ_{ms} in *Aplysia* intestine, i.e., they cause the serosa to become more negative (Gerencser 1978a, 1981a). The experiments described herein suggest that the coupling of metabolic energy to active Cl^- transport across the *Aplysia californica* intestine may be effected by a mechanism which differs significantly from those involved in the activation of the Cl^- independent Na^+ pump in this tissue, namely, a Cl^--stimulated ATPase.

Acknowledgments. The author wishes to acknowledge his gratitude to his technologists, students, and collaborators for their able contributions to the studies reviewed and performed herein. These studies were supported by NIAMD Training Grant (1-T01-AM-05697-02 and -03), NIH Grants (No. AM17361 and RCDA No. AM00367), D.S.R. Seed Award (No. 229K15), Whitehall Foundation Grant (No. 78-156 ck-1), and D.S.R. Award (No. 122101010).

References

Ahearn GA, Maginnis LA, Song YK, Tornquist A (1977) Intestinal water and ion transport in freshwater malacostracan prawns (Crustacea). In: Jungreis AM, Hodges TK, Kleinzeller AM, Schultz SG (eds) Water relations in membrane transport in plants and animals. Academic, London New York, pp 129–142

Akera T, Temma K, Takeda K (1983) Cardiac actions of vanadium. Fed Proc 42:2984–2988

Al-Bazzaz FJ, Al-Awqati Q (1979) Interaction between sodium and chloride transport in canine tracheal mucosa. J Appl Physiol 46:111–119

Alvarado R, Crane RK (1962) Phlorizin as a competitive inhibitor of the active transport of sugars by hamster small intestine in vitro. Biochim Biophys Acta 52:170–172

Amelsvoort JMM van, DePont JJHHM, Bonting SL (1977) Is there a plasma membrane-located anion-sensitive ATPase? Biochim Biophys Acta 466:283–301

Ando M (1975) Intestinal water transport and chloride pump in relation to sea-water adaptation of the eel, *Anguilla japonica*. Comp Biochem Physiol 52A:229–233

Ando M, Utida S, Nagahama H (1975) Active transport of chloride in eel intestine with special reference to sea water adaptation. Comp Biochem Physiol 51A:27–32

Armstrong W McD (1975) Electrophysiology of sodium transport by epithelial cells of the small intestine. In: Csaky TZ (ed) Intestinal absorption and malabsorption. Raven, New York, pp 45–66

Armstrong W McD, Suh TK, Gerencser GA (1972) Stimulation by anoxia of active chloride transfer in isolated bullfrog intestine. Biochim Biophys Acta 225:647–662

Armstrong W McD, Wojtkowski W, Bixenman WR (1977) A new solid-state microelectrode for measuring intracellular chloride activities. Biochim Biophys Acta 465:165–170

Armstrong W McD, Bixenman WR, Frey KF, Garcia-Diaz JF, O'Regan MG, Owens JH (1979a) Energetics of coupled Na^+ and Cl^- entry into epithelial cells of bullfrog small intestine. Biochim Biophys Acta 551:207–219

Armstrong W McD, Garcia-Diaz JF, O'Doherty J, O'Regan MG (1979b) Transmucosal Na^+ electrochemical potential difference and solute accumulation in epithelial cells of the small intestine. Fed Proc 38:2722–2728
Auffret CA, Hanke DE (1981) Improved preparation and assay and some characteristics of Cl^--ATPase activity from *Limonium vulgare*. Biochim Biophys Acta 648:186–191
Baldwin GF, Kirschner LB (1976a) Sodium and chloride regulation in *Uca* adapted to 175% seawater. Physiol Zool 49:158–171
Baldwin GF, Kirschner LB (1976b) Sodium and chloride regulation in *Uca* adapted to 10% seawater. Physiol Zool 49:172–180
Barry RJC, Smyth DH, Wright EM (1965) Short-circuit current and solute transfer by rat jejunum. J Physiol (London) 181:410–431
Bidder AM (1950) The digestive mechanism of the European squids: *Loligo vulgaris, Loligo forbesii, Alloteuthis media*, and *Alloteuthis subulata*. Q J Microsc Sci 91:1–43
Binder HJ, Rawlins CL (1973) Electrolyte transport across isolated large intestinal mucosa. Am J Physiol 225:1232–1239
Bornancin M, de Renzis G, Maetz J (1977) Branchial Cl transport, anion-stimulated ATPase and acid-base balance in *Anguilla anguilla* adapted to freshwater: Effects of hyperoxia. J Comp Physiol 117:313–322
Bornancin M, de Renzis G, Naon R (1980) $Cl^- - HCO_3^-$-ATPase in gills of the rainbow trout: evidence for its microsomal localization. Am J Physiol 238:R251–R259
Bullough DA, Jackson CG, Henderson PJF, Beechey RB, Linnett PE (1982) The isolation and purification of the elvapeptins: a family of peptide inhibitors of mitochondrial ATPase activity. FEBS Lett 145:258–262
Burg M, Green N (1973) Function of the thick ascending limb of Henle's loop. Am J Physiol 224:659–664
Candia OA (1972) Ouabain and sodium effects on chloride fluxes across the isolated bullfrog cornea. Am J Physiol 223:1053–1057
Chien W, Stevens CE (1972) Coupled active transport of Na and Cl across forestomach epithelium. Am J Physiol 223:997–1003
Clarkson TW, Cross AC, Toole SR (1961) Electrical potentials across isolated small intestine of the rat. Am J Physiol 200:1233–1235
Degnan KJ, Karnaky KJ, Zadunaisky JA (1977) Active chloride transport in the in vitro opercular skin of a teleost *Fundulus heteroclitus*, a gill-like epithelium rich in chloride cell. J Physiol (London) 271:155–191
DePew EF, Towle DW (1979) Bicarbonate stimulated ATPase in plasma membrane fractions of fiddler crab (*Uca minax*) gill. Mar Biol Lett 1:59–67
DePont JJHHM, Bonting SL (1981) Anion-sensitive ATPase and $(K^+ + H^+)$-ATPase. In: Bonting SL, DePont JJHHM (eds) Membrane transport. Elsevier/North Holland Biomed Press, Amsterdam, pp 209–222
Duffey ME, Thompson SM, Frizzell RA, Schultz SG (1979) Intracellular chloride absorption by small intestine of the flounder, *Pseudopleuronectes americanus*. Fed Proc 38:1059
Epstein FH, Maetz J, de Renzis G (1973) Active transport of chloride by the teleost gill: inhibition by thiocyanate. Am J Physiol 244:1295–1299
Ericson A-C, Spring KR (1982) Coupled NaCl entry into *Necturus* gallbladder epithelial cells. Am J Physiol 243:C140–C145
Farmer LL, Evans DH (1981) Chloride extrusion in the isolated perfused teleost gill. J Comp Physiol 141:471–476
Field M, Karnaky KJ, Smyth PL, Bolton JE, Kinter WB (1978) Ion transport across the isolated intestinal mucosa of the winter flounder, *Pseudopleuronectes americanus*. I. Functional and structural properties of cellular and paracellular pathways for Na and Cl. J Membr Biol 41: 265–275
Frizzell RA, Dugas M, Schultz SG (1975) Sodium chloride transport by rabbit gallbladder: direct evidence for a coupled NaCl influx process. J Gen Physiol 65:769–795
Frizzell RA, Koch MJ, Schultz SG (1976) Ion transport by rabbit colon. I. Active and passive components. J Membr Biol 27:297–316
Frizzell RA, Field M, Schultz SG (1979) Sodium-coupled chloride transport by epithelial tissues. Am J Physiol 236:F1–F8

Fromter E, Gebler B (1977) Electrical properties of amphibian urinary bladder. III. The cell membrane resistance and the effect of amiloride. Pfluegers Arch 371:99–108
Fuchs W, Larsen EH, Lindemann B (1977) Current-voltage curve of sodium channels and concentration dependence of sodium permeability in frog skin. J Physiol (London) 267:137–166
Gerencser GA (1978a) Enhancement of sodium and chloride transport by monosaccharides in *Aplysia californica* intestine. Comp Biochem Physiol 61A:203–208
Gerencser GA (1978b) Electrical characteristics of isolated *Aplysia californica* intestine. Comp Biochem Physiol 61A:209–212
Gerencser GA (1979a) Metabolic dependence of active sulfate transport in *Aplysia californica* intestine. Comp Biochem Physiol 63A:519–522
Gerencser GA (1979b) Stimulation of sulfate transport by exogenous sugars in *Aplysia californica* intestine. Comp Biochem Physiol 64A:251–255
Gerencser GA (1981a) Effects of amino acids on chloride transport in *Aplysia* intestine. Am J Physiol 240:R61–R69
Gerencser GA (1981b) Electrical transport characteristics of *Aplysia californica* intestinal epithelium. Comp Biochem Physiol 68A:225–230
Gerencser GA (1981c) Review: intestinal potentials. Comp Biochem Physiol 69A:15–21
Gerencser GA (1982a) Membrane physiology of molluscs. In: Podesta RB, Timmers SF (eds) Membrane physiology of invertebrates. Dekker, New York
Gerencser GA (1982b) Paracellular transport characteristics of *Aplysia juliana* intestine. Comp Biochem Physiol 72A:721–725
Gerencser GA (1983a) Invertebrate epithelial transport. Am J Physiol 244:R127–R129
Gerencser GA (1983b) Electrophysiology of chloride transport in *Aplysia* (mollusk) intestine. Am J Physiol 244:R143–R149
Gerencser GA (1983c) Na^+ absorption in *Aplysia* intestine: Na^+ fluxes and intracellular Na^+ and K^+ activities. Am J Physiol 244:R412–R417
Gerencser GA (1984a) Thiocyanate inhibition of active chloride absorption in *Aplysia* intestine. Biochim Biophys Acta 775:389–394
Gerencser GA (1984b) Electrogenic and electrically coupled chloride transport across molluscan intestine. In: Gerencser GA (ed) Chloride transport coupling in biological membranes and epithelia. Elsevier, Amsterdam, pp 183–203
Gerencser GA (1984c) Concentrations and activity coefficients of Na^+, K^+ and Cl^- in *Aplysia californica* enterocytes. Comp Biochem Physiol 77A:717–720
Gerencser GA (1984d) Transapical Na^+ electrochemical potential difference and solute accumulation in *Aplysia californica* enterocytes. Comp Biochem Physiol 78A:137–140
Gerencser GA (1984e) Modifiers of active chloride transport across *Aplysia californica* intestine. Comp Biochem Physiol 78A:607–611
Gerencser GA (1985) Transport across the invertebrate intestine. In: Gilles R, Gilles-Baillien MG (eds) Transport processes, iono- and osmoregulation, vol 2. Springer, Berlin Heidelberg New York, pp 251–264
Gerencser GA (1986a) Inhibition of a chloride pump by acetazolamide in the intestine of *Aplysia californica*. J Exp Biol 125:391–393
Gerencser GA (1986b) Properties and functions of Cl^--stimulated ATPase. Trends Life Sci 1:1–18
Gerencser GA, Hong SK (1975) Roles of sodium and potassium ions on *p*-aminohippurate transport in rabbit kidney slices. Biochim Biophys Acta 406:108–119
Gerencser GA, Hong SK (1977) Ion transport in *Aplysia juliana* intestine: stimulation by exogenous sugars. Comp Biochem Physiol 58A:275–280
Gerencser GA, Lee SH (1983) Cl^--stimulated adenosine triphosphatase: existence, location and function. J Exp Biol 106:143–161
Gerencser GA, Lee SH (1985a) Cl^-/HCO_3^--stimulated ATPase in intestinal mucosa of *Aplysia*. Am J Physiol 248:R241–R248
Gerencser GA, Lee SH (1985b) ATP-dependent chloride transport in plasma membrane vesicles from *Aplysia* intestine. Biochim Biophys Acta 816:415–417
Gerencser GA, Loughlin GM (1983) Extracellular transport properties of *Aplysia californica* intestine. Comp Biochem Physiol 74A:701–704
Gerencser GA, White JF (1980) Membrane potentials and chloride activities in epithelial cells of *Aplysia* intestine. Am J Physiol 239:R445–R449

Gerencser GA, Hong SK, Malvin G (1977 a) Metabolic dependence of transmural potential difference and short-circuit current in isolated *Aplysia juliana* intestine. Comp Biochem Physiol 56A:519 – 523

Gerencser GA, Mustafa T, Hong SK (1977 b) The relationship between intracellular Na^+ and energy charge potential in the kidney slice of the rabbit, *Oryctolagus cuniculus*. Comp Biochem Physiol 58A:329 – 331

Graf J, Giebisch G (1979) Intracellular sodium activity and sodium transport in *Necturus* gallbladder epithelium. J Membr Biol 47:327 – 355

Greger R, Schlatter E (1984) Mechanisms of chloride transport in vertebrate renal tubule. In: Gerencser GA (ed) Chloride transport coupling in biological membranes and epithelia. Elsevier, Amsterdam, pp 271 – 346

Gunter-Smith PJ, White JF (1979) Responce of *Amphiuma* small intestine to theophylline: effect on bicarbonate transport. Am J Physiol 236:E775 – E783

Hanrahan J, Phillips JE (1983) Mechanism and control of salt absorption in locust rectum. Am J Physiol 244:R131 – R142

Harvey WR, Nedergaard S (1964) Sodium independent active transport of potassium in the isolated midgut of *Cecropia* silkworm. Proc Natl Acad Sci USA 51:757 – 762

Harvey WR, Haskell JA, Nedergaard S (1968) Active transport by the *Cecropia* midgut. III. Potential generated by active K transport. J Exp Biol 48:1 – 12

Hill BS, Hanke DE (1979) Properties of the chloride ATPase from *Limonium* salt glands: activation by, and binding to, specific sugars. J Membr Biol 51:185 – 194

House CR, Green K (1965) Ion and water transport in isolated intestine of the marine teleost, *Cottus scorpius*. J Exp Biol 42:177 – 189

Hubel KA (1968) The ins and outs of bicarbonate in the alimentary tract. Gastroenterology 56:647 – 651

Katz AI, Epstein FH (1971) Effect of anions on adenosine triphosphatase of kidney tissue. Enzyme 12:499 – 507

Kerstetter T, Kirschner LB (1972) Active chloride transport by the gills of rainbow trout (*Salmo gairdneri*). J Exp Biol 56:263 – 272

Klyce SD, Wong RKS (1977) Site and mode of adrenaline action on chloride transport across the rabbit corneal epithelium. J Physiol (London) 266:777 – 799

Komnick H, Schmitz M, Hinssen H (1980) Biochemischer Nachweis von HCO_3^- und Cl^--abhängigen ATPase-Aktivitäten in Rectum von anisopteren Libellenlarven und Hemmung der rectalen Chloridaufnahme durch Thiocyanat. Eur J Cell Biol 20:217 – 227

Lahlou B, Fossat B (1971) Mécanisme de transport de l'eau et du sel à travers la vessie urinaire d'un poisson téléostéen en eau douce, la truite arc-en-ciel. C R Acad Sci 273:2108

Lawrence AL, Lawrence D (1967) Sugar absorption in the intestine of the chiton, *Cryptochiton stelleri*. Comp Biochem Physiol 22:341 – 357

Lawrence AL, Mailman DS (1967) Electrical potentials and ion concentrations across the gut of *Cryptochiton stelleri*. J Physiol (London) 193:535 – 545

Lawrence AL, Mailman DS, Puddy RE (1972) The effect of carbohydrates on the intestinal potentials of *Cryptochiton stelleri*. J Physiol (London) 225:515 – 527

Lee SH (1982) Salinity adaptation of HCO_3^--dependent ATPase activity in the gills of blue crab (*Callinectes sapidus*). Biochim Biophys Acta 689:143 – 154

Lerner J (1978) Ion dependency of transport. In: Lerner J (ed) A review of amino acid transport processes in animal cells and tissues. Univ Maine Orono Press, Orono

Leslie BR, Schwartz JH, Steinmetz PR (1973) Coupling between Cl absorption and HCO_3 secretion in turtle urinary bladder. Am J Physiol 25:610 – 617

Maetz J, Garcia-Romeu F (1964) The mechanism of sodium and chloride uptake by the gills of a fresh water fish, *Carassius auratus*. II. Evidence of NH_4^+/Na^+ and HCO_3^-/Cl^- exchanges. J Gen Physiol 47:1209 – 1227

Maren TH (1977) Use of inhibitors in physiological studies of carbonic anhydrase. Am J Physiol 232:F291 – F297

Mircheff AK, Sachs G, Hanna SD, Labiner CS, Rabon E, Douglas AP, Walling MW, Wright EM (1979) Highly purified basal lateral plasma membranes from rat duodenum: physical criteria for purity. J Membr Biol 50:343 – 363

Morisawa M, Utida S (1976) HCO_3^--activated adenosine triphosphatase in intestinal mucosa of the eel. Biochim Biophys Acta 445:458–463

Nagel W, Garcia-Diaz JF, Armstrong W McD (1981) Intracellular ionic activities in frog skin. J Membr Biol 6:127–134

Nellans HN, Frizzell RA, Schultz SG (1973) Coupled sodium-chloride influx across the brush border of rabbit ileum. Am J Physiol 225:467–475

Nellans HN, Frizzell RA, Schultz SG (1974) Brush border processes and transepithelial Na and Cl transport by rabbit ileum. Am J Physiol 226:1131–1141

Newey H, Armstrong PA, Smyth DH (1965) Uranyl ions and intestinal hexose transfer. Nature (London) 205:389–390

Quay JF, Armstrong W McD (1969a) Sodium and chloride transport by isolated bullfrog small intestine. Am J Physiol 217:694–702

Quay JF, Armstrong W McD (1969b) Enhancement of net sodium transport in isolated bullfrog intestine by sugars and amino acids. Proc Soc Exp Biol Med 131:51–64

Renfro JL (1977) Interdependence of active Na and Cl transport by the isolated urinary bladder of the teleost, *Pseudopleuronectes americanus*. J Exp Zool 199:383–390

Renzis G de (1975) The branchial chloride pump in the goldfish (*Carassius auratus*): relationship between Cl^-/HCO_3^- and Cl^-/Cl^- exchanges and the effect of thiocyanate. J Exp Biol 63:587–602

Renzis G de, Bornancin M (1977) Cl^-/HCO_3^- ATPase in the gills of *Carassius auratus:* its inhibition by thiocyanate. Biochim Biophys Acta 467:192–207

Renzis G de, Maetz J (1973) Studies on the mechanism of chloride absorption by the goldfish gill: relation with acid-base regulation. J Exp Biol 59:339–358

Rose RC, Schultz SG (1970) Sugar and amino acid effects on the electrical potential profile across rabbit ileum. Biochim Biophys Acta 211:376–378

Sachs G, Mitch WE, Hirschowitz BI (1965) Frog gastric mucosal ATPase. Proc Soc Exp Biol Med 119:1023–1027

Sachs G, Shah G, Strych A, Cline G, Hirschowitz BI (1972) Properties of ATPase of gastric mucosa. III. Distribution of HCO_3^--stimulated ATPase in gastric mucosa. Biochim Biophys Acta 266: 625–638

Sachs G, Spenney JG, Lewin M (1978) H^+ transport: regulation and mechanism in gastric mucosa and membrane vesicles. Physiol Rev 58:106–178

Schultz SG (1977) Sodium-coupled solute transport by small intestine: a status report. Am J Physiol 223:E249–E254

Schultz SG, Curran PF (1968) Intestinal absorption of sodium chloride and water. In: Code CF, Heidel W (eds) Handbook of physiology: alimentary canal, Sect 6, vol III. Am Physiol Soc, Washington DC, pp 1245–1275

Schultz SG, Curran PF (1970) Coupled transport of sodium and organic solutes. Physiol Rev 50:637–718

Schuurmans Stekhoven F, Bonting SL (1981) Transport adenosine triphosphates: properties and functions. Physiol Rev 61:1–76

Silva P, Stoff J, Field M, Fine L, Forrest N, Epstein FH (1977) Mechanism of active chloride secretion by shark rectal gland: role of Na–K-ATPase in chloride transport. Am J Physiol 233:F298–F306

Simon B (1972) The HCO_3^--stimulated ATPase in cat and rabbit pancreatic tissue. In: Sachs G, Heinz F, Ullrich KJ (eds) Gastric secretion. Academic Press, London New York, pp 345–348

Skadhauge E (1974) Coupling of transmural flows of NaCl and water in the intestine of the eel (*Anguilla anguilla*). J Exp Biol 60:535–546

Skou JC (1965) Enzymatic basis for active transport of Na^+ and K^+ across cell membrane. Physiol Rev 45:596–617

Spring KA, Kimura G (1978) Chloride reabsorption by renal proximal tubules of *Necturus*. J Membr Biol 38:233–254

Stobbart RH (1967) The effect of some anions and cations upon the fluxes and net uptake of chloride in the larva of *Aedes aegypti* (L) and the nature of the uptake mechanisms for sodium and chloride. J Exp Biol 47:35–57

Sudou K, Hoshi T (1977) Mode of action of amiloride in toad urinary bladder. J Membr Biol 32:115–132

Turnberg LA, Bieberdorf FA, Morawaki SG, Fordtran JS (1970) Interrelationships of chloride, bicarbonate, sodium and hydrogen transport in the human ileum. J Clin Invest 49:557–567
Ussing HH, Zerahn K (1951) Active transport of sodium as the source of electric current in the short-circuit isolated frog skin. Acta Physiol Scand 23:110–127
Wais U, Knauf H (1975) H^+ transport and membrane-bound HCO_3-ATPase in salivary duct epithelium. Pfluegers Arch Gesamte Physiol Menschen Tiere 361:61–64
Watlington CO, Jessee SD, Baldwin G (1977) Ouabain, acetazolamide, and Cl^- flux in isolated frog skin: evidence for two distinct active Cl transport mechanisms. Am J Physiol 232:F550–F558
White JF (1977) Activity of chloride in absorptive cells of *Amphiuma* small intestine. Am J Physiol 232:E552–E559
White JF (1980) Bicarbonate-dependent chloride absorption in small intestine: Ion fluxes and intracellular chloride activities. J Membr Biol 53:95–107
White JF, Armstrong W McD (1971) Effect of transported solutes on membrane potentials in bullfrog small intestine. Am J Physiol 22:194–201
Williams D, Phillips JE, Prince WT, Meredith J (1978) The source of short-circuit current across locust rectum. J Exp Biol 77:107–122

Chapter 3

NaCl Transport in the Vertebrate Intestine

J.A. Groot and *R. Bakker*

Contents

Department of Animal Physiology, University of Amsterdam, Kruislaan 320, 1098 SM Amsterdam, The Netherlands

Advances in Comparative and Environmental Physiology
Vol. 1. Edited by R. Greger

1 Introduction

1.1 Developments in Knowledge of Intestinal Transport Function

We will summarize very briefly the developments in our knowledge and in concepts which had and will continue to have, a large contribution to further understanding of the transport processes in the intestinal epithelium of vertebrates.

1. New methods to approximate the osmotic water permeability have brought the insight that just small differences in osmolarity suffice for transepithelial water transport.
2. The introduction of the model for transcellular Cl^- transport, underlying fluid secretion, has generated fruitful experiments, and much evidence in favor of the model. The double exchanger model consisting of Na/H and Cl/anion antiporters has been accepted as an important means for neutral NaCl absorption. In addition, it has been discovered that K^+ transport in the colon is transcellular and requires active uptake mechanisms in the apical as well as in the basolateral membrane.
   Much work has been done in understanding the intracellular regulatory mechanisms for transcellular and paracellular ion transport. The results show the involvement of all known intracellular messengers. Their specific individual functions and their interactions are under active investigation.
3. The recognition of neuronal and hormonal effects in intestinal preparations (the secretion/absorption tonus) has clarified some of the experimental variations between animals and has explained some of the conflicting results obtained in different laboratories. In this respect the recognition of the heterogeneity of the large intestine should also be mentioned.
4. The sophistication of experimental techniques has improved our understanding of the electrophysiological behavior of the intestine and have led, for instance, to the quantitative analyses of the influence of the series resistance in transport studies. In conjunction, we will draw attention to the influence of transport number effects on intestinal transport.
5. The application of new techniques like patch clamping and the use of fluorescent probes in studies with epithelial cell lines have just started and will add precious pieces to our knowledge.

1.2 Salt and Water Transport in the Intestine

In most vertebrates the gut is the only organ involved in the uptake of NaCl and water. Although it has been shown clearly that water absorption can take place from the intestinal lumen to the blood against its thermodynamic potential, suggesting active transport, it is now generally agreed that water transport across the intestinal epithelium is secondary to the transport of solutes (Curran and Solomon 1957; Diamond 1964). The most important solute is presumably NaCl.

The intestine not only absorbs water but it can also secrete large volumes of fluid. Again, the secretory movement of water is thought to be secondary to the transport of NaCl.

The uptake of water and salt is necessary to compensate for obligatory losses associated with respiratory, thermoregulatory, and renal functions which, of course, differ with environmental conditions. Except for the water and salt uptake through the integument (gills and skin) of aquatic lower vertebrates all uptake is performed by the intestinal tract.

However, not only oral uptake but also transepithelial secretions constitute the load for intestinal absorptive processes. As an example we will discuss briefly some figures for man.

A standard daily diet of an adult healthy man contains 2 to 3 liters of water. Another 7 liters is added to the intestinal content by salivary, gastric, billiary, pancreatic, and intestinal secretions. Most of the fluid is reabsorbed so that the normal fluid load to the human colon is strongly reduced to approximately 2 liters per day. This volume is further reduced to the normal fecal water excretion of 150 ml per day.

The maximal absorptive capacity of the colon is considered to be about 5 liters per day (Phillips and Giller 1973; Debongnie and Phillips 1978). Thus, a normal colon can compensate for small-intestinal malabsorption or secretion and, until the fluid load is increased to above 5 liters per day, as for instance in enterotoxin-induced secretory diarrhea, no increased fecal water loss can be detected. However, a small decrease in net colon absorption will result in a significant increase in fecal water content.

The activity of the large intestine in water conservation is even more impressive in larger animals like the pony where a daily load of about 20 liters has been found (Argenzio et al. 1974). According to these authors about 12 liters of fluid may be secreted by the colon itself. Thus, about 30 liters is absorbed daily, an amount equivalent to the animals extracellular fluid volume.

With the exception of the gastric secretion, the principal cation in the secreted as well as in the absorbed fluids is Na^+.

Another example of the importance of intestinal function in salt and water homeostasis can be found in euryhaline fish. In seawater the animal has to gain water to counter the osmotic losses through the gills and the skin. The problem is how the fish can take up water from a solution which has at least three time the osmolarity of the animal. In freshwater the same animal has to get rid of the water entering from the outside down an even larger osmotic gradient. The main factor in adaptation of euryhaline teleosts to water of different salinities is the intestinal NaCl absorption (cf. Skadhauge 1980; Groot et al. 1983; Lahlou 1983; Kirsch et al. 1985).

Because of the importance of the NaCl transport for water transport we will start with a discussion of the coupling between these two transports. This brings us first to the concept of unstirred layers.

1.3 Unstirred Layers

This concept states that even in a very well stirred solution there is a region immediately adjacent to a membrane where no effective stirring occurs and in which solutes only move by diffusion. The actual situation in which the stirring effect gradually disappears on approaching the membrane is usually modeled as a layer with a finite thickness with no stirring at all between the membrane and the perfectly mixed bulk solution. The thickness of the unstirred layer is an operational quantity and is, of course, strongly dependent on stirring rate and geometry of the membrane.

Whether in the steady state the concentration of a solute in the bulk phase differs from the concentration directly adjacent to the membrane depends on its transport through the membrane. If the membrane is impermeable to the solute, its concentration will reach the concentration of the bulk solution after a diffusional delay. If, however, the solute is transported by passive or active transport mechanisms, the concentration at the membrane will be lower than in the bulk solution.

The intestinal epithelium with its mucosal folds, villi, and crypts is notorious for its mucosal unstirred layers (see Winne 1983; Thompson and Dietschy 1980 and references therein). In addition to the unstirred layer at the mucosal side, one has to consider unstirrable regions in the intestinal preparation, e.g., the lateral intercellular spaces, the lamina propria, and the other tissue layers at the serosal side that may be present in experimental situations.

Following Barry and Diamond (1984), the effects due to the presence of an unstirred layer can be divided into three categories: (1) gradient dissipation effects, leading to underestimates of permeability coefficients and K_m's, (2) sweeping away effects in which a transepithelial water flow sweeps solute away from the epithelium at one side and sweeps solute toward the epithelium at the other side, and (3) current-induced transport number effects. The reader is referred to their review for a mathematical discussion of these phenomena. We will use this concept in the next sections.

2 Intestinal Water Transport

2.1 Solute-Linked Water Flow

Already in 1984 Heidenhain drew attention to the observation that water absorption could occur in the absence of a driving gradient across the intestinal wall. Perfusion studies, designed to measure the luminal osmolarity at which no net water flow across the intestinal wall occurs have shown that the intestinal lumen may be 20 – 200 mOsmol l^{-1} higher than plasma osmolarity to stop net water flow (see Skadhauge 1980).

Employing a continuously perfused preparation of rat ileum with the blood supply intact, Curran and Solomon (1957) concluded that water transport followed passively on absorption of solute. Replacing NaCl with LiCl or KCl reduced fluid absorption to negligible amounts. From comparison of the fluxes and the transepithelial potential gradient Curran and Solomon concluded that both Na^+ and Cl^- were actively absorbed.

2.2 The Concept of Local Osmosis

Curran and co-workers argued that the mechanism linking salt and water transport could be a hypertonicity in a confined space in the epithelium (Curran and McIntosh 1962).

Diamond (1964) suggested that this space may be localized to the lateral intercellular space and he (Diamond and Bossert 1967) produced a model called the "standing osmotic gradient model" to quantify the hypothesis. The model has evolved to a simpler system with the following features: Solutes, principally Na^+ and accompanying anions, are taken up from the luminal side into the cells by selective transport processes. This creates and maintains an osmolarity within the cells that is slightly higher than in the luminal solution, so that water follows from the lumen into the cells. Na^+ is transported from the cells across the basolateral membrane against its electrochemical potential by the Na^+/K^+ pump. The transport of Na^+ and accompanying anions into the lateral intercellular space creates and maintains a higher osmolarity in the lateral intercellular space than in the cells and water leaves the cells. This gives rise to an increase in the hydrostatic pressure which will drive the solution through the basement membrane so that it can be taken up in the capillaries, again as a consequence of favorable osmotic and hydrostatic pressure differences (Starling principle).

The concept of local osmosis has been under debate. Diamond (1979) has given a summary of the controversies on the validity and the need for modification of the original standing gradient theory.

Many of the controversies are due to difficulties in measuring the osmotic water permeabilities and the resulting use of permeability coefficients which may be several orders of magnitude to low.

2.3 Osmotic Water Permeability of Epithelial Cell Membranes

The measurement of the osmotic water permeability generally involves the measurement of the steady state water flow (J_v) induced by a driving force, experimentally generated by the addition of a solute in a concentration C with a reflection coefficient σ to one side of the epithelium. The osmotic water permeability follows from $P_{osm} = J_v/\sigma C$ ml $cm^{-2} s^{-1}$ $mOsmol^{-1}$. Due to the unavoidable presence of unstirred layers at both sides of the epithelium, the water flow generated by the osmotic gradient reduces the concentration of all solutes at the side to which water flows and, at the other side, it will concentrate all solutes with reflection coefficients greater than 0.

This is called the "sweeping away" effect, resulting in the reduction of the gradient of the added solute and in the creation of opposing gradients of initially symmetrically distributed solutes. Thus, the water flow induces a "solute polarization" which reduces the driving force for J_v and thereby a decrease in the water flow until a steady state is reached. The correction for unstirred layer effects have led to an updating of the osmotic water permeability. For instance, from 1964 to 1979 estimates of P_{osm} for epithelium of rabbit gallbladder have increased from 0.5×10^{-6} to 1×10^{-3} cm s^{-1} Osmol^{-1} as a result of the improvement of the time resolution of the measurements and of corrections for sweeping away effects (see Diamond 1979).

These values apply to the total epithelium and because of the presence of a paracellular shunt, it is not possible to estimate the water permeability of the cell membranes.

Recently, the application of a quantitative light microscope technique by Spring and co-workers has made it possible to follow the change of individual cell volumes of *Necturus* gallbladder after a change in the osmolarity of the bathing solution. This technique made it possible to calculate the water permeability of individual cell membranes. The maximum values found are 1×10^{-3} cm s^{-1} Osmol^{-1} for the apical membrane and 2.2×10^{-3} cm s^{-1} Osmol^{-1} for the basolateral membrane (Persson and Spring 1982). These values are sufficiently high so that normal fluid absorption by *Necturus* gallbladder could be accomplished by an osmotic pressure gradient of 2 mOsmol across the apical membrane and 1 mOsmol across the basolateral membrane.

There are no data available for water permeabilities of individual cell membranes of intestinal epithelium. Calculation of the epithelial water permeability of goldfish intestine from the rate of shrinkage of the mucosal folds in response to mucosal hypertonicity has produced a value of 0.25×10^{-3} cm s^{-1} Osmol^{-1} (Siegenbeek van Heukelom et al. 1981). This is an underestimate because no correction has been made for the supply of water from the serosal side to the fold. Naftalin and Tripathi (1983) found values of similar magnitude in rabbit ileum from the rate of tissue swelling as measured with an ingenuous technique.

Recently, osmotic permeability has been studied in vesicle preparations obtained from the apical and the basolateral membranes of rat intestine (van Heeswijk and van Os 1986). The permeabilities of the apical and the basolateral vesicles are equal and are of similar magnitude as found for the apical barrier of the goldfish and rabbit intestine.

The general opinion is that the water permeability of intestinal epithelial cells is comparable with the permeability of gallbladder cells and is so high that very small differences in osmolarity suffice for fluid transport.

2.4 The Countercurrent Mechanism in in Vivo Intestinal Villi

The discussion in previous sections relates to in vitro experiments where the blood flow by definition is absent and plays no part in the transport properties. From analysis of the small intestinal villi of a number of mammals including man the group of Lundgren and Jodal has proposed a model for the in vivo situation. They have found a large hyperosmolarity in the lamina propria of the villi, especially in the tips, which is thought bo be created by the concerted action of active salt transport by the epithelial cells and a countercurrent multiplication based on the hairpinlike arrangement of the vasculature in the villi (for a recent review see Jodal and Lundgren 1986).

They suggest that this hyperosmolar region represents Curran's tissue compartment involved in intestinal water absorption in vivo.

Maintenance of the high osmolarity in the tips of the villi requires a low water permeability or a small osmotic gradient. There is no information available concerning these requirements.

An interesting finding is that the villus-tip hyperosmolarity persists in the secretory state of the small intestine. This strongly suggests that the secretory flux of water into the lumen cannot come from the villi and is additional evidence for the role of the crypts in secretion (see Sect. 9).

As far as we know there are no data available for colon or other vertebrate species to corroborate the generality of this model.

3 Intestinal Epithelium is a Leaky Epithelium

3.1 The Paracellular Pathway

In the previous section we have discussed the coupling of water transport to solute transport. In this context it can be said that one of the most important findings of the last decade is the discovery of the large water permeability of the epithelial cell membranes so that much of the constraints of the standing gradient theory have been relieved.

In the early 1970s another important discovery brightened the life of the epitheliologists. Through the work of Frömter and Diamond (1972) it became clear that epithelia could be divided into two groups, one with spontaneous low transepithelial potential differences and low transepithelial resistances, to which small intestinal epithelia belongs, and the group with high transepithelial resistances and larger transepithelial potential differences, the so-called tight epithelia like frog skin and toad bladder. The epithelium of rabbit colon occupies a position in between. Also from the work of Frizzell and Schultz (1972) it became clear that the characteristics of the small intestinal epithelium cannot be attributed to leaky cell membranes but have to be ascribed to the presence of low-resistance transepithelial paracellular pathways. The anatomical site of this pathway is the tight junction in series with the lateral intercellular space.

The conductance of these shunts accounts for more than 95% of the total transepithelial conductance.

The behavior of the ions in the lateral intercellular space is considered to be fully comparable with the behavior in free solution, whereas in the tight junctions the movement of anions is restricted by their cation selective properties.

It has been proposed that the cation selectivity is due to anionic groups with a pK of about 3.5 (Smyth and Wright 1966) and that the tight junctions hold watery pathways for the transepithelial diffusion of ions in their hydrated form. The diameter of these pathways may be between 8 Å and 15 Å (Schultz 1981 b). The electroneutrality of the pathway is maintained by fixed, structural components of the pathway and is not dependent upon the presence of mobile counterions (Frizzell and Schultz 1972).

The properties of the paracellular pathway are such that it seems reasonable to postulate that the junctions are permeable to water. However, their contribution to total transepithelial water flow is dependent on the relative area of the tight junctions and if the cell diameter is rather large, as in villus cells, the fractional area occupied by the tight junctions is so small that their permeability to water would have to be much larger than the water permeability of the cell membranes to make an essential contribution to the total water flow.

3.2 The Cation Selectivity of the Paracellular Pathway: Its Influence on Transepithelial Potentials

3.2.1 Large Unidirectional Fluxes

The picture is totally different for ion fluxes because the paracellular ion permeability is indeed much larger than the cell membrane permeability. Thus, ion fluxes may occur largely via the paracellular pathway. As a consequence, it is a common observation in leaky epithelia that unidirectional ion fluxes are large compared with net fluxes. Table 1 shows examples of unidirectional ion fluxes through small intestines together with the transepithelial conductance.

Table 1. Examples of unidirectional and net fluxes of Na^+ and Cl^- through the small intestine

	J_{ms}^{Na}	J_{sm}^{Na}	J_{net}^{Na}	J_{ms}^{Cl}	J_{sm}^{Cl}	J_{net}^{Cl}	G	
	(μmol cm^{-2} h^{-1})						(mS cm^{-2})	
Rat mid-intestine	25	21	4	11	15	−4	24	Hardcastle et al. (1984)
Rabbit ileum	10	9	1	8	7	1	25	Powell et al. (1974)
Goldfish	15	12	3	8	5	3	48	Bakker and Groot (1984)
Winter flounder	14	12	2	8	3	5	22	Field et al. (1978)

3.2.2 Dilution Potential

If one replaces NaCl at one side of the epithelium isoosmotically with a non-electrolyte, a diffusion potential develops across the epithelium with a polarity opposite to the polarity in free solution. This artificial transepithelial potential is called a "dilution potential". The dilution potential can be used as a monitor for the ion selectivity of intestinal epithelium (e.g., Groot et al. 1986a).

3.2.3 Transport Number Effects

Let us consider an in vitro preparation of intestine mounted in an Ussing chamber and bathed on both sides with identical Ringer's solutions. If we pass a current through the epithelium, because of the leakiness of the intestinal

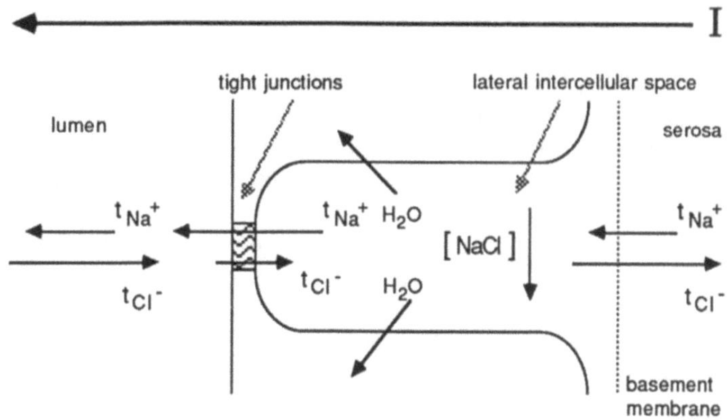

Fig. 1. The transport number effect in the paracellular pathway. In the free solution, electric current (I) is mainly carried by Cl^-, in the tight junction structure mainly by Na^+. Passing a current from the serosal to the mucosal side will lead to NaCl depletion in the lateral intercellular space and to an increase in the mucosal unstirred layer. Water movement will lead to a collapse of the intercellular space

epithelium, most of the current will pass through the paracellular pathway. If the current is passed from the serosal to the mucosal side, the larger mobility of the Cl^- ions as compared with the Na^+ ions in the bathing solutions contrasts with the higher permeability of Na^+ through the tight junctions. Thus, more Na^+ leaves the lateral intercellular space through the tight junctions than is supplied and more Cl^- leaves the intracellular space through the basement membrane than is supplied through the tight junctions. The discontinuity in transport numbers (i.e., the fraction of current carried by each ion) gives rice to local changes in salt concentration in the lateral intercellular space (and the unstirred layers). The loss or gain of NaCl will, in the steady state, be balanced by diffusion and local osmotic flows.

A schematic representation is given in Fig. 1. As a consequence of the solute polarization during current flow, the dimensions of the lateral intercellular space will be influenced. We will discuss the effects on paracellular resistance and water flow further in Section 6.5. In general terms it may be stated that current passage through a compartment whose boundaries differ in transport numbers will lead to salt accumulation or depletion within the compartment (Dibona 1985).

4 Experimental Approaches to Study Ion Transport by Intestinal Epithelium

4.1 Ussing-Type Chambers: Short-Circuit Current and Flux Measurements

Ussing and Zerahn (1951) were the first to use the concept of the short-circuit current as a method to study primary active transport of ions by isolated epithelia, mounted in an experimental chamber between identical solutions. When, after elimination of spontaneous transepithelial potential differences by the short-circuit current, one still finds net transport of ions, the driving force for these transports must be in the tissue itself.

However, there are difficulties with the application of the short-circuit technique to leaky and nonhomogeneous epithelia. First, the unidirectional fluxes may be so large due to passive paracellular transport that an accurate determination of net transport requires simultaneous measurement of both unidirectional fluxes in one piece of tissue.

Second, the ignorance of the series resistance of the tissue (i.e., the resistance of the tissue after the epithelial cells are removed) hamper meaningful measurements of the short-circuit current and comparison with ion fluxes (see Sect. 6.2). Corrections for the displacement of the bathing solution by the tissue as performed by Nellans et al. (1974) reduce the series resistance and makes things even worse.

In addition, the heterogeneity of intestinal preparations is an enormous problem. It has been shown by Marcial et al. (1984) that one may expect large differences between the specific resistances of villus regions and crypt regions, the latter being much more leaky than the villi.

4.2 Microelectrode Techniques

The application of microelectrodes can help in elucidating the properties of the epithelial transport pathways. These measurements are always performed in combination with transepithelial measurements of potential and resistance. The results have to be analyzed in an electrical equivalent circuit (Helman and Thomson 1982) for an epithelium. This requires measurements of membrane resistances and membrane potentials. In addition, one has to know the components of the paracellular resistance and the diffusion potential(s) in the paracellular pathway. We will discuss the electrical equivalent model in Section 6.

By analyzing the changes induced by ion substitutions one may acquire insight in the ion conductances of the barriers in the tissue. Although conventional microelectrodes cannot detect directly the operation of electrically silent ion movements, indirect effects may be pronounced and cause difficulties in interpretation. For instance, the acceleration or inhibition of electroneutral NaCl cotransport by manipulation of extracellular Na^+ or Cl^- may be detected if the change in intracellular ion activity is so large that the cell potential is altered.

In Section 7.3.4 we will discuss the anomalous hyperpolarization of the cell potential upon mucosal omission of Cl^- which is a good example of this problem.

Direct information about ion activities and changes in ion activity upon experimental manipulations like ion substitutions or use of drugs can be obtained by the use of ion-selective microelectrodes. Because the signal from an ion-sensitive microelectrode is the sum of a diffusion potential across the ion exchanger in the tip of the electrode and the membrane potential, one has to measure the membrane potential with a second electrode or the second barrel of double-barrel electrodes. The small diameter of enterocytes hamper the use of double-barrel electrodes. Needless to say that continuous, simultaneous measurements with two microelectrodes in nearby cells during control and experimental situations require good equipment, a skillful and patient experimenter, and cooperative tissue. The literaure shows that the intestinal epithelium of lower vertebrates seems to be more cooperative than mammalian intestine. It has been suggested that this is a consequence of better intercellular communication through gap junctions between the epithelial cells (Curtis et al. 1984).

Application of ion-sensitive electrodes has made it possible to calculate the equilibrium potential for the various ions and thus their electrochemical gradient across the cell membrane. A limitation of the use of ion-selective microelectrodes is that they only signal ion activities; they do not tell whether activity changes occur because of perturbation of influx or efflux or because of a change in volume due to water flow. There may even be no change in activity despite a large salt flux if the water flow compensates the ion movements. Considering the high osmotic permeability this will happen normally. Thus, it is important to have insight in the changes in cell volume. However, as far as we know a combination of simultaneous electrophysiological measurements and cell volume measurements as have been performed by Guggino (1986) in *Amphiuma* kidney have not been done in intestinal epithelium. A reasonable alternative is to combine chemical analysis of ion contents, tissue water, and extracellular space with ion-sensitive electrode analysis (Zuidema et al. 1985).

4.3 Membrane Vesicles from the Apical and Basolateral Side of Intestinal Epithelium Membranes

Plasma membrane vesicles, either from the apical membrane (brush border membrane vesicles) or from the basolateral membrane can be prepared from intestinal epithelium. Brush border membrane vesicles have been used in studies of the sodium cotransport system since the early 1970s (Hopfer et al. 1973; Murer and Hopfer 1974). The use of basolateral membrane vesicles is of earlier date (Mircheff et al. 1980), and very recently the vesicle technique has been applied to the apical membrane of rat colon (Binder et al. 1986; Foster et al. 1986a).

An important advantage of the use of membrane vesicles is the possibility to manipulate the membrane environment. This gives the opportunity to vary ion gradients and membrane potentials. Moreover, the preparation can be stirred efficiently so that unstirred layer problems are minimized.

There are however, some limitations in the use of vesicle preparations. The method of preparation may have effects on the properties of the membrane. The membranes may lose properties that are dependent on the structural integrity of the membranes and cytoskeleton or that are dependent on metabolic substrates. For instance, the study of cAMP-dependent processes in membrane vesicles require the presence of ATP in the vesicles. Other limitations are related to the large surface/volume relation of the vesicles which are about 200 nm in diameter. Because of the small volume, relatively fast diffusing solutes like ions through channels will equilibrate between the outside solution and the vesicle interior within seconds. To follow initial uptake kinetics one needs a faster measuring procedure than the conventional filtration method which has a time resolution of seconds. In addition, the filtration procedure may lead to the loss of fast diffusing tracers. Fast-mixing, stopped-flow techniques in which optical scattering is used to follow vesicle volume changes have been used to measure the osmotic permeability of brush border membrane vesicles as well as basolateral membrane vesicles of rat small intestine (van Heeswijk and van Os 1986). The time resolution of the method is 2 – 3 ms.

The use of voltage-sensitive fluorescent dyes has made it possible to probe the transmembrane potential and to gain insight in relative ion permeabilities of the vesicles (Wright 1984). The fluorescent signal can be calibrated with K^+ equilibrium potentials using valinomycin to make the membranes highly permeable to K^+. As yet, the high relative Cl^- permeability observed in rabbit jejunal brush border membranes seems to contrast with membrane potentials in the intact cells. Thus, although the use of membrane vesicles has important advantages, it is not yet clear what they can tell us about the properties of the original membranes.

5 Plasticity of Transport Functions in Intestinal Epithelium

Of course, the transport properties of epithelial cell membranes are not static entities. There should be coordination between uptake and exit processes, especially for fast transported solutes. Otherwise the cells would be flooded or totally depleted within a short time. An elucidative example given by Schultz (1981 a) and based on the paradigm of the Koefoed-Johnson-Ussing model (1958) may be paraphrased here. It illustrates the large amount of ions passing the epithelium as compared to the amount of ions present in the cells. Consider 1 cm^2 of epithelium consisting of cells with a thickness of 60 μm. Neglecting the volume of the paracellular pathway, the intracellular volume would be 6×10^{-3} cm^3. The intracellular activities of Na^+ and K^+ are usually ca. 10 and 100 μEq cm^{-3}, respectively. Thus, the Na^+ and K^+ contents of this cell layer will be 0.06 and 0.6 μEq cm^{-2}. If the transepithelial net transport of Na^+ is, e.g., 3.6 μEq cm^{-2} h^{-1} then the amount of Na^+ entering and leaving the cell layer every minute is equal to the total intracellular Na^+ content. Assuming a coupling ratio for the Na/K pump of 2 : 3, the amount of K^+ entering and leaving the cells per minute is 15% of the total intracellular K^+.

5.1 Homocellular Regulation

In goldfish intestine it has been found that addition of glucose to the mucosal solution may triple the Na^+ entry rate (Bakker 1986), leading to an increase of the Na-pump rate and therefore to increased K^+ uptake. However, in spite of increased K^+ uptake, the intracellular K^+ activity does not increase in goldfish (Zuidema et al. 1986) nor in *Amphiuma* (White et al. 1986). The driving force for K^+ efflux does not change sufficiently to explain this finding but there is now good evidence that activation of the sugar and amino acid Na-cotransport systems in the apical membrane can induce an increase of the K^+ conductance in the basolateral membrane (Brown and Sepulveda 1985) and may be also in the apical membrane (Bakker 1986).

Schultz (1981a) and Diamond (1982) have stressed the importance of coordination of functions between the two membranes of the polar epithelial cells which Diamond called "crosstalk" and Schultz "intracellular feedback" or "homocellular regulation". The underlying mechanisms are still not elucidated.

5.1.1 Calcium Activity as Coupler Between Apical and Basolateral Membranes

A candidate for the role of intracellular messenger for regulation of the K^+ conductance may be Ca^{2+} (Chase 1984; Brown and Sepulveda 1985). Recently, Sepulveda and Mason (1985) and Morris et al. (1986) using the patch-clamp method have shown the existence of a Ca^{2+}-activated K^+ channel in the basolateral membrane of isolated rabbit and rat enterocytes, respectively. In the case of the rat enterocytes the K^+ channel has a large conductance (250 pS) that is voltage and Ca^{2+}-activated and is blocked by Ba^{2+}. The Ca^{2+} effect was maximal at 10^{-7} M, a concentration that may be below resting Ca activity in the enterocyte. However, these are the first measurements in enterocytes using this powerful technique and we will have to wait for the interesting extensions of the experiments. If Ca^{2+} plays a role in the regulation of the membrane conductance then the question is: what regulates the Ca^{2+} activity? Chase (1984) suggested that a rise in intracellular Na^+ inhibits the Na/Ca exchanger which presumably is present in the basolateral membrane of the enterocytes (Ghijsen et al. 1983). Thus, in the case of an increase in Na^+ influx, as for instance during Na^+-sugar cotransport, an increase of the intracellular Na^+ activity can be detected (Zuidema et al. 1986), which may affect the intracellular Ca^{2+}.

5.1.2 The Cell Volume Regulatory Response as Coupler Between Apical and Basolateral Membrane

Another factor involved in the coordination between apical and basolateral transport may be the cell volume (Lau et al. 1984, 1986). These authors favor the hypothesis that the increase in conductance of the basolateral membrane of *Necturus* small intestine in response to stimulation of apical Na cotransport is a volume regulatory response to cell swelling. Volume regulatory responses have been studied in *Necturus* gallbladder (Spring and Ericson 1982; Larson and Spring 1984). If this tissue is placed in a hypotonic solution, a swelling occurs that is followed by a so-called volume regulatory decrease caused by a loss of KCl through the basolateral membrane. (In contrast, the shrinkage of the cells due to hypertonic solution results in a rapid volume regulatory swelling because the cells take up NaCl. The mechanism for uptake is the parallel operation of Na^+/H^+ and Cl^-/HCO_3^- exchangers which seem to be dormant under control conditions.)

Goldfish intestinal epithelial cells, however, maintain their KCl content and do not show a volume regulatory decrease (Groot 1981) nor do bullfrog and hamster intestinal cells readjust their volume (Armstrong et al. 1975; Dinda et al. 1972).

5.2 Extracellular Regulation

In Sections 8 – 10 we will come back to the regulation of transport functions in the context of regulation of absorption and secretion by neurohumoral agents. Here, we will just summarize that the presence of extracellular humoral and neuronal factors is signaled by receptor-mediated membrane phenomena leading to a change in the concentration of intracellular second messengers like cAMP, cGMP, Ca^{2+}, and probably metabolites from membrane phospholipid hydrolysis. It is now generally accepted that these messengers play a role in the activation of specific protein kinases leading to phosphorylation and thereby altering the function of substrate proteins. Some of the substrate proteins may be membrane-bound ion channels or may be ion exchange carrier systems (see Fig. 10).

6 The Equivalent Electrical Circuit; Spontaneous Short-Circuited Leaky Epithelia

Before discussing the facts that have led to our present understanding of NaCl transport we will elaborate on the equivalent electrical circuit for leaky epithelia. We emphasize the use of current in the analysis of electrophysiological measurements and we will use this approach to suggest a consequence of spontaneous short circuiting in leaky epithelia via the transport number effect.

The evaluation of electrophysiological measurements requires the use of an equivalent electrical circuit as a model.

In the simple equivalent circuit of Fig. 2 both cell membranes are represented by an electromotive force (emf) in series with a resistor, while the paracellular pathway is represented by a resistor only (assuming the absence of ion gradients across the tight junctions, see however, Sect. 6.4). The emf of a membrane is determined by the diffusion (Nernst) potentials for the various ions and their relative conductances and by rheogenic ion translocation mechanisms.

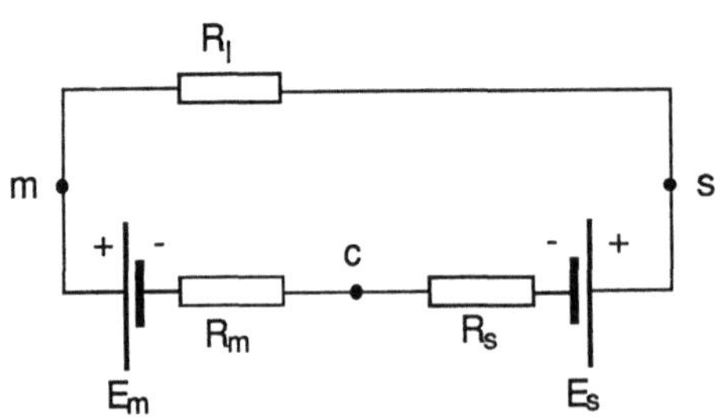

Fig. 2. Equivalent electrical circuit for epithelia. *m* mucosal; *c* cellular; *s* serosal compartment; R_m, R_s, and R_l lumped resistances of the mucosal and serosal membrane and the paracellular pathway, respectively. E_m and E_s are the lumped electromotive forces across the respective membranes, representing the zero-current potential differences

6.1 Open-Circuit Conditions

Although under open-circuit conditions the *transepithelial* current is zero, in *leaky* epithelia the paracellular pathway almost completely short-circuits the series array of the two membrane emf's causing an intraepithelial current through the cell and returning through the paracellular pathway:

$$I_{tc} = \frac{E_s - E_m}{R_t} , \qquad (1)$$

where $R_t = R_m + R_s + R_l$.

(By convention: a positive value for any E indicates an orientation of the emf as shown in Fig. 2; a positive value for I_{tc} indicates a transcellular current from mucosa to serosa; since under open-circuit conditions I_{pc} equals I_{tc}, a positive value for I_{pc} indicates a serosa to mucosa paracellular current; a potential difference Ψ_{ij} equals $\Psi_j - \Psi_i$.)

The transepithelial pd (Ψ_{ms}) simply equals the voltage drop across R_l due to I_{tc}:

$$\Psi_{ms} = R_l * I_{tc} . \qquad (2)$$

Due to I_{tc}, the pd across a membrane differs from its emf by the voltage drop across its resistance:

$$\Psi_{cm} = E_m + R_m * I_{tc} \qquad (3)$$

$$\Psi_{cs} = E_s - R_s * I_{tc} \qquad (4)$$

$$(\Psi_{cs} = \Psi_{cm} + \Psi_{ms}) .$$

Thus, Ψ_{ms} is a measure for the current generated by the cell, and when the ratio's R_m/R_l and R_s/R_l are known, the membrane emf's E_m and E_s can be obtained by extrapolation from the measured membrane pd's Ψ_{cm} and Ψ_{cs}. This is illustrated in Fig. 3, where Eqs. (1) through (4) are represented geometrically.

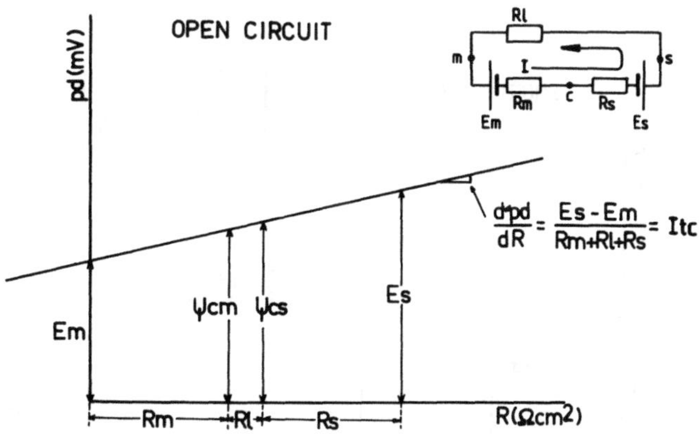

Fig. 3. Geometrical representation of Eqs. (2), (3), and (4); open-circuit conditions (Bakker 1985)

6.2 Short-Circuit Conditions

After Ussing and Zerahn (1951) introduced the short-circuit technique to distinguish between active and passive ion transport in a tight epithelium like frog skin, the short-circuit method was also applied to leaky epithelia with a complex geometry, e.g., intestine.

Leaky epithelia are short-circuited to a large extent by their paracellular pathway. Complete short-circuiting can be achieved by a zero-resistance external current meter. Although, in fact, a short-circuiting device sends a current through the epithelium to abolish Ψ_{ms}, this is of course equivalent to introducing an infinite conductor across the epithelium. Under ideal conditions (see, however, Sect. 6.2.1) the measured current equals the transcellular current:

$$I_{sc} = I_{tc} = \frac{E_s - E_m}{R_m + R_s} \ . \tag{5}$$

The pd across the membranes:

$$\Psi_{cm} = \Psi_{cs} = E_m + R_m * I_{tc} = E_s - R_s * I_{tc} \ . \tag{6}$$

Figure 4 shows that E_m and E_s can be determined by extrapolation from the measured values for I_{sc}, Ψ_{cm}, R_m, and R_s.

It should be noted that in leaky epithelia the transcellular current is only slightly larger under short-circuit conditions as compared to open-circuit conditions. From Eqs. (1) and (5) (see also Figs. 3 and 4):

$$\frac{I_{tc,sc}}{I_{tc,oc}} = \frac{R_m + R_s + R_l}{R_m + R_s} \ . \tag{7}$$

It is obvious that due to the large conductance of the paracellular pathway a small deviation from zero of the clamped transepithelial pd has a large effect on the paracellular current, i.e., the paracellular ion fluxes.

Therefore, perfect short-circuiting is obligatory in order to draw valid inferences on active and passive ion transport from radioisotopic ion fluxes.

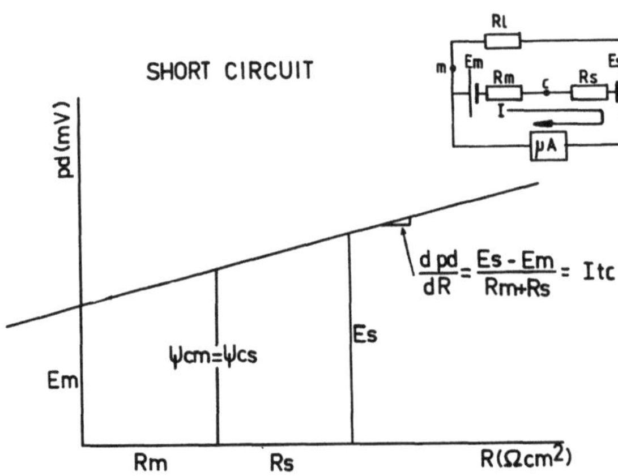

Fig. 4. Geometrical representation of Eqs. (5) and (6); short-circuit conditions (Bakker 1985)

6.2.1 Incomplete Short-Circuiting Due to R_{series}

Rehm (1975) already reviewed the problems encountered in short-circuiting leaky epithelia, but we will illustrate this point since the short-circuit technique is still widely used in intestinal transport research.

Although in theory all transcellular current flows through the short-circuiting device, in practice, I_{sc} will be smaller than I_{tc} whenever a resistance in series with the epithelial cell layer is present, since part of the transcellular current will then be shunted through the paracellular pathway (Fig. 5).

Although the resistance of the solutions bathing the tissue can be minimized or corrected for, the subepithelial tissue in general represents a considerable series resistance. Especially in leaky epithelia like the intestine the mucosal surface amplification by folds and/or villi increases the subepithelial tissue resistance relative to the paracellular resistance. Even in goldfish intestine, stripped completely from its muscle layers, R_{series} accounts for 24% of R_{ms} (Albus et al. 1983).

In unstripped or partially stripped intestinal preparations R_{series} may be equal to or greater than R_l. Gebhardt and Nell (1979) reported that R_{series} accounted for 71, 43, 59, and 14% of the transepithelial resistance of unstripped rat jejunum, stripped rat ileum, stripped rabbit ileum, and stripped rat colon, respectively. Recently, Fromm et al. (1985) reported for unstripped rat jejunum 80% and 83% using impedance and microelectrode measurements, respectively. Fromm et al. concluded that in conventional in vitro studies on intact rat jejunum measured net Na^+ fluxes as well as short-circuit currents will be underestimated by a factor of about 5. Tai and Tai (1981) found a value of 33% for rat ileum and calculated that due to the subepithelial tissue resistance the error in net ion fluxes of Na^+ and Cl^- was 40%.

Thus, whenever R_{series} is significant, I_{sc} grossly underestimates I_{tc}, and the measured net ion fluxes may be erroneous.

Moreover, since the part of I_{tc} which returns through the paracellular pathway depends on the ratio R_{series}/R_l the measured I_{sc} may vary at a fixed I_{tc} due to changes in R_l!

In other words: although the tissue is short-circuited, the pd across the epithelial cell layer will, in general, not equal zero and will affect ion fluxes. Changes in transepithelial pd and resistance may not be reflected in the short-circuit current while ion fluxes do change.

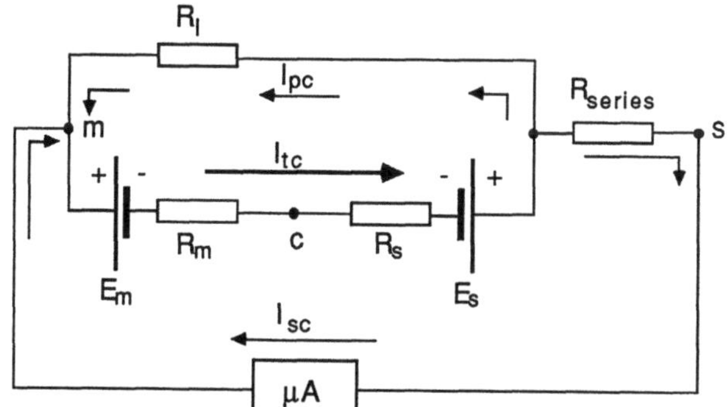

Fig. 5. Effect of series resistance (R_{series}) on currents under short-circuit conditions. A part of the transcellular current (I_{tc}) is shunted through the paracellular pathway (I_{pc}), and is therefore not measured as short-circuit current (I_{sc})

The relation between the short-circuit current and the transcellular current is given by:

$$I_{sc'} = \frac{R_l}{R_l + R_{series}} * I_{tc'} .^{1} \quad (8)$$

Another problem in short-circuiting intestinal preparations is the heterogeneity of the tissue. Marcial et al. (1984) showed that in guinea-pig ileum the crypt cells represent 73% of the paracellular conductance, while only contributing 12% to the surface area. Therefore, it is likely that in short-circuiting mammalian intestine the crypt cells may be reasonably well short-circuited, while the pd across the villus cells (assuming a substantial R_{series} for the villus cells) may not be very different from the open-circuit situation.

6.3 The Electrical Equivalent of the Paracellular Pathway

So far, the paracellular pathway was represented by a single resistor R_l. However, the paracellular pathway is comprised of the tight junction resistance (R_{tj}) in series with the resistance of the lateral intercellular spaces (R_{lis}).

The resistance of the lateral intercellular spaces can be estimated by:

$$R_{lis} = (d * p)/(L * w_i) \quad \text{(Boulpaep and Sackin 1980)} \quad (9)$$

with d = interspace depth (cm)
L = total linear extent ($cm\ cm^{-2}$)
p = resistivity of solution (Ωcm)
w_i = interspace width (cm).

The linear extent L (in $cm\ cm^{-2}$ *mucosal* surface) can be obtained by dividing 1.155 by the edge (in cm) of a cell's hexagonal face (Schafer and DiBona 1984); a mucosa to serosa amplification factor should be used to obtain L for 1 cm^2 serosal area.

6.3.1 Fish Intestinal Epithelia

For the intestinal epithelium of goldfish and flounder L can be estimated at 3×10^4 $cm\ cm^{-2}$ ($\sim 5\ cm^2$ mucosal surface cm^{-2} serosa, cell diameter $\sim 3.5\ \mu m$). Thus, with $L = 3 \times 10^4\ cm^{-1}$, $d = 65\ \mu m$, $p = 65\ \Omega cm^2$, and $w_i > 0.05\ \mu m$, R_{lis} is calculated to be smaller than $4\ \Omega cm^2$. Since $R_l > 15\ \Omega cm^2$, the ratio R_{lis}/R_{tj} is $\leqslant 0.25$.

It should be noted that the small value for R_{lis} mainly results from the huge linear extent, which means that the *specific* tight junction resistance $R_{tj}/{*}L$

---

[1] $$I'_{tc} = \frac{R_m + R_s}{R_m + R_s + R'} \cdot I_{tc} ,$$

with $$R' = \frac{R_l \cdot R_{series}}{R_l + R_{series}} .$$

Since $R_l \ll (R_m + R_s)$, the effect of R_{series} on I_{tc} is negligible compared to the effect on I_{sc}.

(Marcial et al. 1984) is large. In other words, fish intestinal epithelia are *leaky* due to a large linear extent of relatively *tight* junctions!

As will be discussed in Section 6.5 paracellular current flow may lead to changes in lis diameter due to solute polarization caused by the difference in transport numbers for Na^+ and Cl^- of the tight junctions vs free solution (transport number effect). In fact, an estimate for the contribution of R_{lis} to the paracellular resistance may be obtained by imposing a mucosa to serosa current, which will lead to an opening of the lateral intercellular spaces if the tight junctions are cation-selective. The absence of a significant decrease in resistance in goldfish intestine (Bakker et al., in prep.) indicates that under normal conditions R_{lis} is not comparable to R_{tj}. However, as will be shown in Section 6.5.2 even values for R_{lis}/R_{tj} as small as 0.1 to 0.2 may lead to transport number effects in the lis.

6.3.2 Mammalian Intestine

The finding of Marcial et al. (1984) that the crypt cells in guinea-pig ileum represent 73% of the paracellular conductance, while only contributing 12% to the surface area, means that the paracellular pathway in the crypts may function as the paracellular route for the villus cells as well. The implications of an intraepithelial current from mucosal solution through the villus cells and returning through the paracellular pathway of the crypts will be discussed in Section 6.6.2.

6.4 Implications of Local Differences in NaCl Concentration for the Equivalent Electrical Circuit

6.4.1 An Intraepithelial Hypertonic Compartment

As discussed in Section 2.3 the water permeabilities of the intestinal cell membranes are large and therefore only small osmoting gradients (1 – 2%) are required to achieve normal rates of absorption. However, even a small difference in NaCl concentration will generate a diffusion potential across the cation-selective tight junctions (E_l). This can be exemplified as follows: with a permeability ratio of the junctions $P_{Na}/P_{Cl} > = 4$, as found in flounder (Krasny et al. 1982) and goldfish intestine (Bakker and Groot 1984), 1% NaCl concentration difference will cause an E_l of 0.2 mV (serosa-negative).

Therefore, values of E_l up to 1 mV may not be unreasonable. We will show that even small values for E_l should be considered in evaluating electrical measurements in *leaky* epithelia.

6.4.2 Inclusion of a Diffusion Potential Across the Tight Junctions in the Equivalent Circuit

When E_l is introduced in the equivalent circuit (Fig. 6) it can be shown that Ψ_{ms} (open-circuit) no longer reflects the transcellular current:

$$\Psi_{ms} = R_l * I_{tc,oc} + E_l \qquad \text{[cf. Eq. (2)]} , \qquad (10)$$

with

$$I_{tc,oc} = \frac{E_s - E_l - E_m}{R_t} \qquad \text{[cf. Eq. (3)]} , \qquad (11)$$

$$I_{sc} = I_{tc,sc} + \frac{E_l}{R_l} \qquad \text{[cf. Eq. (5)]} . \qquad (12)$$

Since in leaky epithelia $(R_m + R_s) \gg R_l$. I_{tc} and E_l/R_l may be equally important in determining I_{sc}. If transcellular NaCl transport is electroneutral, I_{tc} might be zero, while E_l induces a non-zero I_{sc}. This may be the reason that the interpretation of the observed serosa-negative pd across seawater teleosts has been ambiguous (Bakker 1985). A reliable estimate for E_l is a prerequisite for any conclusion on the transcellular current.

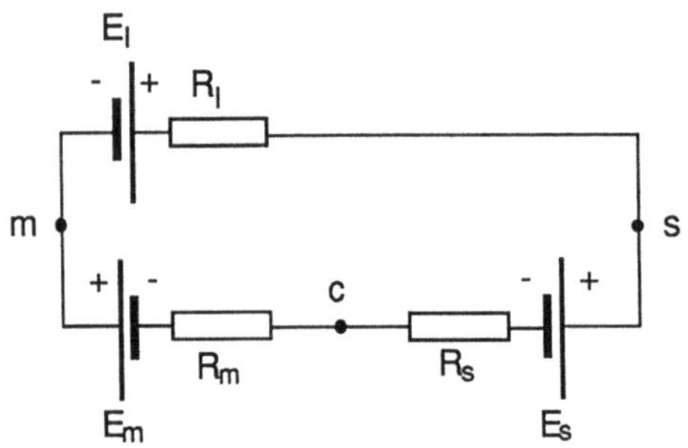

Fig. 6. Equivalent electrical circuit including a diffusion potential across the tight junctions (E_l). Symbols as in Fig. 2

6.5 Transport Number Effects in the Paracellular Pathway

6.5.1 Transport Number Effects Due to an Applied Current

Although under ideal short-circuit conditions the current through the paracellular pathway is zero, normally, under open-circuit conditions and in vivo the transcellular current returns through the paracellular pathway (see also Sect. 6.1). When under experimental conditions current is passed through a leaky epithelium > 90% of the current traverses the paracellular route. Therefore, current passage can be used to study the effects of paracellular current.

Barry and Hope (1969) showed that current-induced water flow may be principally caused by pseudoelectroosmosis, i.e., salt polarization effects due to transport number effects (see Barry and Diamond 1984 for a review). The subsequent phenomena resulting from direct current passage across a leaky epithelium with cation-selective tight junctions are well described by Bindslev et al. (1974) for frog gallbladder. The changes in transepithelial pd due to serosa to mucosa current passage is comprised of three distinct transients: the first transient is related to the electrical capacity of the epithelium, the second transient is the polarization potential due to the transport number effect (i.e., the difference in transport number for Na^+ and Cl^- between the tight junctions and the lis, see also Sect. 3.3), the third transient is caused by a change in resistance of the epithelium.

Frömter (1972) showed in *Necturus* gallbladder that the resistance incease during serosa to mucosa current flow is associated with a collapse of the intercellular spaces. The *decrease* in resistance following mucosa to serosa current passage, however, may not be exclusively due to an opening of the spaces, but may also be ascribed to a breakdown of the tight junctions, an effect maybe similar to the "blistering" observed in toad urinary bladder during serosa hypertonicity (Bindslev et al. 1974; DiBona 1985).

Although current passage in intestinal epithelia has been widely used to study the relation between the transepithelial pd and ion fluxes [Frizzell and Schultz (1972) for rabbit ileum. Field et al.(1978) for flounder intestine], surprisingly little information is available on the time-dependent effects of current passage on transepithelial pd and resistance and water flow. Bindslev et al. (1974) mentioned that they observed effects in frog jejunum which were similar to the effects in frog gallbladder. Krasny et al. (1982) reported resistance changes due to current passage for flounder intestine. Experiments in our laboratory with goldfish intestine showed an increase in resistance upon serosa to mucosa current passage only when the tight junctions are cation-selective.

Earlier, Bindslev et al. (1974) showed for frog gallbladder that the observed effects of current on transepithelial pd and resistance depend on the ion selectivity of the tight junctions: the observed transients in pd and resistance due to mucosa to serosa and serosa to mucosa current, respectively, could be reversed by making the tight junctions *an*ion-selective by treatment with La^{3+}.

The accumulation of NaCl in the intercellular space resulting from mucosa to serosa current will, depending on the water permeability of the cell membranes, induce a water flow. Recently, Naftalin and Tripathi (1985) measured current-induced water fluxes in rabbit ileum. However, since no transients in pd were observed and the change in water flow upon current passage was without delay, they concluded that the observed change in water flow should be ascribed to electroosmosis. However, although rabbit ileum epithelium has been reported to have cation-selective tight junctions, the cation selectivity of the epithelium was not determined in this study. Van Os et al. (1976) concluded that the current-induced changes in water transport in rabbit gallbladder are caused by salt polarization rather than by electroosmosis.

6.5.2 A Simple Model for the Evaluation of Paracellular Transport Number Effects

In order to decide whether the above described effects could play a role in in vitro intestinal preparations under open-circuit conditions (a situation in which, as in the in vivo situation, a paracellular current is present), we will introduce a simple model of the paracellular pathway which can be used to estimate the salt polarization in the intercellular spaces or the induced water flow, respectively.

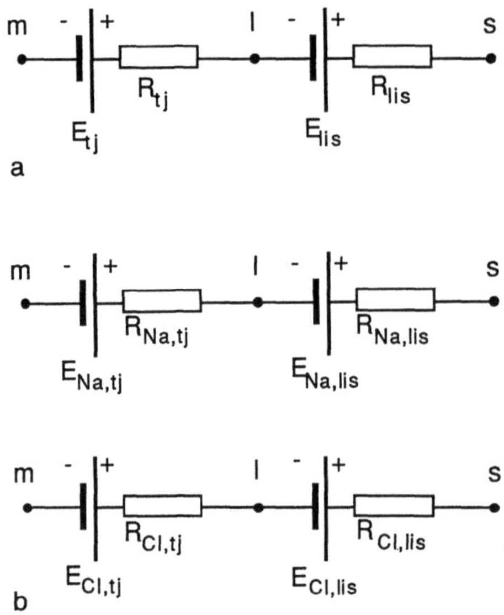

Fig. 7. *a* Equivalent electrical circuit for the paracellular pathway. *m* mucosal; *l* lateral intercellular space; *s* serosal compartment. E_{tj}, R_{tj} and E_{lis}, R_{lis} are the electromotive force and the resistance of the tight junctions and the lateral intercellular spaces, respectively. *b* Equivalent electrical circuit of the paracellular pathway describing the movement of Na^+ and Cl^- under *steady state* conditions. The emf's and resistances represent the Nernst potentials and the diffusional resistances for Na^+ and Cl^- across the appropriate boundaries

The paracellular pathway is represented by the equivalent electrical circuit shown in Fig. 7a. The transepithelial pd (Ψ_{ms}) equals the sum of the two emf's (E_{tj}, E_{lis}) and the voltage drop due to the paracellular current (I_{pc}) across the two resistors (R_{tj}, R_{lis}):

$$\Psi_{ms} = E_{tj} + E_{lis} + I_{pc} * (R_{tj} + R_{lis}) \ . \tag{13}$$

In the *steady state* the NaCl concentration in the lis (at l in Fig. 7a) does not change and therefore the net fluxes of Na^+ and Cl^- through the tight junction should equal the net fluxes of Na^+ and Cl^- through the lis. Since the net flux of an ion is equivalent to the current carried by that ion, the equivalent electrical circuit of Fig. 7a can *under steady state conditions* be split into a Na^+ limb and a Cl^- limb (we will neglect the contribution of the other ions for simplicity) as is illustrated in Fig. 7b [see Schultz et al. (1977) for details on equivalent electrical circuits describing steady state conditions]. Ψ_{ms} for the two limbs is given by:

$$\Psi_{ms} = E_{Na,tj} + E_{Na,lis} + I_{Na} * (R_{Na,tj} + R_{Na,lis}) \ , \tag{14}$$

$$\Psi_{ms} = E_{Cl,tj} + E_{Cl,lis} + I_{Cl} * (R_{Cl,tj} + R_{Cl,lis}) \ . \tag{15}$$

Assuming the absence of mucosal and serosal unstirred layers, the contribution of $E_{tj} + E_{lis}$ (note that $E_{tj} + E_{lis}$ can be identified with the so-called polarization potential) to the change in Ψ_{ms} due to an (applied) I_{pc} can be derived from Eqs. (13) through (15) when the transport numbers of Na^+ and Cl^- in the tight junctions and the lis, and R_{tj} and R_{lis} are known. Moreover, the relative contribution of Na^+ and Cl^- (i.e., the "overall" transport numbers for Na^+ and Cl^-: t'_{Na} and t'_{Cl}) to I_{pc} in the presence of salt polarization can be derived. From $E_{tj} + E_{lis}$ at a known I_{pc} the change in local NaCl concentration can be calculated. In Fig. 8 the polarization potential ($E_{tj} + E_{lis}$) and the "overall" transport numbers, t'_{Na} and t_{Cl}, are plotted as a function of the ratio R_{lis}/R_{tj}. It is apparent that already

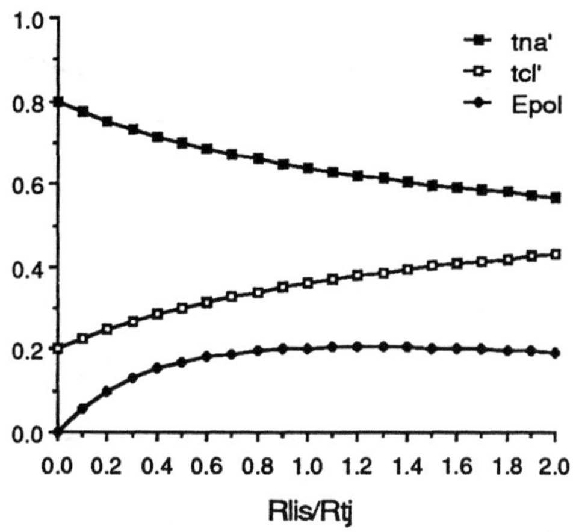

Fig. 8. The effect of R_{lis}/R_{tj} on (1) the "overall" transport numbers t'_{Na} (■——■) and t'_{Cl} (□——□) for Na^+ and Cl^- and (2) the polarization potential $E_{tj}+E_{lis}$ (●——●), expressed as a fraction of the voltage drop across $R_{tj}+R_{lis}$ due to the paracellular current: $(E_{tj}+E_{lis})/(I \cdot (R_{tj}+R_{lis}))$

at a value for R_{lis}/R_{tj} of 0.2 the polarization potential contributes ca. 10% to Ψ_{ms}. This means that whenever a change in I_{tc} causes a change in Ψ_{ms}, I_{pc} will induce a non-negligible polarization potential. A possible means of discerning polarization potentials is by careful analysis of the transient changes in Ψ_{ms}. The need of including changes in E_l when analyzing electrophysiological measurements has been stressed already in Sect. 6.4.

In Fig. 9a the polarization potential $(E_{tj}+E_{lis})$ and the change in local NaCl concentration are plotted as a function of R_{lis}/R_{tj} for $R_{tj} = 20\ \Omega cm^2$ and $I_{pc} = -100\ \mu A\ cm^2$ [values which are comparable to data from flounder intestine (Halm et al. 1985a) and *Tilapia* (Bakker et al., in prep.)]. Of course the use of the equivalent electrical circuit implies that we assume that the movement of Na^+ and Cl^- is exclusively determined by electrodiffusion, and that NaCl gradients are not dissipated by water flow. However, Ikonomov et al. (1985) showed that in *Necturus* gallbladder no changes in Cl^- activity in the lis occurred due to transepithelial current passage. They concluded that the lack of an effect on the NaCl concentration in the lateral intercellular space is not due to the absence of transport number effects, but that the large water permeability of the basolateral cell membrane prevents a measurable change. We should therefore realize that the values for the local NaCl concentration given in Fig. 9a are valid only at a very low water permeability of the basolateral cell membrane.

If, however, the water permeabilities of the cell membranes in intestinal epithelium are comparable to *Necturus* gallbladder, and therefore sufficiently large to prevent measurable NaCl gradients, we may calculate the water transport induced by the paracellular current. In other words, the calculated values for the NaCl concentration in Fig. 9a may be converted into values for isoosmotic fluid transport. Figure 9b shows water flow as a function of the cation selectivity of the tight junctions for $I_{pc} = +100\ \mu A\ cm^{-2}$ and $-100\ \mu A\ cm^{-2}$, respectively. It is feasable that even for small currents the induced water flow may contribute considerably to the total water flow.

When in addition to electroneutral NaCl transport rheogenic mucosa to serosa Cl^- transfer and/or serosa to mucosa K^+ transfer causes a mucosa to

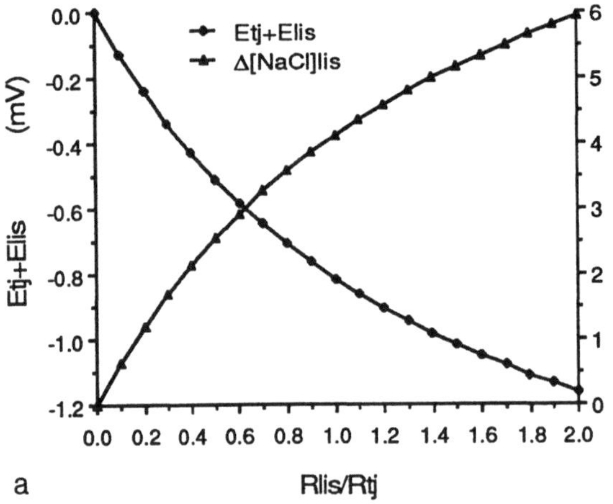

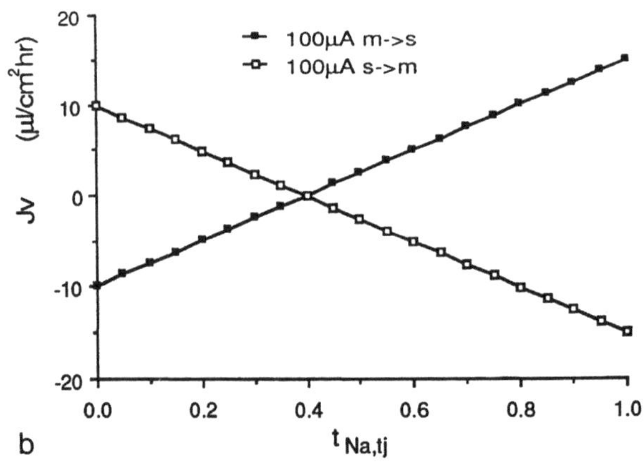

Fig. 9. *a* The polarization potential $E_{tj}+E_{lis}$ (●——●) in mV) and the change in lateral intercellular space NaCl concentration (▲——▲ in % of bath NaCl concentration) as a function of R_{lis}/R_{tj} for $R_{tj} = 20\ \Omega cm^2$ and $I_{pc} = -100\ \mu A\ cm^{-2}$. *b* Current-induced water flow J_v due to salt polarization for a 100 μA cm^{-2} paracellular current from mucosa to serosa (■——■) and serosa to mucosa (□——□) for varying cation selectivity of the tight junctions ($t_{Na,\,tj}$)

serosa paracellular current ($I_{pc}<0$), as found for instance in flounder intestine (Frizzell et al. 1984), cation-selective tight junctions help to increase absorption due to the transport number effect. As marine teleosts are continuously threatened by desiccation and have to rely entirely on their intestine for water uptake, it is tempting to speculate that the observed serosa-negative pd in combination with cation-selective tight junctions increase water uptake by this mechanism. The reported modulation of tight junction Cl^- conductance (Bakker and Groot 1984; Rao et al. 1984; would provide a means for adjusting the water uptake.

In mammalian intestine rheogenic Na^+ transfer through the villus cells ($I_{tc}<0$) may amplify intravillus NaCl accumulation due to the transport number effect, while the returning current through the paracellular pathway of the crypt cells ($I_{pc} > 0$) may lead to NaCl accumulation in the unstirred crypt lumen. Net absorption of water could be modulated by regulation of crypt cell tight junction Cl^- conductance or could result from a difference in cell membrane water permeability between villus and crypt cells (note that a difference in water permeability between crypt and villus may also result from a difference in surface area).

7 Sodium and Chloride Absorption in the Small Intestine

There is much evidence that in intestinal epithelia which possess villi and crypts the villi are concerned with absorption and the crypts with secretion. Elegant optical studies by Welsh et al.(1982) have established that at least in rabbit colon the crypts are the site of secretion of fluid and electrolytes. On the other hand, the maintenance of the high osmolarity in the villus tip during secretion (Jodal and Lundgren 1986) can only be explained if the villi do not participate in secretion. This is, however, not to say that epithelia without crypts cannot secrete. In this section we will discuss the mechanisms of NaCl absorption in small intestine which presumably are located in the villus cells.

7.1 Active NaCl Absorption

Since the in vivo experiments of Curran and Solomon (1957), leading to the postulate that both Na^+ and Cl^- are actively transported from the lumen to the blood, the active absorption of Na^+ and Cl^- has been confirmed by Ussing chamber experiments (Field et al. 1971; Nellans et al. 1974; Powell et al. 1974). The difference between the net Cl^- flux and the net Na^+ flux was, however, not equal to the short-circuit current, indicating a net anion secretion (or a net cation absorption). This so-called residual flux is ascribed to HCO_3^- secretion or a similar process leading to alkalinization of the luminal fluid (see, however, Sect. 6.2).

7.2 Active Na Transport

In considering the mechanism of active transport it is important to know in which membrane the uphill transport takes place. Since the intracellular sodium concentration is lower than in the mucosal and serosal solutions and since the cell potential is negative, the influx of Na^+ through the apical membrane is downhill. The Na^+ influx mechanisms can be divided into three categories (see also Fig. 10): (1) a diffusive influx, determined by the Na^+ permeability and the electrochemical potential difference of Na^+, (2) sugar and amino acid transports driven by the electrochemical potential of Na^+, and (3) electroneutral transport of Na^+ linked to uphill cotransport of anions or countertransport of cations.

With regards to the first mechanism, we can only say that there is still much uncertainty about the sodium permeability in the apical membrane. In the rat

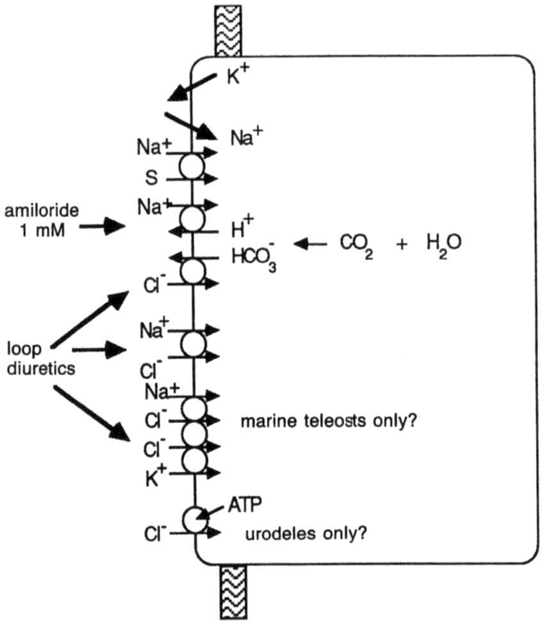

Fig. 10. Overview of relevant brush border transport mechanisms. The *arrows* at the top of the figure represent diffusive movement of K^+ and Na^+. *S* (solute): amino acids and sugars. These three transport processes and the active transport of Cl^- (*bottom* of the figure) are electrogenic. All other cotransporters and countertransporters are electroneutral. The *heavy arrows* indicate inhibitive action

small intestine it is estimated as less than 10% of the permeability of K^+ (Okada et al. 1975). From studies in goldfish intestine values of $1-5 \times 10^{-6}$ cm s^{-1} have been calculated (Bakker 1986; Zuidema et al. 1985). The nonelectrolyte cotransport systems in the small intestine contribute to the electrogenic Na^+ influx as has been shown by brush border vesicle studies (Murer and Hopfer 1974; Wright 1984), as well as by electrophysiological studies (Albus et al. 1983; Gunter-Smith et al. 1982). The influx depends on the non-electrolyte concentration; at a saturating glucose concentration the Na^+ influx through the apical membrane of goldfish intestine increases three fold (Bakker 1986).

The nature of the electroneutral processes will be discussed in Section 7.3.3.

The active step in transepithelial Na^+ transport is located in the basolateral membrane. In most, but not all, epithelia (see Sect. 8.4.3) the ouabain-sensitive, rheogenic NaK-ATPase can only be found in this membrane (DiBona and Mills 1979). Because of its stoichiometry the NaK pump is rheogenic and may contribute to the basolateral emf and thus to the cell potential.

7.3 Active Chloride Transport

7.3.1 Chloride is Accumulated in the Enterocytes

As compared to the sodium pump, the existence of a chloride pump has never been generally accepted. In contrast to the situation for sodium, where a qualitative analysis suffices to gain an idea about the localization of the active step, the assignment of the site for the active step in the transepithelial Cl^- transport requires exact knowledge of the intracellular Cl^- activity and the cell potential to decide whether Cl^- is accumulated in the cell or not. Accumulation would indicate that the active step is in the apical membrane. If chloride is not accumulated, the active step must be sought in the basolateral membrane. From the chemical analysis of the cell concentration of chloride, values between 40 and 70 mM have been reported (see Groot 1981). Equilibrium of Cl^- would require cell potentials of about -15 to -30 mV. Although this strongly suggests that Cl^- is accumulated, definitive conclusions could only be made after the successful application of Cl^--sensitive microelectrodes. It has been shown that the intracellular Cl^- activity in rabbit gallbladder is above its equilibrium value (Duffey et al. 1978). This is also the case in enterocytes of lower vertebrates (Armstrong et al. 1979; Duffey et al. 1979; White 1977, 1986; Zeuthen et al. 1978; Zuidema et al. 1985). Thus, the general finding that Cl^- is accumulated in the epithelial cells requires the presence of an energized Cl^- transport mechanism in the apical membrane.

7.3.2 Na and Cl Fluxes are (Partially) Coupled and Can Be Inhibited by cAMP

Earlier, from flux studies in gallbladder and intestine it has been concluded that a coupled Na^+ and Cl^- transport may be operative in the apical membrane. For intestine, experiments of Nellans et al. (1973, 1974) have shown that in the rabbit ileum part of the unidirectional sodium flux from mucosa to serosa as well as 50% of the net sodium flux is dependent on the presence of Cl^- in the bathing solutions. In addition, in the absence of Na^+, the unidirectional Cl^- flux from mucosa to serosa is reduced and the net Cl^- transport is totally abolished. Short-term influx experiments, in which the apical barrier is exposed to tracers for only 30 to 45 s revealed a reduction of about 20% of the sodium and chloride fluxes if the mucosal solution was chloride-free or sodium-free, respectively. In addition, after preincubation with cAMP or theophylline (an inhibitor of the conversion of cAMP to 5′-AMP), uptake of Na^+ and Cl^- was reduced by the same amount and the ion replacements were without effect.

In earlier experiments of Field (1971) it has been shown that cAMP or theophylline abolished the net Na^+ flux and induced secretion of Cl^-.

Thus, it appears that in the apical membrane the downhill movement of Na^+ can be coupled to an uphill transport of Cl^- which can be inhibited by cAMP. Such a transport system is to be classified as a secondary active transport because the energy is indirectly dependent on metabolic reactions via the sodium gradient (i.e., not only sugar and amino acid transport make use of a Na-coupled transport, but also Cl^-).

Important support for this concept has been obtained from Cl^- activity measurements during sodium-free incubation of the gallbladder (Duffey et al. 1978). In the absence of sodium the membrane potential and the Cl^- equilibrium potential were almost equal, whereas in the presence of sodium the Cl^- activity was 2.5 times its equilibrium value.

7.3.3 NaCl Cotransport or Na^+/H^+ and Cl^-/OH^- Exchangers

It is becoming evident now that the proposed mechanism for mucosal Cl^- accumulation may differ in various intestinal epithelia and there is serious doubt whether a NaCl cotransport mechanism really exists. Actually, the first proposal for coupled NaCl transport came from in vivo studies in human ileum (Turnberg et al. 1970). It was suggested that NaCl transport may occur via a double-exchanger system consisting of a Na^+/H^+ exchanger and a Cl^-/anion exchanger. The latter anion may be bicarbonate or OH^-. If both systems operate simultaneously and with equal turnover rates, a neutral uptake of NaCl will occur in exchange for H_2O and CO_2.

Using brush border membrane vesicles of rat small intestine, Murer et al. (1976) demonstrated the presence of a sodium/proton exchange mechanism and Liedtke and Hopfer (1977) found evidence for Cl/anion exchange.

In a later study Liedtke and Hopfer (1982a) sought evidence for the presence of a directly coupled NaCl cotransport in the same preparation. Their experimental approach was to measure the rate of self-exchange of Na^+ and Cl^- at varying concentrations of Cl^- and Na^+, respectively. They found that the self-exchange rate for Cl^- in the physiologically relevant concentration range was independent from the concentration of Na^+ and vice versa. They concluded that with NaCl concentrations above 25 mM no evidence for a directly coupled cotransport could be found.

Thus, the phenomenon of coupled NaCl transport may originate from the simultaneous operation of the two exchangers, which in the steady state, are coupled by the constancy of intracellular pH. There may be a slack in the coupling of the two exchangers, depending on the buffer capacity in the cell (see Knickelbein et al. 1985). It may be that the source of intracellular H^+ and HCO_3^- is the hydration of CO_2, mediated by carbonic anhydrase. Moreover, a CO_2 gradient across the epithelium may constitute a driving force for NaCl uptake.

In this respect it is interesting that in vivo studies by Charney and Feldman (1984) have shown that intestinal sodium and chloride absorption in rat ileum and rat distal colon is markedly affected by systemic pH and P_{CO_2} (carbon dioxide tension). Respiratory and metabolic acidosis increased net Na^+ and Cl^- absorption in rat ileum and rat distal colon. The respiratory compensation of metabolic acidosis (leading to lower bicarbonate concentration) reduces the net Na^+ and Cl^- absorption. Similar changes in net sodium and chloride transport were also found in rabbit ileum, but not in distal colon. This is interesting because it may be correlated with the absence of transport mechanisms.

Recently, Charney et al. (1986) have reinvestigated the role of carbonic anhydrase in the changes of intestinal electrolyte transport caused by altered systemic pH and P_{CO_2} in rat ileum and colon. From the effects of acetazolamide on the response of electrolyte transport to systemic acid-base alteration they concluded that carbonic anhydrase may play a role in the colon, but not in the ileum. So, other explanations for the inhibitory action of acetazolamide have to be considered (see Nellans et al. 1975).

In conclusion, the correlation of systemic pH and NaCl transport is suggestive for the participation of a Na^+/H^+ exchanger, but a detailed analysis of the relationship between systemic pH and enterocyte pH has to be performed before definitive conclusions can be drawn.

In a recent review, Powell et al. (1985) have compared the kinetic constants of the NaCl-coupled uptake mechanism as obtained from influx measurements into intact epithelium with the constants obtained for sodium/proton exchange and anion exchange in brush border membrane vesicles. The concentrations at which half-maximal transport occurs via the apparent directly coupled NaCl transport is in the range of 4.5 to 29 mM and for influx via the exchangers K_m values from 3.5 to 16 mM were found. These values are not significantly different and cannot be used to distinguish between directly coupled transport or the double exchanger. Moreover, comparison of inhibitors was not conclusive either; furosemide and bumetanide, loop diuretics which are considered specific inhibitors of directly coupled NaCl transport, inhibit the anion exchanger in vesicles as well (Liedtke and Hopfer 1982b), whereas disulfonic stilbenes like DIDS (4,4′-diisothiocyanostilbene-2,2′-disulfonic acid) which are considered to block anion exchangers, are nonselective (Smith et al. 1986). In addition, there seems to be species differences in susceptibility to the inhibitors.

Thus, although it may well be that the process of NaCl coupling in the apical membranes of enterocytes is by the dual exchanger mechanism, no definitive proof can be given for the nonexistence of the cotransporter and the mechanisms may even be present together or may change in activity (see Sect. 5.1.2).

Further work in this field is also necessary to unravel the mechanism of the inhibition of NaCl uptake by cAMP. The question is: on which limb does cAMP exert its inhibitory action? It may well be that such studies can help to clarify the issue of transport of NaCl via direct coupling or via the double exchanger model.

7.3.4 Na/K/Cl Cotransport in Marine Teleosts

So far we have discussed two proposed mechanisms for neutral uptake of NaCl from the lumen into the enterocytes. A third mechanism, revealed in the winter flounder (*Pseudopleuronectes americanus*) and probably common in marine teleosts, is a coupled Na/K/Cl transport similar to the neutral, loop diuretics-sensitive, cotransport mechanism described in avian erythrocytes, Ehrlich ascites tumor cells, amphibian renal-diluting segments, mammalian renal thick ascending limb, and rectal glands of sharks (see Palfrey and Rao 1983).

In isolated intestine of the winter flounder, under short-circuit conditions, a net Na^+ and Cl^- transport has been observed. Sodium-free solutions abolish the net Cl^- flux and Cl^--free solutions abolish the Na^+ flux (Field et al. 1978). In addition, loop diuretics like furosemide, bumetanide, and piretanide inhibit net transepithelial Cl^- transport as well as the mucosal uptake of Cl^- and Na^+ in short-term uptake experiments (Field et al. 1978, 1980; Frizzell et al. 1979b). Moreover, it has been shown that in the absence of Na^+ furosemide has no effect on the Cl^- uptake and in the absence of Cl^- furosemide has no effect on the Na^+ uptake. Thus, furosemide seems to inhibit a coupled NaCl transport. In addition, this coupled transport could be blocked by combined addition of theophylline and dibutyryl cAMP. In the presence of these drugs furosemide had no further effect. Later work in Field's group (Rao et al. 1984) has shown that the inhibitory effect on the cotransport system is by cGMP. cAMP has no inhibitory effect on this mechanism, but modifies the ion selectivity of the epithelium (see Sect. 9.5).

These results are in excellent agreement with the concept of a coupled NaCl transporter in the apical membrane of rabbit ileum. In addition, however, it has been found that both net Cl^- transport and short-term uptake of Cl^- are inhibited by removal of K^+ from the mucosal solution, whereas the influx of Rb^+, used as a substitute for K^+, required the presence of Na^+ and Cl^- (Musch et al. 1982).

The transepithelial fluxes of Rb^+ under short-circuit conditions show a net secretion which could be blocked by Ba^{2+} in the mucosal solution. (Ba^{2+} is known as a blocker of conductive K^+ channels.) The active step for this secretory transport is in the basolateral membrane, presumably in the NaK pump because net Rb^+ secretion is inhibited by ouabain. The important finding is that in the presence of Ba^{2+} a net Rb^+ absorption can be observed. This absorption can be blocked by furosemide or Na^+-free or Cl^--free solutions. In conclusion, Ba^{2+} unmasks a Rb^+ transport mechanism that requires Na^+ and Cl^- and is sensitive to furosemide. All evidence suggest that this coupled Na/K/Cl transport is the mechanism involved in Cl^- accumulation in enterocytes of marine teleosts.

In electrophysiological experiments with winter flounder intestine it has been shown that the mucosal addition of bumetanide or mucosal Cl^--free solutions or Na^+-free solutions all caused a hyperpolarization of the cell potential (see Halm et al. 1985 a, b). These maneuvers result in an inhibition of the coupled Na/K/Cl transport.

The explanation for the hyperpolarization may be found in the fact that upon inhibition of the cotransporter, the Cl^- activity in the cell decreases and attains equilibrium. Secondly, the basolateral membrane of these enterocytes appears to be highly conductive to Cl^-, whereas the apical membrane is primarily conductive to K^+. Thus, as the Cl^- activity falls, the basolateral membrane emf will hyperpolarize and a steady state will be reached in which the cell potential will be determined by the potassium equilibrium potential across the apical membrane.

Interestingly, the anomalous hyperpolarization of the cell potential induced by mucosal Cl^--free solutions has also been observed in early postnatal developing cells of guinea pig colon (Henin and Smith 1976) and in goldfish intestine (Zuidema et al. 1985). This may be an indication for the general existence of basolateral Cl^- conductance. However, in experiments performed in our laboratory we could not find effects of furosemide, bumetanide, or piretanide on ion fluxes or electrical potentials in the goldfish (Bakker and Groot 1984). We propose that the intracellular accumulation of Cl^- may be maintained by the double exchanger. The presence of the exchangers is indicated by the cellular pH response upon Na^+ replacement (acidification) and Cl^- replacement (alkalinization) (Groot et al. 1982). Intracellular alkalinization may increase the K^+ conductance (Reuss et al. 1982) and may lead to a hyperpolarization.

The difference between goldfish and winter flounder led us to initiate a study in an euryhaline fish with respect to the Cl^- transport. In freshwater *Tilapia* (*Oreochromis mossambicus*) we could not detect an effect of bumetanide.

However, in seawater-adapted *Tilapia* bumetanide hyperpolarizes the cell potential and reduces the serosa-negative transepithelial potential (Bakker and Groot, in prep.). This suggests that seawater adaptation may activate a bumetanide-sensitive transporter in an euryhaline fish. Similarly, Lau (1985) has observed that in the seawater-adapted European flounder the Cl^- accumulation is sensitive to piretanide, but not in the freshwater-adapted animals. In conclusion: the Na/K/Cl cotransporter seems to be electroneutral and may have a stoichiometry of $2Cl^-$, 1 Na^+, and 1 K^+. We suggest that the transporter is a prerequisite to solve osmotic problems in marine fish.

Although the inhibitory effect of furosemide on electrolyte transport in mammalian small intestine has been documented (Humphreys 1976; Guandalini et al. 1982) and even a cellular hyperpolarization induced by furosemide has been described (Dinno and Huang 1977), we are not aware of studies concerning the question whether the diuretic-sensitive NaCl transport in mammalian intestine is involved in K^+ uptake and whether the Na/K/Cl cotransporter may exist in the apical membrane of mammalian small intestine.

7.3.5 Electrogenic Cl^- Transport

Finally, a fourth mechanism has been proposed for the Cl^- uptake across the luminal membrane of *Amphiuma* intestine (White 1986; White et al. 1986). In this tissue net Cl^- transport is electrogenic and is not accompanied by net Na^+ absorption nor is the Cl^- accumulation in the cell dependent on luminal Na^+. It is suggested that the luminal Cl^- uptake mechanism is a primary active transport process. An interesting finding is that sugars and amino acids reduce the ability of the cells to maintain an outwardly directed electrochemical potential for Cl^- which may be due to depletion of energy reserves, although the author does not exclude the possibility of increased backflux due to an increase of luminal Cl^- conductance.

7.3.6 Serosal Cl^- Transport

Little is known about the exit of Cl^- through the basolateral membrane. As discussed in Section 7.3.4 there is evidence for a conductive pathway which in the goldfish is estimated to be 20×10^{-6} cm s^{-1} (Zuidema et al. 1985). This is large enough to account for Cl^- movement through the basolateral membrane. In the winter flounder the permeability may be to small (Halm et al. 1985a). In analogy to a neutral KCl transporter as proposed by Reuss (1983) for *Necturus* gallbladder, a similar cotransport may operate in intestinal cells as a nonconductive pathway for Cl^- efflux.

Although a disulfonic stilbene-sensitive anion exchange has been found at the basolateral side of rabbit ileum (Langridge-Smith and Field 1981) and in basolateral membrane vesicles from rat (Grinstein et al. 1980), Cl^- transport is not affected by the stilbenes.

The serosal membrane of *Amphiuma* intestine seems to possess a secondary active transport process in which serosal Na^+ and HCO_3^- are exchanged for intracellular Cl^- (White 1986).

Thus, the study of the Cl^- exit process is still in an early stage and we need much more data to draw a clear picture.

8 Sodium and Chloride Absorption in the Colon

8.1 Functional Heterogeneity of the Large Intestine

Our understanding of electrolyte transport in colon has developed much slower than that of transport in the small intestine. From in vivo studies it is known that the large intestine normally absorbs Na^+ and Cl^- and secretes K^+ and HCO_3^-.

A discussion of the salt transport in the large intestine should take into account the heterogeneity of the transport processes along the length of the colon. It has been shown by a number of laboratories for a number of animals that functional properties of the large intestine, such as the transepithelial potential, the resistance, and the mode of Na^+ and Cl^- transport, differ between proximal and distal parts of the colon (e.g., Fromm and Hegel 1978; von Engelhardt and Rechkemmer 1983; Sandle et al. 1986). In the rabbit colon even four different segments may be distinguished (Clauss et al. 1985). Most in vitro work has been done on distal rabbit colon (for a review see: Schultz 1984), but it is clear now that the rabbit distal colon exemplifies a situation that can be induced in rat distal colon by aldosterone.

8.2 Proximal Colon

Recently, results from Ussing chamber experiments with rabbit proximal colon have revealed that under nonstimulated conditions no net NaCl transport occurs. But a neutral NaCl transport can be induced by α_2 agonists. This transport can be blocked by a high concentration of amiloride (1 mM) which suggests that a Na^+/H^+ exchanger plays a role (Sellin and de Soignie 1987). (The Na^+ conductive pathway and the neutral Na^+/H^+ pathway can be differentiated by the concentration of amiloride needed to inhibit the transport.) The transport is not sensitive to mucosal furosemide or bumetanide and is not inhibited by K^+-free solutions at the mucosal side. Bicarbonate-free solutions or disulfonic stilbenes do not inhibit the net NaCl transport induced by epinephrine. Thus, it is not clear what mechanism is involved in the increase of the mucosa to serosa and net Cl^- transport in proximal colon of rabbit.

In situ experiments with perfused proximal colon of the rat have shown evidence that similar to rabbit proximal colon the NaCl absorption is electrically neutral. In favor of a dual exchange mechanism is that the absorption cannot only be blocked by luminal amiloride at a concentration of 1 mM, but also by acetazolamide and disulfonic stilbenes (Lübcke et al. 1986). Recently, the development of isolation procedures for brush border membrane vesicles from colon revealed the presence of Na^+/H^+ exchangers in the proximal as well as in the distal colon of rat by ^{22}Na uptake (Binder et al. 1986) and by the fluorescence response of a pH-sensitive dye (Foster et al. 1986a).

8.3 Rat Distal Colon

Distal colon from rats which were kept on a normal diet and not treated with mineralocorticoids or glucocorticoids absorbs NaCl in an electroneutral way that can be inhibited by amiloride at 1 mM and acetazolamide. In contrast, sodium depletion or a high K^+ diet or injection of corticoids induces electrogenic Na^+ transport which can be inhibited by amiloride concentrations lower than 1 mM and which is independent of Cl^- (Halevy et al. 1986). The induction of the electrogenic pathway seems to coincide with the loss of the neutral mechanism. However, it has been shown recently that these phenomena are not coupled (Budinger et al. 1986). Moreover, the electrogenic pathway can be induced under in vitro conditions and in these experiments it has also been shown that the neutral NaCl transport still exists after the induction of the low amiloride-sensitive Na transport (Schreiner et al. 1986). In addition to the induction of the Na^+ pathway, the distal colon develops a K^+ secretory pathway (Foster et al. 1986b).

8.4 Rabbit Distal Colon

8.4.1 Na^+ Absorption

The distal part of the colon is much less leaky than the small intestine ($G \approx 5$ mS cm^{-2}) and develops a larger transepithelial potential (serosa-positive). It can transport Na^+ against a large electrochemical potential. This can be inhibited by the pyrazine diuretic amiloride with a half-maximum concentration of ca. 1 µM (Moran et al. 1985). Amiloride induces a hyperpolarization of the apical membrane, a decrease of the apical membrane conductance, and abolishes the transepithelial potential.

Sodium uptake can be increased severalfold by a sodium-deficient diet or aldosterone. Similarly, aldosterone inducible Na^+ transport has been described in the fowl colon (Skadhauge et al. 1985), suggesting the generality of this mechanism.

In recent years the cellular processes that are responsible for electrogenic Na^+ transport in the distal rabbit colon have been studied thoroughly with electrophysiological techniques. It has been shown that the Na^+ entry through the apical membrane behaves as a simple diffusion of a cation that can be described by the Goldman-Hodgkin and Katz flux equation (Thompson et al. 1982), suggesting that Na^+ passes through conductive channels or pores similar to the pores described in the apical membrane of frog skin. Thus, it is concluded that the Na^+ entry is determined by the electrochemical potential difference across the apical membrane, the number of pores in the open state, and the conductance of these pores. The entry of Na^+ through the pores is reversibly blocked by amiloride in a one-for-one way.

An interesting behavior of the Na^+ pores has been described by Turnheim et al. (1983); they found that the number of open pores is larger at a low Na^+ concentration in the luminal fluid and in the cell. This, so-called self-inhibitory mechanism is common to Na^+ pores in frog skin (Fuchs et al. 1977) and in turtle colon (Kirk and Dawson 1985). The mechanism helps to extract Na^+ from the intestinal content.

The effect of aldosterone is to increase the number of pores. The physiological relevance of the aldosterone effect during a Na^+-deficient diet is obvious. It is not clear whether the increased number of Na^+ pores coincides with an increase in NaK pump activity. Edmonds (1976) and Turnheim et al. (1986) did not find evidence for an increase of the biochemical activity (but see Schultz 1984).

To summarize: in the distal colon of a number of animals including toad (Dawson and Curran 1976), turtle (Dawson 1977), fowl (Skadhauge et al. 1985), rabbit (Schultz 1984), pig (Cremaschi et al. 1981), and man (Sandle et al. 1986) Na^+ entry across the apical membrane occurs via a conductive entry process that can be blocked by amiloride. In rat colon the Na^+ entry mechanism is electroneutral or electrogenic depending on the hormonal status of the animal. Mineralocorticoids, but also glucocorticoids and low sodium and high potassium diets, can induce the amiloride-sensitive Na^+ pores and may reduce or even inhibit the electroneutral process. There is evidence that the two types of corticoids do not have the same function, for instance, glucocorticoids stimulate in the small intestine, whereas mineralocorticoids have no effect (cf. Sellin and de Soignie 1985).

8.4.2 Cl^- Absorption

The absorption of Cl^- is not dependent on Na^+ in the luminal fluid, it is electroneutral and it has been proposed that Cl/anion exchange mechanisms in the apical and basolateral membrane are involved (Frizzell et al. 1976). Wills (1985) has recently measured the intracellular Cl^- activity in the surface cells. She concluded that Cl^- is slightly accumulated across the apical membrane, but that the driving force across the basolateral membrane and its conductance are not sufficient for electrogenic Cl^- transport, suggesting the necessity of an electroneutral mechanism in the basolateral membrane. Duffey (1984) has calculated the HCO_3^- concentration from pH measurements with microelectrodes in the surface cells and he concluded that a Cl^- uptake across the apical membrane can be driven by the downhill exit of bicarbonate. Recently, Smith et al. (1986) have used the disulfonic stilbenes, SITS and DIDS, to investigate the anion exchangers. Their results show that DIDS at low concentrations in the serosal fluid can inhibit the mucosa to serosa flux of Cl^-. At higher concentrations, however, DIDS seems to induce synthesis of prostaglandins. The effect of SITS at 1 mM was to reduce both unidirectional fluxes without a change in the net flux. They concluded that the use of the disulfonic stilbenes as specific inhibitors of anion exchange in epithelial tissue is complicated by other effects and precludes their use to demonstrate the role of anion exchangers in the rabbit distal colon. Thus, it seems that the exact mechanism of Cl^- absorption is still an open question.

8.4.3 Transport

The original finding of Frizzell et al. (1976) of a K^+-selective paracellular pathway has led to the hypothesis that all K^+ movement across the colon is by simple paracellular diffusion. Later studies disclosed that absorptive and secretory fluxes are transcellular. In addition, the modulation of the K^+ fluxes by drugs and neurohormonal agents strongly points to transcellular transport.

The absorption of K^+ in rabbit distal colon can be observed in the presence of serosal ouabain, which blocks the serosal uptake. Reduction of the metabolism, however, reduces the absorption of K^+, suggesting that the absorption is active (Wills and Biagi 1982). The active process in rabbit distal colon is presumably related to a K^+-activated ATPase in the apical membrane (Gustin and Goodman 1982). This ATPase has characteristics similar to the K/H-ATPase in gastric mucosa (Foster et al. 1984). The gastric K/H ATPase can be inhibited by a benzimidazole derivative, omeprazole. This drug inhibits the HCO_3^- transport and lowers the intracellular K^+ activity in *Amphiuma* intestine (White 1985). However, the application of this drug in distal rabbit colon did not inhibit the K^+ absorption (Halm and Frizzell 1986).

In at least two species, namely, the turtle (Halm and Dawson 1984) and the guinea pig (Suzuki 1984), the absorptive process in the apical membrane of the colon can be inhibited by mucosal ouabain. This ouabain sensitivity suggest that a Na/K-ATPase in the luminal membrane is involved in these tissues.

The basolateral exit step is downhill and may use a Ba^{2+}-sensitive pore (McCabe et al. 1984). In addition, the apical membrane has a K^+ conductance (cf. Wills 1985) which can also be blocked by Ba^{2+} (Halm and Frizzell 1986). Thus, the potassium accumulation in the surface cells by the Na/K-ATPase in the basolateral membrane and, presumably, by a K/H-ATPase in the apical membrane is counteracted by two leaks. It is not easy to predict in which direction the K^+ will move. Recently, Halm and Frizzell (1986) have shown that in the presence of amiloride (to block the electrogenic Na^+ transport) a spontaneous K^+ secretion can be observed. This can be reversed to net absorption by the cyclooxygenase inhibitor, indomethacine. This drug prevents the synthesis of prostaglandins. We will discuss this point further in Section 9.4.2.

9 Intestinal Secretion

9.1 Electrogenic Secretion of Cl^- is the Motor for Fluid Secretion

With the exception of bicarbonate secretion in the ileum, the mechanisms underlying intestinal secretory processes have been almost ignored until the discovery that addition of vasopressin, theophylline, or cyclic AMP to isolated rabbit ileum increased the short-circuit current (Field et al. 1968). Field (1971) showed that cAMP induces Cl^- secretion by increasing the flux from serosa to mucosa and inhibits net Na^+ absorption by decreasing the mucosa to serosa flux (see Sect. 7.3.2). Analysis of the fluxes and the short-circuit current revealed that the increase in current is primarily due to an electrogenic increase of the Cl^- flux from serosa to mucosa. This basic finding has been confirmed in a number of studies. Moreover, it has been shown that the secretory agent, cholera toxin, increases the cAMP level in the intestinal epithelium. Similarly, prostaglandins induce fluid secretion, an increase in the short-circuit current, and (in some tissues) an increase in the cAMP level.

The capacity of the isolated intestine to secrete electrolytes and fluid has attracted a great deal of research interest in the hope of defining the mechanisms responsible for the losses of salt and water that accompany cholera and other forms of secretory diarrhea.

9.2 The Secretion Model

A number of characteristics of the Cl^- secretory mechanism have been discovered, mainly in rabbit ileum and distal colon, leading to their description in an excellent review by Frizzell et al. (1979a).

In the following summary we also quote results from the work of Heintze et al. (1983). (1) The electrogenic Cl^- secretion is dependent on the presence of sodium in the serosal solution. (2) Ouabain inhibits the Cl^- secretion. (3) Cl^- secretion is inhibited by furosemide in the serosal solution. (4) The exchange of cellular Cl^- with serosal isotopic Cl^- is inhibited by serosal Na^+-free solutions, and by serosal addition of furosemide. (5) The secretion of Cl^- is also inhibited by other loop diuretics like bumetanide and piretanide. (6) Removal of K^+ from the serosal side or increase of K^+ at the serosal side above ca. 12 mM inhibits Cl^- secretion. (7) The transepithelial conductance increases in rabbit distal colon.

The secretion of Cl^- is elicited by cAMP. A role of Ca^{2+} in the secretory process has been proposed on the ground that A23187, a Ca^{2+} ionophore, induces Cl^- secretion. A secretory role for cGMP is also suggested from the finding that the diarrheagenic, heat-stable enterotoxin from an *Escherichia coli* strain causes a striking increase of cGMP (Field 1981) and addition of 8-Br-cGMP had the same effect as the heat-stable toxin (Guandalini et al. 1982). The effects of cGMP and cAMP are nonadditive, suggesting that they may use the same effector. Other intracellular messengers may be involved in secretion induction as well. The presence of Ca^{2+} phospholipid-dependent protein kinase (C-kinase) in rat enterocytes (see de Jonge and Lohmann 1985) has been demonstrated. This kinase can be activated directly with some phorbol esters. The serosal application of 4β-phorbol 12,13-dibutyrate to rat ileum induces a secretory response (Fondacaro and Henderson 1985). The response partially depends on neuronal activity. Recently, Donowitz et al. (1986) have shown that the secretory response of rat distal colon can be blocked by indomethacin, suggesting that the phorbol ester acts via stimulation of the arachidonic acid metabolism. Moreover, these authors suggested that phorbol ester inhibit the Na^+/H^+ exchanger in rat distal colon. This action could not be blocked by indomethacin.

The model for the Cl^- secretion induced by the intracellular messengers consists of Na^+-coupled accumulation of Cl^- across the basolateral membrane and a Cl^- pore in the apical membrane. The action of the messengers via their protein kinases may be the opening of this pore and/or the stimulation of the Cl^-

entry step. The inhibition by ouabain has been considered as evidence that the maintenance of the electrochemical potential difference for Na^+ is the driving force for active Cl^- secretion. The inhibition by serosal omission of K^+ may be an indication for the operation of a Na/K/Cl cotransport mechanism in the basolateral membrane or, alternatively, it may represent the inhibition of the NaK pump. High K^+ concentration may depolarize the cell potential and reduce the driving force for Cl^- efflux through the apical membrane.

In the isolated colon the transepithelial conductance increases during the secretory response. This increase is ascribed to the transcellular pathway because the unidirectional Na^+ fluxes and the mucosa to serosa flux of Cl^- does not change. In rabbit small intestine, however, cAMP induces an increase of the transepithelial resistance. This may be due to a collapse of the intercellular spaces due to the dissipation of the salt accumulation in these compartments (Holman et al. 1979).

In rat ileum, the secretory response is not accompanied by a change in transepithelial resistance (Groot et al. 1986a). We suggest that a conductance increase in the transcellular pathway is balanced by a decrease in the paracellular pathway.

9.3 The Localization of the Secretory Response

Already in 1972 it has been shown by Roggin et al. that the response of rabbit jejunum to cholera toxin is still present after selective damage of the villus cells. This has been confirmed by Hardcastle et al. (1981) for other secretion-inducing drugs and suggests that the crypts are the sites of secretion.

Direct evidence has been obtained by Welsh et al. (1982). They visually observed that in vitro distal rabbit colon secretes fluid from the crypt openings upon stimulation with serosal prostaglandin E_2 (PGE_2). The fluid secretion could be inhibited by serosal furosemide. In addition, microelectrodes were used to follow the changes in cell potential in crypt cells and surface cells after PGE_2 addition. The crypt cells depolarize and the resistance of the apical membrane decreases. Impalement of surface cells showed no change in cell potential or fractional resistance. These cells, however, responded to amiloride with a hyperpolarization, whereas the crypt cells did not respond to amiloride. Recently, Horvath et al. (1986) have extended these observations with the demonstration that after an initial depolarization of the cell potential induced by dibutyryl-cAMP, the potential partially repolarizes again without a detectable change in the fractional resistance. In surface cells they could only detect the hyperpolarization.

In addition, the Cl^- channel blocker A9C (anthracene-9-carboxylic acid) (cf. Wangemann et al. 1986) inhibits active Cl^- secretion and reverses the cAMP-induced depolarization in the crypt cells. These results strongly suggest that the crypt cells are responsible for the cAMP-induced electrogenic Cl^- secretion, while the surface cells are involved in the electrogenic Na^+ absorption.

The only electrophysiological study with small intestine which can be related to the colon studies has been reported recently by Stewart and Turnberg (1986).

In rat ileum the villus cells as well as the crypt cells respond to PGE_2 with a depolarization of 10 – 15 mV. The depolarization is followed by a repolarization toward the prestimulated values. Thus, the Cl^- secretory response in rat ileum may occur in both the villus cells as well as in the crypt cells.

9.4 The Role of Potassium

9.4.1 Repolarization of the Cell Potential

Because the depolarization of the cell will reduce the driving force for Cl^- exit, the secretion would be attenuated. The observation that the initial depolarization is followed by a repolarization is reminiscent of the cell potential change induced by glucose or amino acids in goldfish intestine (Albus et al. 1983; Bakker 1986) and *Necturus* intestine (Gunter-Smith et al. 1982). The repolarization restores the driving force for the sodium-coupled solute uptake mechanism as well as for the Cl^- exit. It is conceivable that in both cases the repolarization is caused by an increase in the K^+ conductance. There is convincing evidence that this happens in canine trachea (see Welsh, this Vol.).

9.4.2 K Secretion

The opening of K^+ channels in the apical membrane may be the underlying mechanism involved in the stimulation of K^+ secretion in rabbit distal colon as part of the secretory response to cAMP, PGE_2, or A23187 (Smith and McCabe 1984; Halm and Frizzell 1986). Both groups have presented evidence that K^+ secretion and Cl^- secretion are related. However, K^+ secretion can be induced by β-adrenergic stimulation without Cl^- secretion. It is suggested that low cAMP concentrations can stimulate K^+ secretion, whereas higher concentrations are necessary for stimulation of Cl^- secretion. Thus, the induction of Cl^- secretion also stimulates K^+ secretion. Furthermore, the finding that the Cl^- secretion induced by A23187 but not the K^+ secretion can be blocked by cyclooxygenase inhibitors is also evidence for separate regulation of the secretory mechanisms. The cyclooxygenase inhibitors like indomethacin and mefanimic acid prevent the synthesis of prostaglandins from arachidonic acid which may arise from phospholipid breakdown, activated by the A23187-induced rise in cell Ca^{2+}. Whether the prostaglandin is synthesized in the epithelial cells or in the lamina propria (which is more plausible) is not known.

Much work has been done to eludidate the role of Ca^{2+} in the secretory process (for a recent review see Donowitz and Welsh 1986). One of the findings is that calmodulin inhibitors couteract the effects of Ca^{2+}. In the rabbit colon the inhibitors trifluoperazine and W-7 inhibit the K^+ secretion but not the Cl^- secretion (Smith and McCabe 1984; McCabe and Smith 1985) again indicating separate pathways (see also Sect. 9.6).

The relation with the Ca^{2+}/phospholipid-dependent protein kinase C which is also blocked by these drugs has not yet been reported.

From present knowledge it is not possible to conclude whether the K^+ absorptive and the K^+ secretory pathways are located in different cells.

9.5 Modulation of the Paracellular Ion Permeability

An intriguing effect of cAMP in fish intestine is the induction of an increased Cl^- permeability of the paracellular pathway (Bakker and Groot 1984; Rao et al. 1984). The evidence for the localization of the cAMP effect to the paracellular pathway is as follows. The addition of cAMP (8-Br-cAMP) to the bathing solution of goldfish intestine mounted in an Ussing chamber induces a small increase of the transepithelial potential of about 0.5 mV (serosa-positive) and a concomitant decrease of the transepithelial resistance from 20 to 15 Ωcm^2. The fractional resistance of the cell membranes as measured with microelectrodes does not change significantly. The cell potential hyperpolarizes transiently but after repolarization does not differ from the control level, while the transepithelial potential remains at a higher level. Both unidirectional Cl^- fluxes increase two to four times, while the net flux remains unchanged. The Na^+ fluxes do not change.

The increase of the unidirectional Cl^- fluxes, together with the increase of the transepithelial conductance, suggest that cAMP increases the paracellular Cl^- conductance. This has been confirmed by studying the effects of cAMP on dilution potentials. Both the mucosal and the serosal dilution potentials are strongly reduced by cAMP. In addition, we applied the transport number effect to solve the question whether the increase of the Cl^- conductance is in the traditional paracellular pathway consisting of the tight junctions and the lateral intercellular space (Bakker et al., in prep.). As discussed in Section 6.5, due to the transport number effect across the tight junctions, a current flow from serosa to mucosa induces an increase of the transepithelial resistance. However, in the presence of forskolin (a drug which directly stimulates adenylate cyclase and thus causes an increase of intracellular cAMP), the transepithelial resistance does not increase. This suggests that the difference between the transport numbers of Na^+ and Cl^- in the tight junctions is decreased by forskolin. In conjunction with the observed increase in transepithelial conductance, this can only mean that the Cl^- conductance of the tight junctions can be modulated by cAMP. Other intracellular mediators like Ca^{2+} (probed by the ionophores A23187 and ionomycine) and (8-Br-)cGMP have no effects on the Cl^- permeability.

A similar increase of the transepithelial conductance and unidirectional Cl^- fluxes has been observed in the winter flounder (Rao et al. 1984). In these animals cGMP affects the Na/K/Cl carrier, whereas cAMP affects the Cl^- conductance in the paracellular pathway.

We propose that the capacity to modulate the Cl^- conductance of the tight junctions has a function in the regulation of transepithelial water transport. In the winter flounder intestine both cyclic nucleotides would act as anti-absorptive agents.

It has been suggested that a selectivity change of the paracellular pathway also occurs in rabbit ileum (Powell 1974; Naftalin and Simmons 1979). However, in this tissue the lateral intercellular spaces collapse so that their resistance increases. This leads to a decrease in cation selectivity of the epithelium. We (Groot et al. 1986a) have used rat ileum to test the hypothesis that the paracellular pathway is involved in the response to cAMP. We found evidence for cAMP-induced increase of Cl^- conductance in the apical membrane and K^+ conductance in the basolateral membrane. Because the cellular responses are much more pronounced than in fish intestine, it is difficult to find effects that can unequivocally be ascribed to the paracellular pathway (but see Sect. 9.6).

9.6 Epithelial Cell Lines as Model in Stimulus Secretion Coupling

In the last few years one colonic tumor cell line has been extensively characterized as a model in the study of stimulus-secretion induction. Work from Dharmsathaporn et al. (1984, 1985, 1986) and Weymer et al. (1985) with the T_{84} cell line has shown that these cells can be grown to confluence in a monolayer with tight junctions and desmosomes between the individual cells. Extracellular chemical messengers like vasoactive intestinal peptide (VIP), PGE_2, and carbachol induce Cl^- secretion with many features in common with the colonic secretion described in Section 9.2. To summarize: cAMP opens a Cl^- channel in the apical membrane and a K^+ channel in the basolateral membrane. The K^+ channel can be blocked by Ba^{2+}. This causes inhibition of Cl^- secretion. The secretion of Cl^- is dependent on Na^+ and K^+ at the serosal side and can be blocked by serosal bumetanide. Carbachol induces a transient increase of the intracellular Ca^{2+} activity, as deduced from Quin-2/AM fluorescence of dispersed cells. There is evidence that Ca^{2+} opens a separate K^+ channel in the basolateral membrane which is not sensitive to Ba^{2+} but can be blocked by quinidine (see Fig. 11).

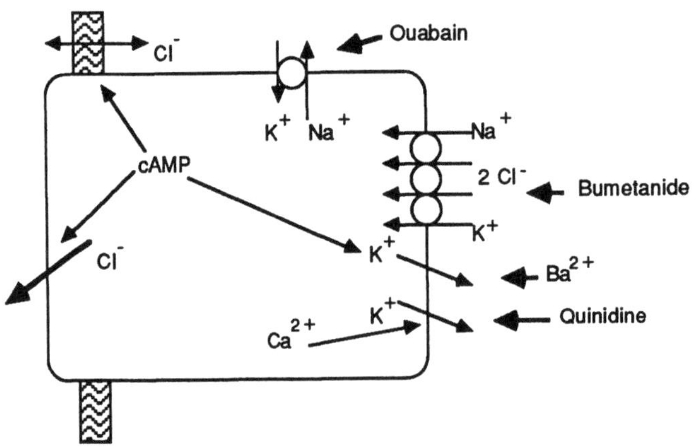

Fig. 11. Cellular model for regulation of Cl^- and K^+ channels in T_{84} colon tumor cells. Cl^- is accumulated by a basolateral, bumetanide-sensitive Na/K/Cl transporter. K^+ is accumulated by the Na/K pump and the Na/K/Cl transporter. cAMP acts via the opening of a Cl^- channel and a Ba^{2+}-sensitive K^+ channel and, presumably, has effect on the Cl^- conductance in the tight junction. Ca^{2+} opens a quinidine-sensitive channel

VIP and PGE_2 induce an increase of the cAMP level and a rise of both unidirectional Cl^- fluxes leading to net secretion. Moreover, they cause a pronounced increase of the transepithelial conductance. The net secretion of Cl^- can be inhibited by bumetanide, but the unidirectional Cl^- fluxes remain high, suggesting an effect on the paracellular pathway.

Two other cell lines have been used in transport studies, namely, the human colonic carcinoma cell line Caco-2 (Grasset et al. 1985) and the human colonic adenocarcinoma cell line HT29 (Augeron et al. 1986).

The importance of the work with these cell lines is that the effect of neurohormonal agents can be studied without interaction of neuronal or enterochromaffin cells in the tissue. Of course, it can be that the cell lines show peculiarities that are not present in intestinal cells or that the cells have lost receptors or transport mechanism as compared with colonic cells.

10 Regulation of Ion Transport in the Intestine

In the previous sections we have shown that intracellular mediators are involved in the regulation of cellular transport processes and in the regulation of the paracellular permeability. A summary is given in Fig. 12. For a well-documented review on this subject the reader should consult the article by de Jonge and Lohmann (1985).

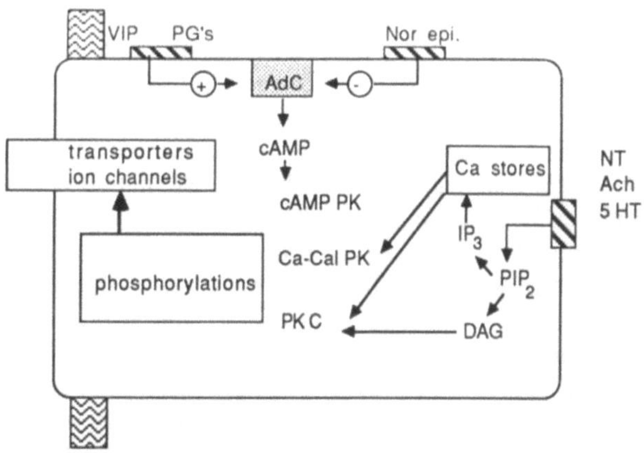

Fig. 12. Stimulus/effect coupling. The *hatched rectangles* represent receptors for *VIP* = vasoactive intestinal peptide, *PG* = prostaglandin, which stimulate *AdC* = adenylate cyclase. *Nor epi* = norepinephrine, which inhibits *AdC*. *cAMP* activates cAMP-dependent protein kinase (*cAMP PK*). *NT* = neurotensin; *Ach* = acetylcholine; *5 HT* = serotonin, which stimulates phospholipase C-mediated hydrolysis of phosphatidylinositol 4,5-biphosphate (*PIP_2*), leading to soluble diacylglycerol (*DAG*) and inositol triphosphate (*IP_3*). The intracellular effect of *IP_3* is the release of Ca^{2+} from intracellular stores. One of the effects of Ca^{2+} is the potentiation of protein kinase C (*PK C*) which further requires *DAG* and membrane-bound phospholipids to become active. Another effect of Ca^{2+} is the activation of calmodulin (*Cal*) kinases (*Ca-Cal PK*). Other possible functions of Ca^{2+}, directly or via calmodulin are not shown. Activation of the kinases leads to phosphorylation of their substrates (e.g., ion channels and transporters in the membrane) which activates or inactivates the function of these membrane proteins

A large number of extracellular chemical messengers are known to affect intestinal transport under experimental conditions (for a review see Turnberg 1983).

However, most of the experimental procedures do not differentiate between effects directly on the epithelial cells or via neuronal cells. It is now known that besides the classical acetylcholine (which stimulates secretion) and noradrenaline (which opposes the acetylcholine effect) many more chemical messengers have been identified in the enteric nervous system (Schultzberg et al. 1980).

The enteric nervous system may be divided into three plexuses, although the development of these plexuses differ between vertebrate classes. After the usually performed stripping procedure, the plexus of Auerbach (between the longitudinal and the circular muscle layer) will be removed. However, the submucosal plexus between the muscularis propria and the muscularis mucosae may still be present and the third plexus within the lamina propria of the mucosa is definitely present. Hubel (see Hubel 1985) has shown that the ion transport in isolated intestine mounted in an Ussing chamber can be modulated by electrical stimulation. The neurotoxic agent tetrodotoxin prevents the effect of stimulation. Thus, the enteric nervous system is not only involved in regulation of motor activity and circulation, but also in the regulation of the epithelial transport function (Cooke 1986; Wood 1984).

Even in goldfish intestinal epithelium which can be thoroughly stripped (Albus et al. 1979) spontaneous oscillations of cell potential and transepithelial potential can sometimes be observed. These oscillations can be blocked by tetrodotoxin. In addition, the effect of low concentrations of forskolin or theophylline is inhibited by the neurotoxin, but the effects of VIP and serotonin cannot be blocked by tetrodotoxin. This suggests that the toxin does not act on the epithelial cells (Groot 1985; Groot and Bakker, in prep.). These results show that a functional neuronal network is still present after stripping and that not all effects of experimentally added chemical messengers need to be ascribed to a direct effect on the epithelial cells.

Recently (Bridges et al. 1986; Andres et al. 1985), it has been shown that the mucosal plexus in rat distal colon is a neatly organized network around the crypts and that there are ganglia in this plexus. Stimulation of this plexus leads to inhibition of net Na^+ and Cl^- absorption.

The presence of the neuronal network in the experimental preparations makes it extremely difficult to draw conclusions on the effects of Ca^{2+} omission or Ca^{2+}channel blockers because it is well known that the release of neurotransmitters is dependent on extracellular Ca^{2+}. Moreover, cAMP-generating agents may give rise to release of transmitters (Carey et al. 1985; Groot et al. 1986b) which secondarily stimulate the epithelial cells.

The occurrence of secondary effects via the neuronal structures adds to the secondary effects of stimulation of the arachidonic acid metabolism which generate prostaglandins. This makes it difficult to delineate the precise route from extracellular chemical messenger via intracellular messenger to the transport pathway under regulation. The demonstration of receptors on the epithelial cells is a prerequisite for the interpretation of transmitter function.

Acknowledgments. Part of this work was supported by the foundation for Biological Research in the Netherlands (ZWO-BION).

References

Albus H, Groot JA, Siegenbeek van Heukelom J (1979) Effects of glucose and ouabain on transepithelial electrical resistance and cell volume in stripped and unstripped goldfish intestine. Pfluegers Arch 383:55 – 66

Albus H, Bakker R, Siegenbeek van Heukelom J (1983) Circuit analysis of membrane potential changes due to electrogenic sodium dependent sugar transport in goldfish intestine. Pfluegers Arch 398:1 – 9

Andres H, Bock R, Bridges RJ, Rummel W, Schreiner J (1985) Submucosal plexus and electrolyte transport across rat colonic mucosa. J Physiol (London) 364:301 – 312

Argenzio RA, Lowe JE, Pickard DW, Stevens CE (1974) Digesta passage and water exchange in the equine large intestine. Am J Physiol 226:1035 – 1042

Armstrong WMcD, Byrd BJ, Cohen ES, Cohen SJ, Hamang PH, Myers CJ (1975) Osmotically induced electrical changes in isolated bullfrog small intestine. Biochim Biophys Acta 401:137 – 151

Armstrong WMcD, Bixenman WR, Frey KF, Garcia JF, O'Regan HG, Owens JL (1979) Energetics of coupled Na and Cl entry into epithelial cells of bullfrog small intestine. Biochim Biophys Acta 551:207 – 219

Augeron C, Maoret JJ, Laboisse CL, Grasset E (1986) Permanently differentiated cell clones isolated from the human colonic adenocarcinoma cell line HT29: possible models for the study of ion transport and mucus production. In: Alvarado F, Os CH van (eds) Ion gradient coupled transport. Elsevier, Amsterdam, pp 363 – 366

Baerentsen HJ, Christiansen O, Thomsen PG, Zeuthen T (1982) Steady states and the effect of ouabain on *Necturus* gallbladder epithelium: a model analysis. J Membr Biol 68:215 – 225

Bakker R (1985) Electrophysiology of fish intestine. In: Gilles R, Gilles-Baillien M (eds) Transport processes, iono- and osmoregulation. Springer, Berlin Heidelberg New York Tokyo, pp 240 – 250

Bakker R (1986) Feedback on potassium permeability by sodium-glucose cotransport in fish intestine. In: Alvarado F, Os CH van (eds) Ion gradient coupled transport. Elsevier, Amsterdam, pp 245 – 248

Bakker R, Groot JA (1984) cAMP mediated effects of ouabain and theophylline on paracellular ion selectivity. Am J Physiol 246:G213 – G217

Barry PH, Diamond JM (1984) Effects of unstirred layers on membrane phenomena. Physiol Rev 64:763 – 872

Barry PH, Hope AB (1969) Electroosmosis in membranes: effects of unstirred layers and transport numbers. Biophys J 9:729 – 757

Binder HJ, Stange G, Murer H, Stieger B, Hauri HP (1986) Sodium-proton exchange in colon brush border membranes. Am J Physiol 251:G382 – G390

Bindslev N, Tormey JMcD, Wright EM (1974) The effects of electrical and osmotic gradients on lateral intercellular spaces and membrane conductance in a low resistance epithelium. J Membr Biol 19:357 – 380

Boulpaep EL, Sackin H (1980) Electrical analysis of intraepithelial barriers. In: Boulpaep EL (ed) Current topics in membranes and transport. Academic Press, London New York, pp 169 – 196

Bridges RJ, Rack M, Rummel W, Schreiner J (1986) Mucosal plexus and electrolyte transport across the rat colonic mucosa. J Physiol (London) 376:531 – 542

Brown PD, Sepulveda FV (1985) Potassium movements associated with amino acid and sugar transport in enterocytes isolated from rabbit jejunum. J Physiol (London) 363:271 – 285

Budinger ME, Foster ES, Hayslett JP, Binder HJ (1986) Sodium and chloride transport in the large intestine of potassium-loaded rats. Am J Physiol 251:G249 – G252

Carey HV, Cooke HJ, Nemeth PR, Zafirof DH, Wood JD (1985) Nerve mediated action of forskolin on guinea-pig ileal mucosa. Experientia 41:1156 – 1158

Charney AN, Feldman GM (1984) Systemic acid-base disorders and intestinal electrolyte transport. Am J Physiol 247:G1 – G12

Charney AN, Wagner JD, Birnbaum GJ, Johnstone JN (1986) Functional role of carbonic anhydrase in intestinal electrolyte transport. Am J Physiol 251:G682 – G687
Chase HS (1984) Does calcium couple the apical and basolateral membrane permeabilities in epithelia? Am J Physiol 247:F869 – F876
Clausen C, Lewis SA, Diamond JM (1979) Impedance analysis of a tight epithelium using a distributed resistance model. Biophys J 26:291 – 317
Clauss W, Schäfer H, Horch I, Hörnicke H (1985) Segmental differences in electrical properties and Na transport of rabbit caecum, proximal and distal colon in vitro. Pfluegers Arch. 403: 278 – 282
Cooke HJ (1986) Neurobiology ot the intestinal mucosa. Gastroenterology 90:1057-1081
Cremaschi D, James PS, Meyer G, Smith MW (1981) Electrophysiology of pig distal colon measured during early post-natal development. J Physiol (London) 314:137 – 149
Curran PF, McIntosh JR (1962) A model system for biological water transport. Nature (London) 193:347 – 348
Curran PF, Solomon AK (1957) Ion and water fluxes in the ileum of rats. J Gen Physiol 41:143 – 168
Curtis RL, Trier JS, Frizzell RA, Lindem NM, Madara JL (1984) Flounder intestinal absorption cells have abundant gap junctions and may be coupled. Am J Physiol 246:C77 – C83
Dawson DC (1977) Na and Cl transport across the isolated turtle colon: parallel pathways for transmural ion movement. J Membr Biol 37:213 – 233
Dawson DC, Curran PF (1976) Sodium transport by the colon of *Bufo marinus*: Na uptake across the mucosal border. J Membr Biol 28:295 – 307
Debongnie JC, Phillips SF (1978) Capacity of the human colon to absorb fluid. Gastroenterology 74:698 – 703
Dharmsathaporn K, Pandol SJ (1986) Mechanism of chloride secretion induced by carbachol in a colonic epithelial cell line. J Clin Invest 77:348 – 354
Dharmsathaporn K, McRoberts JA, Mandel KG, Tisdale LD, Masui H (1984) A human colonic cell line that maintains vectorial electrolyte transport. Am J Physiol 246:G204 – G208
Dharmsathaporn K, Mandell KG, Masui H, McRoberts JA (1985) Vasoactive intestinal polypeptide-induced chloride secretion by a colonic epithelial cell line: direct participation of a basolaterally localized Na,K,Cl cotransport system. J Clin Invest 75:462 – 471
Diamond JM (1964) The mechanism of isotonic water transport. J Gen Physiol 48:15 – 42
Diamond JM (1979) Osomotic water flow in leaky epithelia. J Membr Biol 51:195 – 216
Diamond JM (1982) Transcellular crosstalk between epithelial cell membranes. Nature (London) 300: 683 – 685
Diamond JM, Bossert WH (1967) Standing-gradient osmotic flow. A mechanism for coupling of water and solute transport in epithelia. J Gen Physiol 50:2061 – 2083
DiBona DR (1985) Functional analysis of tight junction organization. Pfluegers Arch 405: S59 – S66
DiBona DR, Mills JW (1979) Distribution of Na pump sites in transporting epithelia. Fed Proc 38:134 – 143
Dinda PK, Beck M, Beck IT (1972) Effect of changes in the osmolarity of the luminal fluid on intracellular concentration of solutes in hamster jejunum. Can J Physiol Pharmacol 50:72 – 82
Dinno MA, Huang KC (1977) Effect of glucose and diuretics on intracellular potentials of mouse intestinal mucosa. Proc Soc Exp Biol Med 155:71 – 78
Donowitz M, Welsh MJ (1986) Ca and cyclic AMP in regulation of intestinal Na, K, and Cl transport. Annu Rev Physiol 48:135 – 150
Donowitz M, Cheng HY, Sharp GWG (1986) Effects of phorbol esters on sodium and chloride transport in rat colon. Am J Physiol 251:G509 – G517
Duffey ME (1984) Intracellular pH and bicarbonate activities in rabbit colon. Am J Physiol 246:C558 – C561
Duffey ME, Turnheim K, Frizzell RA, Schultz SG (1978) Intracellular chloride activities in rabbit gallbladder: direct evidence for the role of the sodium gradient in energizing uphill chloride transport. J Membr Biol 42:229 – 245
Duffey ME, Thompson SM, Frizzell RA, Schultz SG (1979) Intracellular chloride activities and active chloride absorption in the intestinal epithelium of the winter flounder. J Membr Biol 50:331 – 341
Edmonds CJ (1976) Effects of aldosterone on the colon. In: Robinson JWL (ed) Intestinal ion transport. MTP Press, London, pp 269 – 279

Engelhardt W von, Rechkemmer G (1983) Absorption of inorganic ions and short chain fatty acids in the colon of mammals. In: Gilles-Baillien M, Gilles R (eds) Intestinal transport. Springer, Berlin Heidelberg New York, pp 26–45
Field M (1971) Ion transport in rabbit ileal mucosa. II. Effects of cyclic 3′,5′-AMP. Am J Physiol 221:992–997
Field M (1981) Secretion of electrolytes and water by mammalian small intestine. In: Johnson LR (ed) Physiology of the gastrointestinal tract. Raven, New York, pp 963–989
Field M, Plotkin GR, Silen W (1968) Effects of vasopressin, theophylline and cyclic adenosine monophosphate on short circuit current across isolated rabbit ileal mucosa. Nature (London) 217:469–470
Field M, Fromm D, McColl I (1971) Ion transport in rabbit ileal mucosa. I. Na and Cl fluxes and short circuit current. Am J Physiol 220:1388–1396
Field M, Karnaky KJ, Smith PL, Bolton JE, Kinter WB (1978) Ion transport across the isolated intestinal mucosa of the winter flounder, *Pseudopleuronectes americanus*. J Membr Biol 41: 265–293
Field M, Smith PL, Bolton JE (1980) Ion transport across the isolated intestinal mucosa of the winter flounder, *Pseudopleuronectes americanus*: II Effects of cyclic AMP. J Membr Biol 55:157–163
Fondacaro JD, Henderson LS (1985) Evidence for protein kinase C as a regulator of intestinal electrolyte transport. Am J Physiol 249:G422–G426
Foster ES, Hayslett JP, Binder HP (1984) Mechanism of active potassium absorption and secretion in the rat colon. Am J Physiol 246:G611–G617
Foster ES, Dudeja PK, Brasitus TA (1986a) Na^+ H^+ exchange in rat colonic brush border membrane vesicles. Am J Physiol 250:G781–G787
Foster ES, Sandle GI, Hayslett JP, Binder HJ (1986b) Dietary potassium modulates active potassium absorption and secretion in rat distal colon. Am J Physiol 251:G619–G626
Frizzell RA, Schultz SG (1972) Ionic conductances of extracellular shunt pathway in rabbit ileum. J Gen Physiol 59:318–346
Frizzell RA, Koch MJ, Schultz JS (1976) Ion transport by rabbit colon I. Active and passive components. J Membr Biol 27:297–316
Frizzell RA, Field M, Schultz SG (1979a) Sodium coupled chloride transport by epithelial tissues. Am J Physiol 236:F1–F8
Frizzell RA, Smith PL, Vosburgh E, Field M (1979b) Coupled sodium chloride influx across brush border of flounder intestine. J Membr Biol 46:27–39
Frizzell RA, Halm DR, Musch MW, Stewart CP, Field M (1984) Potassium transport by flounder intestinal mucosa. Am J Physiol 246:F946–F951
Fromm M, Hegel U (1978) Segmental heterogeneity of epithelial transport in rat large intestine. Pfluegers Arch 378:71–83
Fromm M, Schulzke JD, Hegel U (1985) Epithelial and subepithelial contributions to transmural electrical resistance of intact rat jejunum, in vitro. Pfluegers Arch 405:400–402
Frömter E (1972) The route of passive ion movement through the epithelium of *Necturus* gallbladder. J Membr Biol 8:259–301
Frömter ER, Diamond JM (1972) Route of passive ion permeation in epithelia. Nature (London) 235:9–13
Fuchs W. Larsen EH, Lindemann B (1977) Current voltage curve of sodium channels and concentration dependence of sodium permeability in frog skin. J Physiol (London) 267:137–166
Gebhardt U, Nell G (1979) Measurement ot the "true" epithelial resistance of the intestinal mucosa. Gastroenterol Clin Biol 3:175
Ghijsen WEJM, De Jong MD, Os CH van (1983) Kinetic properties of Na/Ca exchange in basolateral plasma membranes of rat small intestine. Biochim Biophys Acta 730:85–94
Grasset E, Bernabeu J, Pinto M (1985) Epithelial properties of human colonic carcinoma cell line Caco-2: effect of secretagogues. Am J Physiol 248:C410–C418
Grinstein S, Turner RJ, Silverman M, Rothstein A (1980) Inorganic anion transport in kidney and intestinal brush border and basolateral membranes. Am J Physiol 238:F452–F460
Groot JA (1981) Cell volume regulation in goldfish intestinal mucosa. Pfluegers Arch. 392:57–66
Groot JA (1985) Modulation of transepithelial Cl^- permeability. In: Gilles R, Gilles-Baillien M (eds) Transport processes, iono- and osmoregulation. Springer, Berlin Heidelberg New York Tokyo, pp 206–217

Groot JA, Dekker K, Riel JW van, Zuidema T (1982) Intracellular ion concentrations and pH of stripped mucosa of goldfish (*Carassius auratus*) intestine in relation to Cl^- transport. 4th Conf ESCPB, Bielefeld, pp 132 – 133

Groot JA, Albus H, Bakker R, Siegenbeek van Heukelom J, Zuidema T (1983) Electrical phenomena in fish intestine. In: Gilles-Baillien M, Gilles R (eds) Intestinal transport. Springer, Berlin Heidelberg New York, pp 321 – 340

Groot JA, Bakker R, Dekker K, Oosterhuis WP (1986a) Modulation of transepithelial ion permeability by neurohumoral agents; comparison of fish and rat intestine. In: Alvarado F, Os CH van (eds) Ion gradient coupled transport. Inserm Symp no 26. Elsevier, Amsterdam, pp 411 – 414

Groot JA, Bakker R, Bijlsma P, Oosterhuis WP (1986b) The response of stripped rat ileum to forskolin is partially mediated by neuronal cells. 7th Meet EITG Univ Goteborg, Dep Physiol, p 27

Guandalini S, Rao MC, Smith PL, Field M (1982) cGMP modulation of ileal ion transport: in vitro effects of *Escherichia coli* heat stable enterotoxin. Am J Physiol 243:G36 – G41

Guggino WB (1986) Functional heterogeneity in the early distal tubule of the *Amphiuma* kidney: evidence for two modes of Cl^- and K^+ transport across the basolateral cell membrane. Am J Physiol 250:F430 – F440

Gunter-Smith PJ, Grasset E, Schultz SG (1982) Sodium coupled amino acid and sugar transport by *Necturus* small intestine: an equivalent electrical circuit analysis of a rheogenic cotransport system. J Membr Biol 66:25 – 39

Gustin MC, Goodman DBP (1982) Characterization of the phosphorylated intermediate of the K-ouabain-insensitive ATPase of the rabbit colon brush border membrane. J Biol Chem 257:9629 – 9633

Halevy J, Budinger ME, Hayslett JP, Binder HJ (1986) Role of aldosterone in the regulation of sodium and chloride transport in the distal colon of sodium depleted rats. Gastroenterology 91:1227 – 1233

Halm DR, Dawson DC (1984) Potassium transport by turtle colon: active secretion and active absorption. Am J Physiol 246:C315 – C322

Halm DR, Frizzell RA (1986) Active K transport across rabbit distal colon: relation to Na absorption and Cl secretion. Am J Physiol 251:C252 – C267

Halm DR, Krasny EJ, Frizzell RA (1985a) Electrophysiology of flounder intestinal mucosa. I. Conductance properties of the cellular and paracellular pathways. J Gen Physiol 85:843 – 864

Halm DR, Krasny EJ, Frizzell RA (1985b) Electrophysiology of flounder intestinal mucosa. II. Relation of the electrical potential profile to coupled NaCl absorption. J Gen Physiol 85:865 – 883

Hardcastle J, Hardcastle PT, Redfern JS (1981) Action of 5-hydroxytryptamine on intestinal ion transport in the rat. J Physiol (London) 320:41 – 55

Hardcastle J, Hardcastle PT, Noble JM (1984) The involvement of calcium in the intestinal response to secretagogues in the rat. J Physiol (London) 355:465 – 478

Heeswijk MPE van, Os CH van (1986) Osmotic water permeability of brush border and basolateral membrane vesicles from rat renal cortex and small intestine. J Membr Biol 92:183 – 193

Heidenhain R (1894) Neue Versuche über die Aufsaugung im Dünndarm. Pfluegers Arch 56: 579 – 631

Heintze K, Stewart CP, Frizzell RA (1983) Sodium dependent chloride secretion across rabbit descending colon. Am J Physiol 244:G357 – G365

Helman SI, Thomson SM (1982) Interpretation and use of electrical equivalent circuits in studies of epithelial tissues. Am J Physiol 243:F519 – F531

Henin S, Smith MW (1976) Electrical properties of pig colonic mucosa measured during early postnatal development. J Physiol (London) 262:169 – 187

Holman GD, Naftalin RJ, Simmons NL, Walker M (1979) Electrophysiological and electron-microscopical correlations with fluid and electrolyte secretion in rabbit ileum. J Physiol (London) 290:367 – 386

Hopfer U, Nelson K, Perotto J, Isselbacher KJ (1973) Glucose transport in isolated brush border membranes from rat small intestine. J Biol Chem 248:25 – 32

Horvath PJ, Ferriola PC, Weiser M, Duffey ME (1986) Localization of chloride secretion in rabbit colon: inhibition by anthracene-9-carboxylic acid. Am J Physiol 250:G185 – G190

Hubel KA (1985) Intestinal nerves and ion transport: stimuli, reflexes, an responses. Am J Physiol 248:G261 – G271

Humphreys MH (1976) Inhibition of NaCl absorption from perfused rat ileum by furosemide. Am J Physiol 230:1517–1523

Ikonomov O, Simon M, Frömter E (1985) Electrophysiological studies on lateral intercellular spaces of *Necturus* gallbladder epithelium. Pfluegers Arch 403:301–307

Jodal M, Lundgren O (1986) Countercurrent mechanisms in the mammalian gastrointestinal tract. Gastroenterology 91:225–241

Jonge HR de, Lohmann SM (1985) Mechanisms by which cyclic nucleotides and other intracellular mediators regulate secretion. In: Evered D, Whelan J (eds) Microbial toxins and diarrhoeal disease. Ciba Found Symp 112. Pitman, London, pp 116–138

Kirk KL, Dawson DC (1985) Passive cation permeability of turtle colon: evidence for a negative interaction between intracellular sodium and apical sodium permeability. Pfluegers Arch 403:82–89

Kirsch R, Humbert W, Simonneaux V (1985) The gut as an osmoregulatory organ: comparative aspects and special references to fishes. In: Gilles R, Gilles-Baillien M (eds) Transport processes, iono- and osmoregulation. Springer, Berlin Heidelberg New York Tokyo, pp 265–277

Knickelbein RG, Aronson PS, Schron CM, Seifter J, Dobbins JW (1985) Na und Cl transport across rabbit ileal brush border. II. Evidence for $Cl:HCO_3$ exchange and mechanism of coupling. Am J Physiol 249:G236–G245

Koefoed-Johnsen V, Ussing HH (1958) The nature of the frog skin potential. Acta Physiol Scand 42:298–308

Krasny EJ, DiBona AI, Frizzell RA (1982) Regulation of paracellular permselectivity in flounder intestine. Bull MDIBL 22:82–85

Lahlou B (1983) Intestinal transport and osmoregulation in fishes. In: Gilles-Baillien M, Gilles R (eds) Intestinal transport. Springer, Berlin Heidelberg New York, pp 341–353

Langridge-Smith JL, Field M (1981) Sulfate transport in rabbit ileum: characterization of the serosal border anion exchange process. J Membr Biol 63:207–214

Larson M, Spring KR (1984) Volume regulation by *Necturus* gallbladder: basolateral KCl exit. J Membr Biol 81:219–232

Lau KR (1985) The effects of salinity adaptation on intracellular chloride accumulation in the European flounder. Biochim Biophys Acta 818:105–108

Lau KR, Hudson RL, Schultz SG (1984) Cell swelling increases a barium-inhibitable potassium conductance in the basolateral membrane of *Necturus* small intestine. Proc Natl Acad Sci USA 81:3591–3594

Lau KR, Hudson RL, Schultz SG (1986) Effect of hypertonicity on the increase in basolateral conductance of *Necturus* small intestine in response to Na^+ sugar co-transport. Biochim Biophys Acta 855:193–196

Liedtke CM, Hopfer U (1977) Anion transport in brush border membranes isolated from rat small intestine. Biochem Biophys Res Commun 76:579–585

Liedtke CM, Hopfer U (1982a) Mechanism of Cl^- translocation across small intestinal brush border membrane. I Absence of Na–Cl co-transport. Am J Physiol 242:G263–G271

Liedtke CM, Hopfer U (1982b) Mechanism of Cl^- translocation across small intestinal brush border membrane. II Demonstration of $Cl^- - OH^-$ exchange and Cl^- conductance. Am J Physiol 242: G272–G280

Lübcke R, Haag K, Berger E, Knauf H, Gerok W (1986) Ion transport in rat proximal colon in vivo. Am J Physiol 251:G132–G139

Marcial MA, Carlson SL, Madara JL (1984) Partitioning of paracellular conductance along the crypt-villus axis: a hypothesis based on structural analysis with detailed consideration of tight junction structure function relationships. J Membr Biol 80:59–70

McCabe RD, Smith PL (1985) Colonic potassium and chloride secretion: role of cAMP and calcium. Am J Physiol 248:G103–G109

McCabe RD, Smith PL, Sullivan LP (1984) Ion transport by rabbit descending colon: mechanism of transepithelial potassium transport. Am J Physiol 246:G594–G602

Mircheff AK, Os CH van, Wright EM (1980) Pathways for alanine transport in intestinal basal lateral membrane vesicles. J Membr Biol 52:83–92

Moran WM, Hudson RL, Schultz SG (1985) Kinetics of the effect of amiloride on the permeability of the apical membrane of rabbit descending colon to sodium. J Membr Biol 87:55–65

Morris AP, Gallacher DV, Lee JAC (1986) A large conductance, voltage and calcium activated K^+ channel in the basolateral membrane of rat enterocytes. FEBS Lett 206:87–92

Murer H, Hopfer U (1974) Demonstration of electrogenic Na^+ dependent glucose transport in intestinal brush border membranes. Proc Natl Acad Sci USA 71:484–488
Murer H, Hopfer U, Kinne R (1976) Sodium/proton antiport in brush border membrane vesicles isolated from rat small intestine and kidney. Biochem J 154:597–604
Musch MW, Orellana SA, Kimberg LS, Field M, Halm DR, Krasny EJ, Frizell RA (1982) $Na^+-K^+-Cl^-$ co-transport in the intestine of a marine teleost. Nature (London) 300:351–353
Naftalin RJ, Simmons NL (1979) The effects of theophylline and choleragen on sodium and chloride ion movements within isolated rabbit ileum. J Physiol (London) 290:331–350
Naftalin RJ, Tripathi S (1983) Routes of water flow across the intestine and their relationship to isotonic transport. In: Gilles-Baillien M, Gilles R (eds) Intestinal transport. Springer, Berlin Heidelberg New York, pp 14–25
Naftalin RJ, Tripathi S (1985) Passive water flows across the isolated rabbit ileum by osmotic, hydrostatic and electrical gradients. J Physiol (London) 360:27–50
Nellans HN, Frizzell RA, Schultz SG (1973) Coupled sodium chloride influx across the brush border of rabbit ileum. Am J Physiol 225:467–475
Nellans HN, Frizzell RA, Schultz SG (1974) Brush-border processes and transepithelial Na and Cl transport by rabbit ileum. Am J Physiol 226:1131–1141
Nellans HN, Frizzell RA, Schultz SG (1975) Effect of acetazolamide on sodium and chloride transport by in vitro rabbit ileum. Am J Physiol 228:1808–1814
Okada Y, Sato T, Inouye A (1975) Effects of K^+ and Na^+ on membrane potentials of epithelial cells of rat duodenum. Biochim Biophys Acta 413:104–115
Os CH van, Michels JA, Slegers JFG (1976) Effects of electrical gradients on volume flows across gallbladder epithelium. Biochim Biophys Acta 443:545–555
Palfrey HC, Rao MC (1983) Na/K/Cl cotransport and its regulation. J Exp Biol 106:43–54
Persson BE, Spring KR (1982) Gallbladder epithelial cell hydraulic water permeability and volume regulation. J Gen Physiol 79:481–505
Phillips SF, Giller J (1973) The contribution of the colon to electrolyte and water conservation in man. J Lab Clin Med 81:733–746
Powell DW (1974) Intestinal conductance and permselectivity changes with theophylline and choleragen. Am J Physiol 227:1436–1443
Powell DW, Farris RK, Carbonetto ST (1974) Theophylline, cyclic AMP, choleragen and electrolyte transport by rabbit ileum. Am J Physiol 227:1428–1435
Powell DW, Berschneider HM, Lawson LD, Martens H (1985) Regulation of water and ion movement in intestine. In: Evered D, Whelan J (eds) Microbial toxins and diarrhoeal disease. Ciba Found Symp 112. Pitman, London, pp 14–28
Rao MC, Nash NT, Field M (1984) Differing effects of cGMP and cAMP on ion transport across flounder intestine. Am J Physiol 246:C167–C171
Rehm WS (1975) Ion transport and short-circuit technique. In: Bronner F, Kleinzeller A (eds) Current topics in membranes and transport, vol 7. Academic Press, London New York, pp 217–270
Reuss L (1983) Basolateral KCl co-transport in a NaCl absorbing epithelium. Nature (London) 305:723–726
Reuss L, Cheung LY, Grady TP (1982) Mechanisms of cation permeation across apical cell membrane of *Necturus* gallbladder: effect of luminal pH and divalent cations on K and Na permeability. J Membr Biol 59:211–224
Roggin GM, Banwell JG, Yardley JH, Hendrix TR (1972) Unimpaired response of rabbit jejunum to cholera toxin after selective damage to villus epithelium. Gastroenterology 63:981–989
Sandle GI, Wills NK, Alles W, Binder HJ (1986) Electrophysiology in the human colon: evidence of segmental heterogeneity. Gut 27:999–1005
Schafer JA, DiBona DR (1984) Epithelial permeability. In: Staub NC, Taylor AE (eds) Edema. Raven, New York, pp 95–115
Schreiner J, Bridges RJ, Rummel W (1986) In vitro effects of dexamethasone on sodium transport across rat colon descendens from normal rat. 7th Meet Eur Intest Transport Group, Univ Goteborg, Dep Physiol, p 52
Schultz SG (1980) Sodium-coupled solute transport by small intestine: a status report. Am J Physiol 233:E249–E254
Schultz SG (1981a) Homocellular regulatory mechanisms in sodium-transporting epithelia: avoidance of extinction by "flush through". Am J Physiol 241:F579–F590

Schultz SG (1981b) Salt and water absorption by mammalian small instestine. In: Johnson LR (ed) Physiology of the gastrointestinal tract. Raven, New York, pp 983 – 989

Schultz SG (1984) A cellular model for active sodium absorption by mammalian colon. Annu Rev Physiol 46:435 – 451

Schultz SG, Frizzell RA, Nellans HN (1977) An equivalent electrical circuit model for Na^+ transporting epithelia. J Theor Biol 65:351 – 384

Schultzberg M, Hökfelt T, Nilsson G, Terenius L, Rehfeld JF, Brown M, Elde R, Goldstein M, Said S (1980) Distribution of peptide and catecholamine containing neurons in the gastro-intestinal tract of rat and guinea pig: immunohistochemical studies with antisera to substance P, vasoactive intestinal polypeptide, enkephalins, somatostatin, gastrin/cholecystokinin, neurotensin, and dopamine β-hydroxylase. Neuroscience 5:689 – 744

Sellin JH, Soignie R de (1985) Steroids alter ion transport and absorptive capacity in proximal and distal colon. Am J Physiol 249:G113 – G119

Sellin JH, Soignie R de (1987) Regulation of Na-Cl absorption in rabbit proximal colon in vitro. Am J Physiol 252:G45 – G51

Sepulveda FV, Mason WT (1985) Single channel recordings obtained from basolateral membranes of isolated rabbit enterocytes. FEBS Lett 191:87 – 91

Siegenbeek van Heukelom J, Ham MD van den, Albus H, Groot JA (1981) Microscopical determination of the filtration permeability of the mucosal surface of the goldfish intestinal epithelium. J Membr Biol 63:31 – 39

Skadhauge E (1980) Water transport in the vertebrate intestine. In: Gilles R (ed) Animals and environmental fitness. Pergamon Press, New York, pp 79 – 90

Skadhauge E, Clauss W, Arnason SS, Thomas DH (1985) Mineralocorticoid regulation of lower intestinal ion transport. In: Gilles R, Gilles-Baillien M (eds) Transport processes, iono- and osmoregulation. Springer, Berlin Heidelberg New York Tokyo, pp 118 – 133

Smith PL, McCabe RD (1984) A23187-induced changes in colonic K and Cl transport are mediated by separate mechanisms. Am J Physiol 247:G695 – G702

Smith Pl, Sullivan SK, McCabe RD (1986) Concentration dependent effects of disulfonic stilbenes on colonic chloride transport. Am J Physiol 250:G44 – G49

Smyth DH, Wright EM (1966) Streaming potentials in the rat small intestine. J Physiol (London) 182:591 – 602

Spring KR, Ericson AC (1982) Epithelial cell volume modulation and regulation. J Membr Biol 69:167 – 176

Stewart CP, Turnberg LA (1986) Electrophysiological response to secretagogues of crypt and villus enterocytes in rat ileum in vitro. 7th Meet EITG, Univ Goteborg, Dep Physiol, p 57

Suzuki Y (1984) Proton secretion in the guinea pig colonic mucosa. In: Case RM, Lingard JA, Young JA (eds) Manchester Univ Press, Manchester, pp 71 – 75

Tai YH, Tai CY (1981) The conventional short-circuiting technique under-short-circuits most epithelia. J Membr Biol 59:173 – 177

Thompson SM, Suzuki Y, Schultz SG (1982) The electrophysiology of rabbit colon: I. Instantaneous transepithelial current voltage relations and the current voltage relations of the Na-entry mechanisms. J Membr Biol 66:41 – 54

Thomson ABR, Dietschy JM (1980) Intestinal kinetic parameters: effects of unstirred layers and transport preparation. Am J Physiol 239:G372 – G377

Turnberg LA (1983) Neurohormonal control of intestinal transport. In: Gilles-Baillien M, Gilles R (eds) Intestinal transport. Springer, Berlin Heidelberg New York, pp 240 – 248

Turnberg LA, Bieberdorf FA, Morawwski SG, Fordtran JS (1970) Interrelationships of chloride, bicarbonate, sodium and hydrogen transport in human ileum. J Clin Invest 49:557 – 567

Turnheim K, Thompson SM, Schultz SG (1983) Relation between intracellular sodium and active sodium transport in rabbit colon: current voltage relations of the apical sodium entry mechanism in the presence of varying luminal sodium concentrations. J Membr Biol 76:229 – 309

Turnheim K, Plass H, Grasl M, Krivanek P, Wiener H (1986) Sodium absorption and potassium secretion in rabbit colon during sodium deficiency. Am J Physiol 250:F235 – F245

Ussing HH, Zerahn K (1951) Active transport of sodium as the source of the electrical current in the short-circuited isolated frog skin. Acta Physiol Scand 23:110 – 127

Wangemann P, Wittner M, Di Stefano A, Englert HC, Lang HJ, Schlatter E, Greger R (1986) Cl^--channel blockers in the thick ascending limb of the loop of Henle. Structure activity relationship. Pfluegers Arch 407 (Suppl 2):S128 – S141

Welsh MJ, Smith PL, Fromm M, Frizzell RA (1982) Crypts are the site of intestinal fluid and electrolyte secretion. Science 218:1219 – 1221

Weymer A, Huott P, Liu W, McRoberts JA, Dharmsathaporn K (1985) Chloride secretory mechanisms induced by prostaglandin E_2 in a colonic epithelial cell line. J Clin Invest 76:1828 – 1836

White JF (1977) Activity of chloride in absorptive cells of *Amphiuma* small intestine. Am J Physiol 232:E553 – E559

White JF (1985) Omeprazole inhibits H^+ secretion by *Amphiuma* jejunum. Am J Physiol 248:G256 – G259

White JF (1986) Modes of Cl^- transport across the mucosal and serosal membranes of urodele intestinal cells. J Membr Biol 92:75 – 89

White JF, Burnup K, Ellingsen D (1986) Effects of sugars and amino acids on amphibian intestinal Cl^- transport and intracellular Na^+, K^+ and Cl^- activity. Am J Physiol 250:G109 – G119

Wills NK (1985) Apical membrane potassium and chloride permeabilities in surface cells of rabbit descending colon epithelium. J Physiol (London) 358:433 – 445

Wills NK, Biagi B (1982) Active potassium transport by rabbit descending colon epithelium. J Membr Biol 64:195 – 203

Winne D (1983) Unstirred layer as a diffusion barrier in vitro and in vivo. In: Skadhauge E, Heintze K (eds) Intestinal absorption and secretion. Falk Symp 36. MTP Press, Lancaster, pp 21 – 38

Wood JD (1984) Enteric neurophysiology. Am J Physiol 247:G585 – G598

Wright EM (1984) Electrophysiology of plasma membrane vesicles. Am J Physiol 246:F363 – F372

Zeuthen T, Ramos M, Ellory JC (1978) Inhibition of active chloride transport by piretanide. Nature (London) 273:678 – 680

Zuidema T, Riel JW van, Siegenbeek van Heukelom J (1985) Cellular and transepithelial responses of goldfish intestinal epithelium to chloride substitutions. J Membr Biol 88:293 – 304

Zuidema T, Kamermans M, Siegenbeek van Heukelom J (1986) Influence of glucose absorption on ion activities in cells and submucosal space in goldfish intestine. Pfluegers Arch 407:292 – 298

Chapter 4

NaCl Transport in the Kidney

F. Lang

Contents

1 Introduction

Most of the work performed within the kidney serves to reabsorb filtered NaCl. Moreover, transport of almost any other substance in the kidney is in some way linked to renal NaCl transport and most factors influencing renal function do

Physiologisches Institut der Universität, A-6010 Innsbruck, Austria

Advances in Comparative and Environmental Physiology
Vol. 1. Edited by R. Greger

modify renal transport of NaCl. A comprehensive review of renal handling of NaCl would thus require an almost complete description of renal physiology, which is clearly beyond the scope of this series.

Instead, an attempt will be made to briefly summarize the current views on the mechanisms of renal sodium chloride transport and their modulation by hormones. Special emphasis will be placed upon recent developments. The review will focus mainly on mammalian kidney. Space does not allow in-depth discussion of controversial issues nor the consideration of a number of highly interesting topics related to renal handling of NaCl, e.g., modulation of renal NaCl transport by factors other than hormones, such as diet, hemodynamics, or drugs. Furthermore, the reader may refer to excellent reviews for comparative aspects of renal function (Beyenbach 1985; Braun 1985; Dantzler 1985; Nishimura 1985; Stoner 1985), on renal metabolism (Wirthensohn and Guder 1986), on morphological correlates of transport (Kaissling 1985; Madsen and Tisher 1986; Pfaller 1982), on intracellular mechanisms of hormonal action (Morel and Doucet 1986), and on complex regulatory functions such as glomerular tubular balance (Häberle and v. Baeyer 1983), tubulo-glomerular feedback (Briggs and Schnermann 1986), and the adaptation of renal NaCl excretion to expansion of extracellular volume (Knox and Haas 1982).

2 Segmental Organization of Renal NaCl Transport

Both sodium and chloride ions are freely filtered in the glomerulum. According to the Gibbs-Donnan distribution the concentration in the glomerular ultrafiltrate (and interstitial fluid) of sodium is some 10% lower and of chloride is some 10% higher than the respective concentration in plasma water.

Figure 1 displays the transepithelial movements of NaCl in the mammalian kidney under normal (hydropenic) conditions. Some two-thirds of filtered NaCl are reabsorbed along the proximal tubule. In the proximal convoluted tubule, the

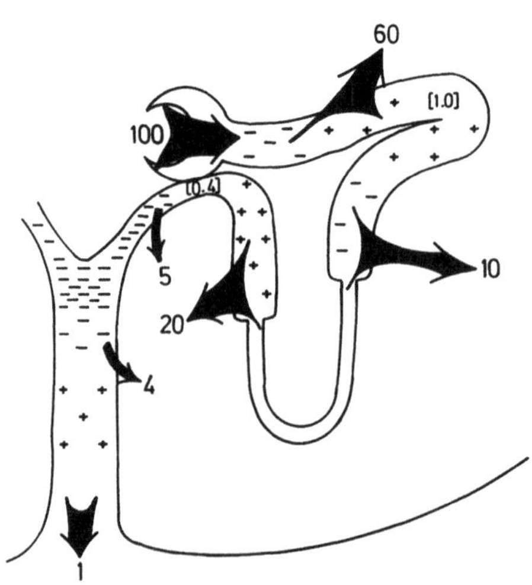

Fig. 1. Segmental NaCl transport (for references see text). Given are the concentration ratios in the tubular fluid over ultrafiltrate (TF/UF) in *parenthesis* [ ] and the transport rates (% of filtered load). TF/UF is slightly higher and the reabsorption rates slightly lower for chloride than for sodium. The *negative* and *positive signs*, respectively, indicate the polarity of the transepithelial potential difference

Table 1. General properties of different tubule segments in mammalian kidney[a]

Mammalian	Rte[a] (Ωcm^2)	Ra, Rb, Rs[a]	Pf[a] ($\mu m\ s^{-1}$)
Proximal tubule	≈10	Ra > Rb ≫ Rs	1000 – 7000
Thin descending limb loop of Henle	700		2500
Thin ascending limb loop of Henle	<10		≈20
Thick ascending limb loop of Henle	10 – 35	Ra ≈ Rb ≈ Rs	<10
Distal convoluted tubule	10 – 400		1000 (300)[b]
Connecting tubule	30		10
Cortical collecting duct	100 – 900	Ra ≈ Rb ≈ Rs	200 – 1000 (10)[b]
Medullary collecting duct			
× Outer stripe	500	Ra > Rb, Rs	200 (10)[b]
× Inner stripe	500	Ra ≫ Rb, Rs	
Papillary collecting duct	1000 – 2000		60 – 600 (40)[b]

[a] Rte, Ra, Rb, and Rs are transepithelial, luminal cell membrane, basolateral cell membrane, and shunt resistance. Pf is the water permeability (for references see Abramow and Orci 1980; Bello-Reuss 1986; Berry 1983; Boulpaep 1979; Frömter 1979, 1982, 1984; Greger 1985; Knepper and Burg 1983; Koeppen and Giebisch 1983; Koeppen 1985, 1986; Lang et al. 1986a; Lapointe et al. 1984; Malnic and Giebisch 1972; O'Neil and Boulpaep 1982; Reif et al. 1984; Rocha and Kudo 1982; Roy and Jamison 1985; Sansom and O'Neil 1986; Ullrich et al. 1984).
Transepithelial and cellular membrane resistances are one order of magnitude higher in amphibian than in mammalian tubules, but the relations of the various resistances are similar to those in the mammalian kidney (Bello-Reuss 1986; Guggino et al. 1982b; Lang et al. 1986a; Oberleithner et al. 1985).
[b] In the absence of ADH.

reabsorption is isotonic and the osmolarity of luminal fluid remains virtually constant. At the end of the proximal convoluted tubules, the luminal sodium concentration is equal to the sodium concentration in the ultrafiltrate. Due to predominant bicarbonate reabsorption (see Sect. 3.1), chloride reabsorption lags behind and the end proximal luminal chloride concentration is some 10 to 20% higher than the chloride concentration in the ultrafiltrate (Higashihara et al. 1979; Le Grimellec 1975; Marchand et al. 1976; Vari and Ott 1982; Windhager and Giebisch 1961; Wong et al. 1986a, b).

Reabsorption of solutes is not isotonic in the loop of Henle (Bennett et al. 1967, 1968; de Rouffignac and Morel 1969; de Rouffignac 1969; Gottschalk and Mylle 1959; Gutsche et al. 1980; Jamison et al. 1967, 1973; Landwehr et al. 1967; Marsh 1970; Morgan and Berliner 1969; Pennell et al. 1974): Thin and thick ascending limbs are almost impermeable to water (see Table 1) and in these segments NaCl is reabsorbed without water. As a result, NaCl concentration and osmolarity of the luminal fluid decrease along the ascending limb of Henle's Loop. Interstitial NaCl concentration and osmolarity increase accordingly. Due to the increase of interstitial osmolarity more water than NaCl is reabsorbed from the water-permeable descending limb of Henle's Loop. Thus, luminal osmolarity increases along the descending limb of Henle's Loop. At the tip of Henle's Loop osmolarity, both in luminal and interstitial fluid, may exceed

1000 mOsmol l^{-1}. Due to the arrangement of the nephron in form of a loop these high interstitial osmolarities are achieved without steep osmotic gradients across the epithelium of a given nephron segment. At the end of the thick ascending limb, luminal osmolarity may be lower than 100 mOsmol l^{-1}. At this nephron site some 80% of the filtered water and more than 90% of the filtered NaCl have been reabsorbed (see Fig. 1). In the absence of ADH the distal tubule and collecting duct are poorly permeable to water (see Table 1) and most of the fluid and NaCl delivered to the early distal tubule are excreted into the final urine. In the presence of ADH, the distal tubule and especially the collecting duct are highly permeable to water. Due to the low luminal osmolarity, more water than solutes leaves the early distal tubule and luminal osmolarity approaches plasma isotonic values al least in the rat (Gottschalk and Mylle 1959). Luminal osmolarity is in large part due to solutes other than NaCl (among others: urea) and the luminal NaCl concentration decreases further along the distal convoluted tubule. At the end of the superficial distal convoluted tubule, some 5% of the filtered water and some 2% of the filtered NaCl are recovered (see Fig.1). NaCl and water reabsorption along the collecting duct leads to excretion of only some 1% of the filtered water and NaCl in the final urine.

The osmolarity profile at the tip of papilla and the thin limbs of Henle's Loop is substantially influenced by urea movements (for references see Roy and Jamison 1985): The tick ascending limb, distal convoluted tubule, and cortical collecting duct are poorly permeable to urea. Fluid reabsorption in the distal convoluted tubule and cortical collecting duct thus increase the luminal urea concentration. In the papillary collecting duct the luminal urea concentration approaches values of up to 500 mmol l^{-1} and the relatively high urea permeability of that segment allows urea to leave the lumen and to enhance osmolarity in the medullary interstitium (see Fig. 2). Urea thus accumulated in the kidney medulla augments water reabsorption in the thin descending limbs of Henle's Loop (see Sect. 3.2). The energy for urea accumulation in the medulla comes basically from fluid reabsorption in the distal convoluted tubules and cortical collecting duct, which are in part the result of NaCl reabsorption in the thick ascending limb.

Fig. 2. Urinary concentrating mechanism. Movements of NaCl (*solid closed arrows*), water (*open arrows*), and urea (*broken arrows*) in the presence of ADH

3 Mechanisms of Renal NaCl Transport

3.1 Proximal Tubule

The systems transporting NaCl in the proximal convoluted are displayed in Fig. 3. Intracellular sodium activity is maintained at 10–15 mmol l^{-1} (see Table 2) by peritubular sodium/potassium ATPase, which extrudes sodium in exchange for potassium (for references see P.L. Joergensen 1980). Both sodium and potassium are transported against their electrochemical gradient (see Table 2). Potassium accumulated by sodium/potassium ATPase may leave the cell via potassium channels in both the luminal and peritubular cell membrane (Gögelein and Greger 1984). The potassium diffusion generates a cell-negative potential of some 60 to 80 mV (Bello-Reuss 1982, 1986; Biagi et al. 1981a, b; Burckhardt et al. 1984a, b; Cardinal et al. 1984; Frömter 1982; Lapointe et al. 1984, 1986; Samarzija et al. 1982; Samarzija and Frömter 1982a, b, c; Völkl et al. 1986).

A major portion of the sodium entry across the luminal cell membrane is due to sodium/hydrogen ion exchange (Murer et al. 1976; Aronson et al. 1982). The transport is electroneutral, the driving force comes from the chemical gradient of both sodium and hydrogen ions (see Table 2). The secreted hydrogen ions react with filtered bicarbonate to form CO_2 via carbonic acid. CO_2 diffuses across the luminal cell membrane into the cell, where it generates hydrogen ions and bicarbonate. The dehydration of carbonic acid to CO_2 in the lumen and the reverse reaction in the cell require the catalytic action of carbonic anhydrase in the

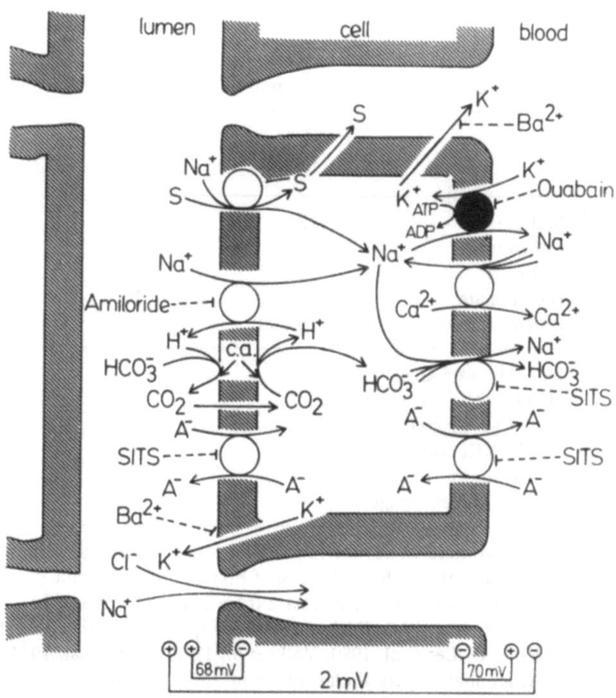

Fig. 3. Cellular mechanisms and inhibitors of NaCl transport in the proximal nephron (for references see text)

luminal cell membrane to catch up with the fast transport rates (Maren 1967; Lang et al. 1978). Hydrogen ions are resecreted into the lumen, whereas bicarbonate leaves the cell mainly across the peritubular cell membrane. Peritubular bicarbonate exit is in large part coupled to sodium (Boron and Boulpaep 1983; Yoshitomi et al. 1985). Most likely, three bicarbonate ions are needed to extrude one sodium ion (see Table 2). Obviously, only a very small fraction of bicarbonate utilizes other transport mechanisms, such as anion exchangers (Boron and Boulpaep 1983; Burckhardt et al. 1984b). In amphibian kidney, sodium/hydrogen ion exchange is not restricted to the luminal cell membrane, but operates as well in the peritubular cell membrane (Boron and Boulpaep 1983). The amphibian proximal tubule shows very little transepithelial hydrogen ion secretion and it is not clear at present, whether peritubular sodium/hydrogen ion exchange occurs as well in mammalian proximal tubules with high rates of transepithelial hydrogen ion transport rates.

A substantial fraction of sodium entry across the luminal cell membrane is coupled to the transport of various substrates (see Table 3). The driving force for these transport systems comes from the steep electrochemical sodium gradient across the luminal cell membrane (see Table 2). For some substrates, the transport is coupled to two or even three sodium ions, and at low filtered loads the driving force allows for almost complete reabsorption of these substances in the proximal tubule. The sodium-coupled transport of neutral substrates carries a positive charge and is thus electrogenic (for references see Frömter 1979, 1982, 1984; Lang et al. 1986a). The cotransport of sodium and anions may be similarly electrogenic, if the number of sodium ions transported is not identical to the charge of the anion transported. Stimulation of sodium-coupled transport may

Table 2. Driving forces in proximal tubules and thick ascending limbs[a]

	ai ($mmol\,l^{-1}$)	ae ($mmol\,l^{-1}$)	$\frac{E \cdot Z \cdot F}{R \cdot T}$	Eo (mV)	Em – Ea (mV)
Sodium	12	115	2.3	+ 60	– 130
Potassium	80	4	– 3.0	– 80	+ 10
Chloride	15	90	– 1.8	– 50	– 20
Bicarbonate	10	20	0.7	– 20	– 50
Calcium	0.0001	0.8	9.0	+ 120	– 190
Na^+/H^+ exchange			3.0	–	–
$HCO_3^- - Na^+$ cotransport (3:1)			4.3	– 58	– 15
Na^+/Ca^{2+} exchange (3:1)			– 2.2	– 60	– 10
NaCl – KCl cotransport[b]			2.8	–	–
KCl symport[b]			1.2	–	–
Cl^-/HCO_3^- exchange			1.1	–	–

[a] Approximate values for ion activities in intracellular (ai) and extracellular (ae) fluid, equilibrium potential (Eo) and the difference of Eo and cell membrane potential (Em = – 70 mV). (Values for Em and intracellular ion activities are from Biagi et al. 1981a, b; Cardinal et al. 1984; Cassola et al. 1983; Cemerikic et al. 1982; Edelman et al. 1978; Frömter 1979, 1982, 1984; Greger et al. 1983b, 1984; Greger 1985; C.O. Lee et al. 1980; Völkl et al. 1986; Wang et al. 1984; Yoshitomi and Frömter 1984, 1985).
[b] At luminal or peritubular ion composition as above, which is not usually the case in vivo.

Table 3. Substrates for sodium-coupled transport in proximal renal tubules (for references see Frömter 1982, 1984; Hoffmann et al. 1976; Hoshi et al. 1976; Kinne 1976; Lang et al. 1986a; Lapointe et al. 1986; Murer and Burckhardt 1983; Samarzija et al. 1982; Samarzija and Frömter 1982a, b, c; Silbernagel 1979; Turner 1984; Ullrich et al. 1977, 1980, 1982; Ullrich 1979; Wright et al. 1981, 1982)

1. Electrolytes
   - H^+ (Na^+/H^+ exchange)
   - Ca^{2+} (Na^+/Ca^{2+} exchange)
   - Phosphate
   - Sulphate
   - Chloride (?)
2. Carbohydrates
   - D-glucose
   - α-methyl-D-glucose
   - D-galactose
   - 3-o-methyl-D-glucose
3. Amino acids
   - L-phenylalanine
   - L-histidine
   - L-aminoisobutyrate
   - L-aminobicyclo-(2,2,1)-heptane-2-carboxylic acid
   - L-lysine
   - L-arginine
   - L-ornithine
   - L-asparate
   - L-proline
   - Glycine
4. Organic acids
   - *Monocarboxylic acids*
     - Lactate
     - Pyruvate
     - Nicotinic acid
     - Picolinic acid
     - Pyrazinoic acid
     - Acetacetate
     - β-hydroxybutyrate
   - *Dicarboxylic acids*
     - Succinate
     - Malate
     - Oxalacetate
     - Fumarate
     - β-ketoglutarate
   - *Tricarboxylic acids*
     - Citrate
   - *Others*
     - Biliary acids (e.g., taurocholate)

lead to profound depolarization of the cell membrane (Frömter 1979, 1982, 1984). The driving force is maintained by a delayed increase of basolateral potassium conductance (Lang et al. 1983, 1986a). Sodium-coupled transport systems are found as well in the peritubular cell membrane (K.E. Joergensen et al. 1983; Kahn et al. 1985; Samarzija and Frömter 1982c; Ullrich et al. 1984). Probably these systems serve primarily to supply the cell with various substrates if luminal delivery is not sufficient. In addition, they allow for secondary active secretion of organic acids into the tubule lumen.

Another peritubular transport system involving sodium is the sodium/calcium exchanger (Gmaj et al. 1979; Jajakumar et al. 1984; Talor and Arruda 1986). This transport system allows for extrusion of calcium at the peritubular cell membrane. The system probably operates close to equilibrium (see Table 1), i.e., the transport direction is reversed at moderate depolarization or at a minimal increase of intracellular sodium activity. This system has been thought to be involved in the regulation of sodium transport (Windhager et al. 1986).

In the straight portion of the proximal tubule, sodium transport is in part due to sodium channels at the luminal cell membrane (Gögelein and Greger 1986a; Kawamura et al. 1975; Schafer et al. 1978).

The contribution of the different transport systems to transcellular sodium transport in the proximal tubule is quite variable, but roughly some two-thirds of the transcellular sodium transport are due to luminal sodium/hydrogen ion exchange, the remaining one-third in large part due to luminal, sodium-coupled substrate transport. At the basolateral cell membrane, probably some 20% of the sodium extrusion is coupled to bicarbonate exit, some 80% has to be transported by sodium/potassium ATPase. The contribution of the other systems mentioned above to transcellular sodium transport is most likely negligible.

Possibly more than 50% of the transepithelial sodium transport and the major part of transepithelial chloride transport are paracellular via tight junctions and thus passive (Frömter et al. 1973; Frömter 1974). This transport could be in part driven by sovent drag, i.e., NaCl is dragged with reabsorbed water (Frömter et al. 1973; Hierholzer et al. 1980; Kokko et al. 1971). However, significant NaCl reabsorption by solvent drag is not generally believed to occur (Jacobson et al. 1982). Passive paracellular chloride reabsorption is further favored by the lumen-negative transepithelial potential difference of up to 4 mV in the early proximal tubule, which is created by electrogenic, sodium-coupled transport systems. However, the conductance of the tight junctions in early proximal tubules is relatively low and paracellular fluxes of NaCl restricted accordingly (Seely 1973; Lang et al. 1986b). In later portions of the proximal convoluted tubules, passive chloride reabsorption is driven by a chemical gradient: In the second two-thirds of the proximal convoluted tubule, the luminal chloride concentration approaches a value of some 10 to 20% above the interstitial chloride concentration, since avid sodium bicarbonate reabsorption in the early proximal tubule leaves chloride behind and the luminal chloride concentration increases due to volume reabsorption (Higashihara et al. 1979; Le Grimellec 1975; Marchand et al. 1976; Vari and Ott 1982; Windhager and Giebisch 1961).

Passive paracellular chloride reabsorption creates a lumen-positive transepithelial potential difference of up to 2 mV in later portions of the superficial proximal convoluted tubules (Frömter et al. 1973). The generation of this potential difference is favored by the decline of electrogenic sodium-coupled transport due to low luminal substrate concentrations and due to lower transport capacity in later segments of the proximal convoluted tubule (for references see Knepper and Burg 1983). The lumen-positive transepithelial potential difference is an important driving force for passive paracellular sodium reabsorption in late proximal convoluted tubules, where the tight junctions are highly conductive to sodium (Frömter et al. 1973; Frömter 1974). In late proximal convoluted tubules of juxtamedullary nephrons, the transepithelial potential difference does not become lumen-positive, since in those segments the paracellular shunt is similarly conductive to bicarbonate and chloride and the concentration differences of these ions across the epithelium do not result in a diffusion potential (Kawamura et al. 1975).

Probably some transcellular chloride transport contributes to proximal tubular chloride reabsorption in mammalian (Alpern et al. 1985; Baum and Berry 1984; Chantrelle et al. 1985; Howlin et al. 1986; Lucci and Warnock 1979; Rector 1983) and amphibian (Kimura and Spring 1979) kidneys. Anion exchangers are identified at both luminal and peritubular cell membranes and may serve at least in part to transport chloride, e.g., in exchange for bicarbonate, or for formate (Karniski and Aronson 1985; Schild et al. 1986; Warnock and

Eveloff 1982). In amphibian kidneys transport of chloride across the basolateral cell membrane is influenced by sodium and bicarbonate (Guggino et al. 1983). Nevertheless, the major portion of chloride reabsorption in the proximal tubules is passive, in large part via the paracellular shunt (Guggino et al. 1983; Frömter 1973). The shunt is preferably conductive to chloride in amphibian proximal tubules (Guggino et al. 1982a; Lang et al. 1986b) and it is slightly less conductive to chloride than to sodium in the mammalian proximal tubule (Frömter et al. 1973).

NaCl concentration in the paracellular space appears to be some 4% higher than the peritubular or luminal NaCl concentration (Sackin 1986a, b). The osmotic gradient across the tight junctions could serve as a driving force for transepithelial water reabsorption.

In addition to the transport systems displayed in Fig. 3, a rather unselective ionic channel has been identified in basolateral cell membranes of the proximal convoluted tubules with so far undefined physiological significance (Gögelein and Greger 1986b).

3.2 Thin Limbs of Henle's Loop

Active transcellular transport appears to be minimal in the thin limbs of Henle's Loop. In the absence of transepithelial driving forces, net transport of NaCl is not detectable in isolated perfused segments of thin descending or ascending limbs of Henle's Loop (Kokko 1970). Accordingly, sodium/potassium-ATPase activity is very low (Garg et al. 1981; Katz et al. 1979). In vivo, the transepithelial potential difference is close to zero (for references see Boulpaep 1979) and transepithelial transport is mainly due to chemical transepithelial gradients. At least in the descending limbs NaCl diffuses across the epithelium mainly through paracellular shunts which are more conductive to sodium than to chloride (Tabei and Imai 1986). Considerable differences occur between the thin limbs of short and long Henle's loops (Imai et al. 1984) and between the descending limbs of different species (Imai 1984). In general, the water permeability is high in the thin descending limb and very low in the thin ascending limb of Henle's Loop (see Table 1). On the other hand, NaCl permeability is much higher in ascending as compared to descending limbs of Henle's Loop (Abramow and Orci 1980; Imai 1977; Imai and Kokko 1974; Kokko 1970, 1982; Rocha and Kokko 1973a). Most likely due to these properties thin limbs of Henle's Loop contribute significantly to the urinary concentrating mechanism (Imai and Kokko 1974; Imai 1977; Jamison et al. 1967, 1973; Marsh 1970; Kokko 1982): Urea does not equilibrate fully across the epithelium and urea concentration remains higher in the interstitial than in the luminal fluid of the descending limbs. Osmotic equilibration in descending limbs is in large part due to water absorption and luminal NaCl concentration exceeds NaCl concentration in surrounding tissue. After the bend of Henle's Loop NaCl permeability allows NaCl to leave the lumen following its chemical gradient. Due to the relatively low luminal urea concentrations, NaCl diffusion may leave a lumen behind, which is slightly hypotonic as compared to the surrounding interstitial fluid. NaCl reabsorption in the thin ascending limb contributes to interstitial hyperosmolarity since it exceeds simultaneous entry of urea.

3.3 Thick Ascending Limb of Henle's Loop

The transport systems in the mammalian thick ascending limb or the amphibian diluting segment are displayed in Fig.4. A cotransport system at the luminal cell membrane allows for entry of NaCl and KCl into the cell (Greger 1981 b, 1985; Greger and Schlatter 1981; Greger et al. 1983 a; Kinne at al. 1986; Oberleithner et al. 1982a, b, 1983 a). The transport is driven by the steep electrochemical gradient for sodium, whereas potassium and chloride are transported against their electrochemical gradient (see Table 1). Since one sodium ion and one potassium ion are transported with two chloride ions, the transport system is electroneutral. A striking feature of the NaCl – KCl cotransport system is its extremely high affinity for sodium. It is half-maximally saturated at luminal sodium concentrations of as low as 3 mmol l^{-1} and for this very reason escaped detection until a few years ago (Greger 1981 a, 1985).

Potassium recirculates via a potassium channel at the luminal cell membrane (Greger and Schlatter 1981; Oberleithner et al. 1983 a). Sodium is extruded at the peritubular cell membrane by sodium/potassium ATPase in exchange for potassium. Chloride leaves the cell across the peritubular cell membrane in part via a chloride channel (Oberleithner et al. 1983 b), and in part via a symport with potassium (Greger 1985).

The chloride diffusion at the peritubular cell membrane and the potassium diffusion at the luminal cell membrane create a lumen-positive transepithelial potential difference of up to 10 mV (Greger 1985; Oberleithner et al. 1982 b). This transepithelial potential difference constitutes a driving force for paracellular cation transport. The paracellular shunt is highly conductive to sodium

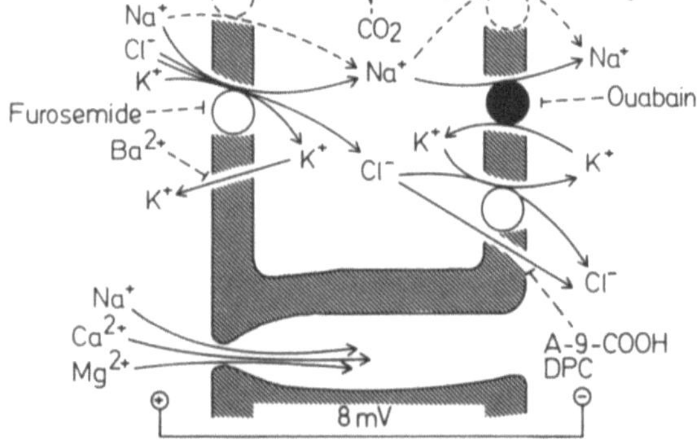

Fig. 4. Cellular mechanisms and inhibitors of NaCl transport in the thick ascending limb (Good 1985; Greger 1985; Oberleithner et al. 1982 b; Schlatter et al. 1983; Wangemann et al. 1986)

(Burg and Green 1973; Burg 1982; Greger 1981 a; Hebert et al. 1981; Rocha and Kokko 1973 b). As a result paracellular passive fluxes contribute up to 50% of the transepithelial sodium reabsorption in the thick ascending limb. On the other hand, the electrical gradient would favor backleak of chloride into the tubule lumen. However, chloride conductance of the paracellular shunt is very low and chloride fluxes restricted accordingly (Burg and Green 1973; Greger 1981 a).

The paracellular shunt is conductive to some cations other than sodium. It allows the reabsorption of calcium and magnesium. The transport of these ions in the thick ascending limb thus depends highly on transcellular transport activity (Greger 1985).

The NaCl – KCl cotransport system accepts ammonium ions in place of potassium. The system allows for accumulation of ammonium ions in the kidney medulla and thus for efficient excretion of ammonia in urine (Good et al. 1984; Knepper and Burg 1983).

During potassium excess an amiloride-sensitive sodium hydrogen ion exchange system is found in the diluting segment of the amphibian kidney operating in parallel to the NaCl-KCl cotransport system (Münich et al. 1986; Oberleithner et al. 1983 c, 1986). Similarly, amiloride-sensitive bicarbonate transport has been observed in thick ascending limbs of rat kidney (Good et al. 1984; Good 1985), whereas rabbit thick ascending limbs apparently do not transport bicarbonate (for references see Greger 1985). The secretion of hydrogen ions leads to acidification of luminal fluid. The transport system alkalinizes the cell. Intracellular alkalosis enhances the potassium conductance at the luminal cell membrane, allowing for increased potassium flux into the lumen (Oberleithner et al. 1983 c, 1985). The formation of hydrogen ions from carbonic acid within the cell leads to generation of bicarbonate, which leaves the cell probably via a sodium-coupled transport system in the peritubular cell membrane (Wang et al. 1987).

The thick ascending limb is not homogeneous. The medullary thick ascending limb is more leaky to sodium, facilitating passive reabsorption of sodium (Burg and Green 1973; Burg 1982; Rocha and Kokko 1973 b). The higher resistance of the cortical thick ascending limb impairs passive sodium reabsorption, on the one hand, but allows for NaCl reabsorption against a steeper transepithelial gradient.

In the diluting segment of *Amphiuma* two cell types have been identified, one with separate conductive pathways and the other with electroneutral cotransport for potassium and chloride in the basolateral cell membrane (Guggino 1986).

3.4 Distal Convoluted Tubule and Collecting Duct

The description of cellular transport mechanisms in nephron segments beyond the thick ascending limb is complicated by the coexistence of different cells in a given segment, the marked axial heterogeneity of the various nephron segments, and differences between the same segments of different animal species. From morphology, several segments can be discriminated which are likely to serve specific transport functions (Madsen and Tischer 1986): The thick ascending limb is followed by the macula densa, the distal convolutes tubule (DCT), the

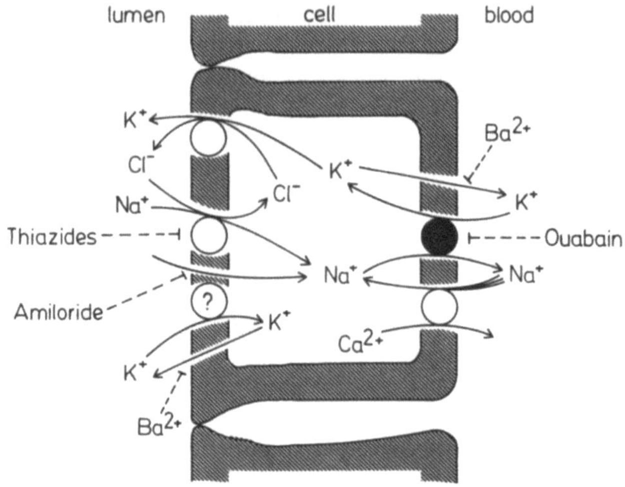

Fig. 5. Cellular mechanisms and inhibitors of NaCl transport in the distal convoluted tubule cells (Ellison et al. 1986; Velazque et al. 1982; Velazque and Wright 1986). The transport systems shown (e.g., NaCl cotransport, potassium pump) in the luminal cell membrane are less well defined than in other nephron segments and they may be operative in different cell types. Potassium uptake across the luminal cell membrane may at least in part involve NaCl – KCl cotransport, which is, however, apparently unimportant for transepithelial NaCl transport. For the sake of clarity the figure does not consider mechanisms of hydrogen ion secretion

connecting tubule (CNT), the initial collecting tubule (ICT), the cortical collecting tubule (CCT), the outer medullary collecting tubule (OMCT), the inner medullary collecting tubule (IMCT), and the papillary collecting tubule (PCT). The cells lining the DCT have a similar appearance as cells in the medullary thick ascending limb. In CNT and collecting tubules principal and intercalated cells can be discriminated. In the rat, intercalated cells of type B prevail in CCT, and intercalated cells of type A prevail in CNT and OMCT. However, as pointed out above, the distribution of the cells along the distal nephron is subject to marked species differences.

In DCT cells (see Fig. 5), sodium transport across the luminal cell membrane is in part electroneutral, possibly mediated by a hydrochlorothiazide inhibitable sodium chloride cotransport system (Costanzo and Windhager 1978; Velazquez and Wright 1986). Another part of the sodium enters the cell via an amiloride-sensitive sodium channel (Velazquez and Wright 1986). The DCT is capable of reabsorbing or secreting potassium. Potassium secretion is probably in part mediated by potassium conductance at the luminal cell membrane. In addition, KCl cotransport may be involved in potassium secretion (Ellison et al. 1986). The carrier responsible for potassium uptake into the cell has not been identified as yet. At the basolateral cell membrane, sodium is expelled in exchange for potassium by the sodium/potassium-ATPase. At the basolateral cell membrane, sodium entry probably serves to extrude calcium. Thus, intracellular calcium activity and transepithelial calcium transport probably depend on low intracellular sodium activity, and thus on sodium transport. In DCT hydrogen ion

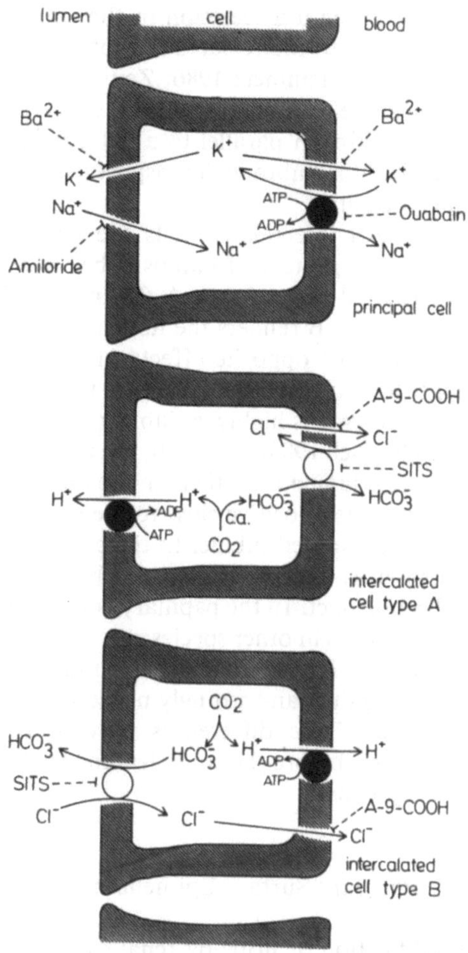

Fig. 6. Transport systems and inhibitors in the collecting duct, i.e., in principal cells (PC) and in intercalated cells type A (ICA) and type B (ICB) (Hunter et al. 1984; O'Neil 1983; Palmer 1986; Steinmetz 1986; Stone et al. 1983; Tago et al. 1986a, b)

secretion occurs in parallel to sodium reabsorption (Malnic et al. 1972). This hydrogen ion secretion may be acccomplished by an ATP-driven hydrogen ion pump and/or a sodium hydrogen ion exchanger. Sodium reabsorption via luminal sodium channels creates a lumen-negative transepithelial potential difference which is blunted by conductive potassium fluxes from cell to lumen. The net result will depend on the transport rates of the different systems.

In the principal cell (see Fig. 6) transepithelial sodium reabsorption involves sodium channels at the luminal cell membrane and sodium/potassium ATPase at the basolateral cell membrane. Potassium thus accumulated leaves the cell via potassium channels in both the luminal and basolateral cell membrane (Koeppen et al. 1983; O'Neil 1981; O'Neil and Boulpaep 1979, 1982; Palmer 1986; Schlatter and Schafer 1987; Stokes 1981, Windhager et al. 1986). Transcellular transport across the principal cells creates a lumen-negative transepithelial potential difference. The sodium channel at the luminal cell membrane is blocked by an increase of intracellular calcium activity (Taylor and Windhager 1979; Windhager et al. 1986).

Hydrogen ion secretion in the connecting tubules and collecting duct is accomplished by intercalated cells of type A: In these cells an ATP-driven hydrogen ion pump (Steinmetz 1986; Zeidel et al. 1986) resides in the luminal cell membrane and electroneutral chloride/bicarbonate exchange (Stone et al. 1983; Tago et al. 1986b) in parallel to a chloride channel (Tago et al. 1986c) at the basolateral cell membrane. Transport across intercalated cells of type A renders the lumen positive.

Type-B intercalated cells are probably involved in bicarbonate secretion. Their arrangement of pumps is believed to be almost opposite to the pumps in intercalated cells of type A (Steinmetz 1986). Transport across the intercalated cells of type B renders the lumen negative.

Given the opposite effects of the different transport systems on the transepithelial potential difference, it is not surprising that the transepithelial potential difference is highly variable in the distal nephron (for references see Koeppen and Giebisch 1983). After all, the potential difference is the result of all transport processes lumped together: In the rabbit, e.g., the transepithelial potential difference is lumen-positive close to the macula densa, but becomes negative along the distal convoluted tubule. In the cortical and outer medullary collecting duct the potential is lumen-negative and becomes lumen-positive in the inner medullary collecting duct. In the papillary collecting duct the potential difference is close to zero. Studies in other species may yield quite different results: The transepithelial potential difference has been found to be lumen-positive in the human cortical collecting duct and strongly negative in the hamster papillary collecting duct for instance. These differences may in part be the result of different dietary conditions before the study leading to different stimulation of the various transport systems.

3.5 Papillary Surface Epithelium

Modification of urine by renal transport systems extends to the renal pelvis, to the epithelium separating "final" urine and renal papilla (Bargman et al. 1984; Bonventre et al. 1980; Sands et al. 1985; Schmidt-Nielsen 1977; Schütz and Schnermann 1972). NaCl transport across this epithelium is obviously accomplished by a bumetanide-sensitive cotransport system in the apical cell membrane (exposed to pelvic urine) and sodium/potassium ATPase at the basolateral cell membrane (Sands et al. 1986). Thus, this epithelium displays some similarities to the thick ascending limb.

4 Hormonal Modulation of Renal Tubular NaCl Transport

4.1 Mineralocorticosteroids

Aldosterone reduces renal excretion of sodium and stimulates potassium excretion as well as urinary acidification. In principal cells of the distal convoluted tubule and collecting duct mineralocorticoids stimulate the parallel synthesis of the luminal sodium channel (Palmer and Frindt 1986), the basolateral sodium/potassium ATPase (El Merinissi and Doucet 1983, 1984; Fimognari et al. 1967; Garg et al. 1981; Horster et al. 1980; Mujais et al. 1986), and of enzymes related to energy supply (Trachewsky et al. 1985). These effects concert to enhance transepithelial sodium reabsorption and potassium secretion and render the lumen negative (Gross et al. 1975; Gross et al. 1986; Gross and Kokko 1977; Hierholzer et al. 1965; Horisberger and Diezi 1984; Koeppen and Helman 1982; Lifschitz et al. 1973; Marver 1984; O'Neil and Helman 1977; Sansom and O'Neil 1986; Schwartz and Burg 1978; Stokes et al. 1981; Tomita et al. 1985; Wingo et al. 1985). Original evidence that enhancement of sodium/potassium ATPase activity is solely the result of increased sodium entry across the luminal cell membrane (Petty et al. 1981) has been misleading: At least in other epithelia stimulated synthesis of sodium/potassium ATPase subunits occurs even in the absence of enhanced sodium entry across the luminal cell membrane (Geering et al. 1982) and at least in other epithelia intracellular sodium does not increase following application of mineralocorticosteroids (Handler et al. 1969; Lipton and Edelman 1971). Under the influence of mineralocorticoids the sodium to potassium coupling ratio of the basolateral sodium pump increases apparently from 3:2 to 3:1 (Sansom and O'Neil 1986). As a result, the potential difference across the basolateral cell membrane increases above the equilibrium potential for potassium, allowing for passive entry of potassium into the cell (Sansom and O'Neil 1986). Following acute administration of aldosterone, enhancement of sodium reabsorption was found to precede alterations of transepithelial potential difference and potassium secretion (Wingo et al. 1985). Besides their more or less direct effect on transport, mineralocorticoids inhibit prostaglandin synthesis in the collecting duct, as discussed below (Sect. 4.14).

In addition to their effects on principal cells, mineralocorticosteroid hormones stimulate hydrogen ion secretion by intercalated cells. This effect is due to enhancement of proton ATPase activity in the luminal cell membrane (Koeppen and Helman 1982).

Furthermore, significant effects of mineralocorticoid hormones have been observed in the diluting segment of the amphibian kidney. In this segment, aldosterone obviously activates a luminal sodium/hydrogen ion exchanger, usually not found in this nephron segment (Weigt et al. 1987). The enhanced activity of this carrier leads to acidification of the luminal fluid and to intracellular alkalosis. The latter enhances luminal potassium conductance and thus facilitates potassium secretion (Oberleithner et al. 1983c). In the rabbit thick ascending limb, aldosterone enhances sodium/potassium ATPase activity (Horster et al. 1980).

Other nephron segments appear to be rather insensitive to mineralocorticoids.

4.2 Glucocorticosteroids

Glucocorticosteroids bind to all segments of the nephron (S.-M.K. Lee et al. 1983). They increase the glomerular filtration rate, stimulate intrarenal gluconeogenesis and ammonia production, lead to phosphaturia and increased acid excretion. In the proximal tubule glucocorticosteroids inhibit sodium-coupled phosphate transport (Durasin et al. 1984; Freiberg et al. 1982; Frick and Durasin 1980) and enhance the transport capacity of sodium/hydrogen ion exchange (Freiberg et al. 1982; Kinsella et al. 1985). Furthermore, they are required for the adaptive increase of sodium/hydrogen ion exchange during chronic metabolic acidosis (Kinsella et al. 1984b). In the thick ascending limb (Doucet et al. 1986; El Merinissi and Doucet 1983, 1984) distal convoluted tubule and collecting duct (El Merinissi et al. 1983; Garg et al. 1985), they enhance sodium/potassium ATPase activity and apparently increase transepithelial sodium transport. The effect in the collecting duct is probably mediated by mineralocorticoid receptors (El Merinissi and Doucet 1983). In the papillary collecting duct chronic administration of glucocorticoids decrease cAMP-phosphodiesterase activity and thus enhances the responsiveness of that segment to ADH (Jackson et al. 1983).

4.3 Sex Hormones

Progesterone is natriuretic mainly due to competitive displacement of mineralocorticoid hormones from their receptors (Trachewsky et al. 1985; Wambach and Higgins 1978; Wambach et al. 1979). In toad urinary bladder, testosterone similarly competes for mineralocorticoid receptors (Rossier et al. 1980). In kidney, androgens lead to hyperplasia of proximal convoluted tubules but apparently do not interfere directly with sodium transport (Rossier et al. 1985).

4.4 Thyroid Hormones

Thyroid hormones stimulate sodium and fluid reabsorption and/or sodium/potassium ATPase activity in the proximal tubule and collecting duct (Lo and Lo 1979). As a result they reduce sodium excretion despite an increase of glomerular filtration rate. Thyroid hormones have been shown to stimulate both sodium-coupled phosphate transport (Espinosa et al. 1984) and sodium/hydrogen ion exchange (Kinsella et al. 1986). On the other hand, the increase of sodium reabsorption is thought to be the consequence of enhanced potassium conductance (Capasso et al. 1985a), because it is mimicked by application of the potassium ionophore valinomycin. In the presence of the ionophore, thyroid hormones do not further increase fluid reabsorption. An increase of potassium conductance hyperpolarizes the cell, thus increasing the driving force for sodium-coupled transport. Furthermore, hyperpolarization by stimulation of electrogenic bicarbonate exit across the basolateral cell membrane leads to intracellular acidosis (Lang et al. 1987) which in turn activates sodium/hydrogen ion exchange (Aronson et al. 1982; Kinsella et al. 1984a). Luminal sodium permeability is apparently unaltered after treatment with thyroid hormones (Capasso et al. 1985b). With considerable delay thyroid hormones stimulate synthesis of sodium/potassium ATPase (Barlet and Doucet 1986a), which is thus believed to be a consequence but not the cause of enhanced sodium transport. The increase of the glomerular filtration rate may be secondary to enhanced fluid reabsorption in proximal tubules, which decreases the delivery of solute to the macula densa and reduces negative tubuloglomerular feedback (Briggs and Schnermann 1986). Interestingly, following prolonged thyroid deprivation the kidneys lose their responsiveness to thyroid hormones (Barlet and Doucet 1986b).

4.5 Antidiuretic Hormone

Besides its stimulatory action on hydraulic permeability mainly in the collecting duct (for references see Handler and Orloff 1981), ADH modifies electrolyte transport in various nephron segments: In mouse medullary thick ascending limb ADH stimulates NaCl transport (Greger 1985; D.A. Hall and Varney 1980; Hebert et al. 1981, 1984; Hebert and Andreoli 1984a, b) by enhancing chloride conductance at the basolateral cell membrane and subsequent stimulation of NaCl−KCl cotransport across the luminal cell membrane (Greger 1985). The effect is mediated by cAMP (V_2 receptors). The enhanced chloride efflux across the basolateral cell membrane depolarizes both cell membranes and thus stimulates potassium efflux across the luminal cell membrane. Cellular loss of chloride and potassium should tend to shrink the cell. This may be the signal stimulating the NaCl−KCl cotransport at the luminal cell membrane (Eveloff and Calamia 1986), since the carrier is involved in volume regulation in other cells (for references see Geck and Heinz 1986). The stimulation of transepithelial NaCl transport is paralleled by an increase of magnesium, calcium, and potassium reabsorption (De Rouffignac et al. 1983).

Despite some stimulation of cAMP (Imbert et al. 1975) ADH appears to exert only poor effects in rabbit ascending limbs (Rocha and Kokko 1974; Sasaki and Imai 1980).

In distal convoluted tubules ADH stimulates via cAMP sodium reabsorption paralleled by enhancement of calcium and magnesium reabsorption (Costanzo and Windhager 1980; Elalouf et al. 1984a).

In principal cells of the collecting duct ADH via cAMP stimulates sodium transport and increases lumen negativity (hyperpolarization). The effect is sustained in rat cortical collecting duct (Reif et al. 1984; Schlatter and Schafer 1987; Tomita et al. 1985), but is only transient in rabbit collecting duct and connecting tubule (Frindt and Burg 1972; Helman et al. 1971), where the hyperpolarization may be followed by depolarization (Frindt and Burg 1972; Imai 1979).

Beyond the collecting duct, ADH influences transport in papillary surface epithelium (Sands et al. 1985, 1986), possibly in a similar way as in thick ascending limbs.

ADH stimulates the formation of PGE_2 in both epithelial and interstitial cells. ADH probably exerts this effect by increasing intracellular calcium activity (V_1 receptors) (Wuthrich and Vallotton 1986). Since PGE_2 inhibits cAMP formation, this effect probably accounts for the transient nature of the hormone effect on transepithelial transport (see Sect. 4.14). Another factor inhibiting ADH-stimulated cAMP formation is adenosine (Torikai 1987).

In addition to its acute effect on transepithelial transport, ADH stimulates hypertrophy of medullary thick ascending limb (Bouby et al. 1985) and enhances the responsiveness to acute application of the hormone (Besseghir et al. 1986).

4.6 Atrial Natriuretic Factor

The natriuretic action of atrial natriuretic factor (ANF) has been ascribed in part to enhancement of the glomerular filtration rate and washout of hypertonicity in the renal medulla by increased blood flow (De Bold et al. 1981; Huang et al. 1985). Accordingly, natriuresis has been abolished by clamping of the renal artery (Sosa et al. 1986). On the other hand, ANF has been shown to produce natriuresis in the absence of altered glomerular filtration rate and renal blood flow (Pollock and Arendshorst 1986; Yukimura et al. 1984). Micropuncture experiments and microcatheterization point to inhibition of sodium transport in the collecting duct and perhaps thick ascending limb (Briggs et al. 1982; Sonnenberg et al. 1982, 1986). Accordingly, ANF has been shown to bind to the medullary collecting duct (Koseki et al. 1986) and to inhibit adenylate cyclase in the loops of Henle and collecting duct (Anand-Srivastave et al. 1986). On the other hand, ANF proved ineffective in isolated perfused segments of thin and thick ascending limbs (Kondo et al. 1986). Although evidence has been presented for inhibition of proximal tubular sodium phosphate cotransport and sodium/hydrogen ion exchanger (Hammond et al. 1985), no effect of ANF on proximal tubular fluid reabsorption has been observed (Baum and Toto 1986). Obviously, additional studies appear necessary to elucidate the mechanisms and prerequisites for natriuretic action of ANF.

4.7 Hypothetical Natriuretic Hormone

A number of studies point to the existence of a natriuretic hormone other than ANF. In contrast to ANF (Pollock et al. 1983), the hormone is thought to inhibit sodium/potassium ATPase (for references see De Wardener and Clarkson 1985; Kramer 1981). A description of the properties of this hypothetical hormone is beyond the scope of this overview.

4.8 Parathyroid Hormone

Besides its well-know inhibitory effect on sodium-coupled phosphate transport (for references see Gmaj and Murer 1986) PTH stimulates sodium/calcium exchange at the basolateral cell membrane (Jayakumar et al. 1984; Scoble et al. 1985), possibly secondarily to increasing intracellular calcium activity (Goligorsky et al. 1986) and inhibits net sodium bicarbonate reabsorption in the proximal nephron (Bank and Aynedjian 1976; Dennis 1976; McKinney and Myers 1980a, b; Puschett and Zurbach 1976). The effects on phosphate and bicarbonate transport have been shown to be mimicked and are thus believed to be mediated by cAMP. Due to its low transport capacity, inhibition of sodium-coupled phosphate transport does not appreciably affect transepithelial sodium reabsorption. Similarly, direct influence of stimulated sodium/calcium exchange on transepithelial sodium transport can be neglected. Inhibition of bicarbonate reabsorption, however, leads to marked reduction of net sodium and fluid reabsorption. Furthermore, the increase of luminal bicarbonate concentration is paralleled by a decline of luminal chloride concentration and reabsorption.

Some controversy exists on the mechanism accounting for impaired bicarbonate reabsorption. Most likely, PTH inhibits the sodium/hydrogen ion exchanger at the luminal cell membrane (Cohn et al. 1983), whereas inhibition of cellular carbonic anhydrase (N. Beck et al. 1975) is probably not relevant (Garg 1976). Some evidence points to an increase of shunt conductance in the presence of PTH, which could impair net bicarbonate reabsorption by increased backleak (Lorentz 1974, 1976).

In cortical thick ascending limb, distal convoluted tubule, and connecting tubule PTH stimulates the formation of cAMP as well as transepithelial calcium and magnesium reabsorption (Bourdeau and Burg 1980; Costanzo and Windhager 1980; Greger et al. 1978; Imai 1981; Shareghi and Stoner 1978; Shareghi and Agus 1982; Suki et al. 1980). Probably this effect is not paralleled by an increase of NaCl transport (Bailly et al. 1984; Bourdeau and Burg 1980; Costanzo and Windhager 1980; Imai 1981). The effect is mimicked and probably mediated by cAMP. However, in addition to its effect on calcium transport, cAMP stimulates sodium transport probably in ADH-sensitive cells (see Sect. 4.5).

4.9 Calcitonin

Calcitonin decreases the fractional excretion of NaCl, calcium, magnesium, and potassium (Elalouf et al. 1986). While the hormone inhibits proximal tubular sodium-coupled phosphate transport, it stimulates, via formation of cAMP, transport for NaCl, calcium, magnesium, and potassium in the medullary thick ascending limb (Elalouf et al. 1984b; Suki and Rouse 1981). In the distal convoluted tubule of the rat calcitonin similarly stimulates reabsorption of sodium, chloride, calcium, and magnesium (Elalouf et al. 1983). In the rabbit distal convoluted tubule, calcitonin reduces the lumen-negative transepithelial potential difference (Imai and Nakamura 1982). The effects on transport remain to be established. In the collecting duct, calcitonin may mimic the effect of ADH, i.e., formation of cAMP and increase of water permeability (Carney et al. 1983; Morel et al. 1982; de Rouffignac and Elalouf 1983).

4.10 Somatotropin

Somatotropin (growth hormone) stimulates sodium-coupled transport in the proximal tubule (Hammerman et al. 1980) with only little effect on renal sodium excretion.

4.11 Insulin

Insulin inhibits proximal tubular sodium reabsorption despite stimulation of sodium-coupled phosphate transport (DeFronzo et al. 1976; Hammerman 1985). Inhibition of parathyrin-stimulated cAMP formation probably accounts for increased phosphate reabsorption (Northrup et al. 1979). Reduced proximal tubular fluid reabsorption probably accounts for calciuria. The inhibitory effect on proximal tubular sodium transport may be surprising, since in other tissues insulin stimulates sodium/hydrogen ion exchange and sodium/potassium ATPase (for references see Moore 1983). In the distal nephron insulin stimulates sodium reabsorption and potassium secretion. As a result, fractional sodium excretion is actually decreased (DeFronzo et al. 1975; Nizet et al. 1971).

4.12 Glucagon

Glucagon stimulates cAMP formation in the thick ascending limb and distal convoluted tubules (Bailly et al. 1980). As a result, reabsorption of calcium, magnesium, potassium, and possibly chloride are enhanced by the hormone (Bailly et al. 1984, 1985). Glucagon-sensitive adenylate cyclase is found as well in the collecting duct of the rat (Bailly et al. 1980). The receptors are probably limited to intercalated cells, since glucagon does not produce antidiuresis (Bailly et al. 1984) in contrast to ADH, which stimulates cAMP formation in principal cells (see Sect. 4.5). The significance of this effect for transepithelial transport remains to be established.

4.13 Angiotensin

Besides its effects on renal vasculature, sympathetic nerve system, and aldosterone release, angiotensin II directly influences renal tubular sodium transport (J.E. Hall et al. 1977; Johnson and Malvin 1977). At concentrations prevailing in plasma (picomolar range) angiotensin II stimulates proximal tubular sodium reabsorption, whereas at several orders of magnitude higher concentrations angiotensin II inhibits sodium reabsorption in the proximal convoluted tubule (Harris and Young 1977; Harris 1979; Harris and Naver 1985; Schuster et al. 1984a; Steven 1974).

4.14 Prostaglandins

PGE_2 is formed along the entire nephron, with the highest synthesis rates in the collecting duct (Farman et al. 1986). In the medullary thick ascending limb, PGE_2 inhibits chloride reabsorption (Hebert and Andreoli 1984b), probably by inhibition of cAMP formation. In the collecting duct PGE_2, $PGF_{2\alpha}$, and PGE_1 similarly inhibit sodium transport (Iino and Imai 1978; Stokes and Kokko 1977) and PGE_2 blunts the stimulatory effect of ADH on hydraulic permeability and sodium transport (Grantham and Orloff 1968; Holt and Lechene 1981; Nadler et al. 1986). Since ADH stimulates the formation of PGE_2 (Wuthrich and Vallotton 1986), prostaglandins are part of a negative feedback limiting the effect of ADH (T.R. Beck et al. 1980; Garcia-Perez and Smith 1984; Hebert and Andreoli 1984b). Inhibition of prostaglandin synthesis has been shown (Barber et al. 1986) to impair natriuresis following denervation of the kidney (see Sect. 4.16). Formation of prostaglandins in the collecting duct is suppressed by mineralocorticoids (Kirschenbaum et al. 1982) and hypertonicity (Sato and Dunn 1986), leading to enhanced sensitivity of that segment to ADH.

4.15 Bradykinin

Bradykinin, which may be formed in the connecting tubule, reduces the ADH-stimulated hydraulic permeability (Schuster et al. 1984b) and sodium reabsorption (Tomita et al. 1985) in the collecting duct. The effect is probably mediated by PGE_2 (Schuster et al. 1984b), which is produced under the influence of bradykinin (Garcia-Perez and Smith 1984; Grenier et al. 1981). ADH, on the other hand, stimulates renal excretion of kallikrein (Fejes-Toth et al. 1980) and of kinins (Kauker et al. 1984). In cultured renal epithelial cells bradykinin has been observed to enhance potassium conductance by increasing intracellular calcium activity (Paulmichl et al. 1987).

4.16 Catecholamines

Denervation of the kidney leads to natriuresis, whereas renal nerve stimulation elicits antinatriuresis in part due to alterations of the proximal sodium, bicarbonate, chloride, phosphate, glucose, and paraaminohippurate transport (Bello-Reuss et al. 1975; Cogan 1986; DiBona 1982). The effects are believed to be mediated by sympathetic nerves: (Nor)epinephrine (Bello-Reuss 1980; Chan 1980a, b), phenylepinephrine, and isoproterenol (Weinman et al. 1982) have been shown to stimulate proximal tubular sodium bicarbonate and fluid reabsorption. α_2-Agonists have been shown to inhibit PTH-induced cAMP formation in the kidney (Guder and Rupprecht 1975), which could in part account for stimulation of sodium and fluid reabsorption (see Sect. 4.8). Another segment increasing NaCl transport in response to renal nerve stimulation is apparently the thick ascending limb (DiBona and Sawin 1982). β-Agonist isoproterenol stimulates the formation of cAMP, which increases NaCl reabsorption in this segment (Morel and Doucet 1986). In the connecting tubule and collecting duct, isoproterenol via cAMP depolarizes the lumen-negative potential (Iino et al. 1981; Imai 1979) probably by increasing chloride reabsorption in intercalated cells (Tago et al. 1986a). Accordingly, β-agonists enhance distal NaCl reabsorption (Greven and Heidenreich 1975). In principal cells, on the other hand, α_2-agonists appear to inhibit ADH-stimulated cAMP generation (N.P. Beck et al. 1972). In cultured epithelioid kidney cells epinephrine has been shown to enhance potassium and chloride conductance (Lang et al. 1986c; Paulmichl et al. 1986). The enhancement of potassium conductance is apparently mediated by α-receptors and an increase of intracellular calcium activity (Paulmichl et al. 1986), whereas the increase of chloride conductance is apparently mediated by β-receptors and stimulation of adenylate cyclase (Lang et al. 1986c). The sensitivity to circulating catecholamines is enhanced following chronic denervation (Szénási et al. 1986).

4.17 Dopamine

Dopamine slightly increases sodium and phosphate excretion at least partially by inhibition of the proximal sodium and phosphate transport (Bello-Reuss et al. 1982; Cuche et al. 1976; Kaneda and Bello-Reuss 1983). The effect is most likely mediated by D_1 receptors and formation of cAMP (Felder et al. 1984, Nakajima et al. 1977) and may modify denervation natriuresis (Jose et al. 1986). Some natriuresis may be accounted for by the vasodilating effect of dopamine (Steinhausen et al. 1986).

4.18 Substance P

Infusion of substance P into the renal artery is followed by marked natriuresis, which apparently occurs in the absence of an altered glomerular filtration rate and renal blood flow (Kramer et al. 1983). The effect is at least in part due to decreased proximal tubular sodium reabsorption (Arendshorst et al. 1976) and is paralleled by increased cAMP excretion (Gullner et al. 1979).

4.19 Histamine

Histamine produces a marked natriuresis, which is at least partially due to an enhanced glomerular filtration rate and renal blood flow (O'Brien and Williamson 1971). However, indirect evidence points to some influence on renal tubular transport (Abboud and Dousa 1983; Banks et al. 1978; Radke et al. 1985).

4.20 Other Hormones

In addition to the hormones listed above, a variety of hormones and mediators alter renal NaCl excretion by influencing renal blood flow and glomerular filtration rate.

In this review, only hormones with apparently direct influence on renal tubular NaCl transport are considered.

References

Abboud HE, Dousa TP (1983) Renal metabolism and actions of histamine and serotonin. Miner Electrolyte Metab 9:246–259

Abramow M, Orci L (1980) On the "tightness" of the rabbit descending limb of the loop of Henle – physiological and morphological evidence. Int J Biochem 12:23–27

Alpern RJ, Howlin KJ, Preisig PA (1985) Active and passive components of chloride transport in the rat proximal convoluted tubule. J Clin Invest 76:1360–1366

Anand-Srivastava MB, Vinay P, Genest J, Cantin M (1986) Effect of atrial natriuretic factor on adenylate cyclase in various nephron segments. Am J Physiol 251:F412–F423

Arendshorst WJ, Cook MA, Mills IH (1976) Effect of substance P on proximal tubular reabsorption in the rat. Am J Physiol 230:1662–1667

Aronson PS, Nee J, Suhm MA (1982) Modifier role of internal H^+ in activating the Na^+-H^+ exchanger in renal microvillus membrane vesicles. Nature (London) 299:161–163

Bailly C, Imbert-Teboul M, Chabardès D, Hus-Citharel A, Montégut M, Clique A, Morel F (1980) The distal nephron of rat kidney: a target site for glucagon. Proc Natl Acad Sci USA 77: 3322–3422

Bailly C, Roinel N, Amiel C (1984) PTH-like glucagon stimulation of Ca and Mg reabsorption in Henle's loop of the rat. Am J Physiol 246:F205–F212

Bailly C, Roinel N, Amiel C (1985) Stimulation by glucagon and PTH of Ca and Mg reabsorption in the superficial distal tubule of the rat kidney. Pfluegers Arch 403:28–34

Bank N, Aynedjian HS (1976) A micropuncture study of the effect of parathyroid hormone on renal bicarbonate reabsorption. J Clin Invest 58:336–344

Banks RO, Fondacaro JD, Schwaiger MM, Jacobson ED (1978) Renal histamine H_1 and H_2 receptors: characterization and functional significance. Am J Physiol 235:F570–F575

Barber JD, Harrington WW, Moss NG, Gottschalk CW (1986) Prostaglandin blockade impairs denervation diuresis and natriuresis in the rat. Am J Physiol 250:F895–F900

Bargman J, Leonard SL, McNeely E, Robertson C, Jamison RL (1984) Examination of transepithelial exchange of water and solute in the rat renal pelvis. J Clin Invest 74:1860–1870

Barlet C, Doucet A (1986a) Kinetics of triiodothyronine action on Na-K-ATPase in single segments of rabbit nephron. Pfluegers Arch 407:27–32

Barlet C, Doucet A (1986b) Lack of stimulation of kidney Na-K-ATPase by thyroid hormones in long-term thyroidectomized rabbits. Pfluegers Arch 407:428–431

Baum M, Berry CA (1984) Evidence for neutral transcellular NaCl transport and neutral basolateral chloride exit in the rabbit proximal convoluted tubule. J Clin Invest 74:205–211

Baum M, Toto RD (1986) Lack of a direct effect of atrial natriuretic factor in the rabbit proximal tubule. Am J Physiol 250:F66–F69

Beck NP, Reed SW, Murdaugh HV, Davis BB (1972) Effects of catecholamines and their interaction with other hormones on cyclic 3′,5′-adenosine monophosphate of the kidney. J Clin Invest 51:939–944

Beck N, Kim KS, Wolak M, Davis BB (1975) Inhibition of carbonic anhydrase by parathyroid hormone and cyclic AMP in rat renal cortex in vitro. J Clin Invest 55:149-156

Beck TR, Hassid A, Dunn MJ (1980) The effect of arginine vasopressin and its analogs on the synthesis of prostaglandin E2 by rat renal medullary interstitial cells in culture. J Pharmacol Exp Ther 215:15–19

Bello-Reuss E (1980) Effect of catecholamines on fluid reabsorption by the isolated proximal convoluted tubule. Am J Physiol 238:F347–F352

Bello-Reuss E (1982) Electrical properties of the basolateral membrane of the straight portion of the rabbit proximal renal tubule. J Physiol (London) 326:49–63

Bello-Reuss E (1986) Cell membranes and paracellular resistances in isolated renal proximal tubules from rabbit and Ambystoma. J Physiol (London) 370:25–38

Bello-Reuss E, Colindres RE, Pastoriza-Muñoz E, Mueller RA, Gottschalk CW (1975) Effects of acute unilateral renal denervation in the rat. J Clin Invest 56:208–217

Bello-Reuss E, Higashi Y, Kaneda Y (1982) Dopamine decreases fluid reabsorption in straight portions of rabbit proximal tubule. Am J Physiol 242:F634–F640

Bennett CM, Clapp JR, Berliner RW (1967) Micropuncture study of the proximal and distal tubule in the dog. Am J Physiol 213:1254–1262

Bennett CM, Brenner BM, Berliner RW (1968) Micropuncture study of nephron function in the rhesus monkey. J Clin Invest 47:203 – 216
Berry CA (1983) Water permeability and pathways in the proximal tubule. Am J Physiol 245: F279 – F294
Besseghir K, Trimble ME, Stoner L (1986) Action of ADH on isolated medullary thick ascending limb of the Brattleboro rat. Am J Physiol 251:F271 – F277
Beyenbach KW (1985) Comparative physiology of the renal proximal tubule. Renal Physiol 8: 222 – 236
Biagi B, Kubota T, Sohtell M, Giebisch G (1981 a) Intracellular potentials in rabbit proximal tubules perfused in vitro. Am J Physiol 240:F200 – F210
Biagi B, Sohtell M, Giebisch G (1981 b) Intracellular potassium activity in the rabbit proximal straight tubule. Am J Physiol 241:F677 – F686
Bonventre JV, Roman RJ, Lechene C (1980) Effect of urea concentration of pelvic fluid on renal concentrating ability. Am J Physiol 239:F609 – F618
Boron WF, Boulpaep EL (1983) Intracellular pH regulation in the renal proximal tubule of the salamander. J Gen Physiol 81:29 – 94
Bouby N, Bankir L, Trinh-Trang-Tan M-M, Minuth WW, Kriz W (1985) Selective ADH-induced hypertrophy of the medullary thick ascending limb in Brattleboro rats. Kidney Int 28:456 – 466
Boulpaep EL (1979) Electrophysiology of the kidney. In: Giebisch G, Tosteson D, Ussing H (eds) Membrane transport in biology. Springer, Berlin Heidelberg New York, pp 97 – 145
Bourdeau JE, Burg MB (1980) Effect of PTH on calcium transport across the cortical thick ascending limb of Henle's loop. Am J Physiol 239:F121 – F126
Braun EJ (1985) Comparative aspects of the urinary concentrating process. Renal Physiol 8:249 – 260
Briggs JP, Schnermann J (1986) Macula densa control of renin secretion and glomerular vascular tone: evidence for common cellular mechanisms. Renal Physiol 9:193 – 203
Briggs JP, Steipe B, Schubert G, Schnermann J (1982) Micropuncture studies of the renal effects of atrial natriuretic substance. Pfluegers Arch 395:271 – 276
Burckhardt B-C, Sato K, Frömter E (1984 a) Electrophysiological analysis of bicarbonate permeation across the peritubular cell membrane of rat kidney proximal tubule, I. Basic observations. Pfluegers Arch 401:34 – 42
Burckhardt B-C, Cassola AC, Frömter E (1984 b) Electrophysiological analysis of bicarbonate permeation across the peritubular cell membrane of rat kidney proximal tubule, II. Exclusion of HCO_3^- effects on other ion permeabilities and of coupled electroneutral HCO_3^- transport. Pfluegers Arch 401:43 – 51
Burg MB (1982) Thick ascending limb of Henle's loop. Kidney Int 22:454 – 464
Burg MB, Green N (1973) Function of the thick ascending limb of Henle's loop. Am J Physiol 224:659 – 668
Capasso G, Lin J-T, De Santo NG, Kinne R (1985 a) Short term effect of low doses of tri-iodothyronine on proximal tubular membrane Na-K-ATPase and potassium permeability in thyroidectomized rats. Pfluegers Arch 403:90 – 96
Capasso G, Kinne-Safran E, De Santo NG, Kinne R (1985 b) Regulation of volume reabsorption by thyroid hormones in the proximal tubule of rat: minor role of luminal sodium permeability. Pfluegers Arch 403:97 – 104
Cardinal J, Lapointe J-Y, Laprade R (1984) Luminal and peritubular ionic substitutions and intracellular potential of the rabbit proximal convoluted tubule. Am J Physiol 247:F352 – F364
Carney S, Morgan T, Ray C, Thompson L (1983) Effect of calcitonin on urine concentration in the rat. Am J Physiol 244:F432 – F435
Cassola AC, Mollenhauer M, Frömter E (1983) The intracellular chloride activity of rat kidney proximal tubular cells. Pfluegers Arch 399:259 – 265
Cemerikic D, Wilcox CS, Giebisch G (1982) Intracellular potential and K^+ activity in rat kidney proximal tubular cells in acidosis and K^+ depletion. J Membr Biol 69:159 – 165
Chan YL (1980 a) Adrenergic control of bicarbonate absorption in the proximal convoluted tubule of the rat kidney. Pfluegers Arch 388:159 – 164
Chan YL (1980 b) The role of norepinephrine in the regulation of fluid absorption in the rat proximal tubule. J Pharmacol Exp Ther 215:65 – 70
Chantrelle BM, Cogan MG, Rector FC, Jr (1985) Active and passive components of NaCl absorption in the proximal convoluted tubule of the rat kidney, Miner Electrolyte Metab 11:209 – 214

Cogan MG (1986) Neurogenic regulation of proximal bicarbonate and chloride reabsorption. Am J Physiol 250:F22 – F26

Cohn DE, Klahr S, Hammerman MR (1983) Metabolic acidosis and parathyroidectomy increase $Na^+ - H^+$ exchange in brush border vesicles. Am J Physiol 245:F217 – F222

Costanzo LS, Windhager EE (1978) Calcium and sodium transport by the distal convoluted tubule of the rat. Am J Physiol 235:F492 – F506

Costanzo LS, Windhager EE (1980) Effects of PTH, ADH, and cyclic AMP on distal tubular Ca and Na reabsorption. Am J Physiol 239:F478 – F485

Cuche J-L, Marchand GR, Greger RF, Lang FC, Knox FG (1976) Phosphaturic effect of dopamine in dogs. J Clin Invest 58:71 – 76

Dantzler WH (1985) Comparative aspects of renal function. In: Seldin DW, Giebisch G (eds) The kidney, vol 1. Raven, New York, pp333 – 364

De Bold AJ, Borenstein HB, Veress AT, Sonnenberg H (1981) A rapid and potent natriuretic response to intravenous injection of atrial myocardial extract in rats. Life Sci 28:89 – 94

DeFronzo RA, Cooke CR, Andres R, Faloona GR, Davis PJ (1975) The effect of insulin on renal handling of sodium, potassium, calcium, and phosphate in man. J Clin Invest 55:845 – 855

DeFronzo RA, Goldberg M, Agus ZS (1976) The effects of glucose and insulin on renal electrolyte transport. J Clin Invest 58:83 – 90

Dennis VW (1976) Influence of bicarbonate on parathyroid hormone-induced changes in fluid absorption by the proximal tubule. Kidney Int 10:373 – 380

DiBona GF (1982) The functions of the renal nerves. Rev Physiol Biochem Pharmacol 94:75 – 181

DiBona GF, Sawin LL (1982) Effect of renal nerve stimulation on NaCl and H_2O transport in Henle's loop of the rat. Am J Physiol 243:F576 – F580

Doucet A, Hus-Citharel A, Morel F (1986) In vitro stimulation of Na-K-ATPase in rat thick ascending limb by dexamethasone. Am J Physiol 251:F851 – F857

Durasin I, Frick A, Neuweg M (1984) Glucocorticoid-induced inhibition of the reabsorption of inorganic phosphate in the proximal tubule in the absence of parathyroid hormone. Renal Physiol 7:115 – 123

Edelman A, Curci S, Samarzija I, Frömter E (1978) Determination of intracellular K^+ activity in rat kidney proximal tubular cells. Pfluegers Arch 378:37 – 45

Elalouf JM, Roinal N, Rouffignac C de (1983) Stimulation by human calcitonin of electrolyte transport in distal tubules of rat kidney. Pfluegers Arch 399:111-118

Elalouf JM, Roinel N, Rouffignac C de (1984a) Effects of antidiuretic hormone on electrolyte reabsorption and secretion in distal tubules of rat kidney. Pfluegers Arch 401:167 – 173

Elalouf JM, Roinel N, Rouffignac C de (1984b) ADH-like effects of calcitonin on electrolyte transport by Henle's loop of rat kidney. Am J Physiol 246:F213 – F220

Elalouf JM, Roinel N, Rouffignac C de (1986) Effects of human calcitonin on water and electrolyte movements in rat juxtamedullary nephrons: inhibition of medullary K recycling. Pfluegers Arch 406:502 – 508

Ellison DH, Velazquez H, Wright FS (1986) Unidirectional potassium fluxes in renal distal tubule: effects of chloride and barium. Am J Physiol 250:F885 – F894

El Merinissi G, Doucet A (1983) Short-term effects of aldosterone and dexamethasone on Na – K-ATPase along the rabbit nephron. Pfluegers Archiv 399:147 – 151

El Merinissi G, Doucet A (1984) Specific activity of Na-K-ATPase after adrenalectomy and hormone replacemant along the rabbit nephron. Pfluegers Arch 402:258 – 263

El Merinissi G, Chabardès D, Doucet A, Hus-Citharel A, Imbert-Teboul M, Le Bouffant F, Montégut M, Siaume S, Morel F (1983) Changes in tubular basolateral membrane markers after chronic DOCA treatment. Am J Physiol 245:F100 – F109

Espinosa RE, Keller MJ, Yusufi ANK, Dousa TP (1984) Effect of thyroxine administration on phosphate transport across renal cortical brush border membrane. Am J Physiol 246:F133 – F139

Eveloff JL, Calamia J (1986) Effect of osmolarity on cation fluxes in medullary thick ascending limb cells. Am J Physiol 250:F176 – F180

Farman N, Pradelles P, Bonvalet JP (1986) Determination of prostaglandin E2 synthesis along rabbit nephron by enzyme immunoassay. Am J Physiol 251:F238 – F244

Fejes-Toth G, Zahajszky T, Filep J (1980) Effect of vasopressin on renal kallikrein excretion. Am J Physiol 239:F388 – F392

Felder RA, Blecher M, Calcagno PL, Jose PA (1984) Dopamine receptors in the proximal tubule of the rabbit. Am J Physiol 247:F499 – F505

Fimognari GM, Fanestil DD, Edelman IS (1967) Induction of RNA and protein synthesis in the action of aldosterone in the rat. Am J Physiol 213:954 – 962

Freiberg JM, Kinsella J, Sacktor B (1982) Glucocorticoids increase the $Na^+ - H^+$exchange and decrease the Na^+ gradient-dependent phosphate-uptake systems in renal brush border membrane vesicles. Proc Natl Acad Sci USA 79:4932 – 4936

Frick A, Durasin I (1980) Proximal tubular reabsorption of inorganic phosphate in adrenalectomized rats. Pfluegers Arch 385:189 – 192

Frindt G, Burg MB (1972) Effect of vasopressin on sodium transport in renal cortical collecting tubules. Kidney Int 1:224 – 231

Frömter E (1974) Electrophysiology and isotonic fluid absorption of proximal tubules of mammalian kidney. In: Guyton AC, Thurau K (eds) MTP Int Rev Sci. Physiol Ser 1. Kidney and urinary tract physiology, vol 6. Butterworth, London; Univ Park Press, Baltimore, pp 1 – 38

Frömter E (1979) Solute transport across epithelia: what can we learn from micropuncture studies on kidney tubules? J Physiol (London) 288:1 – 31

Frömter E (1982) Electrophysiological analysis of rat renal sugar and amino acid transport. I. Basic phenomena. Pfluegers Arch 393:179 – 189

Frömter E (1984) Viewing the kidney through microelectrodes. Am J Physiol 247:F695 – F705

Frömter E, Rumrich G, Ullrich KJ (1973) Phenomenologic description of Na^+, Cl^- and HCO_3^- absorption from proximal tubules of the rat kidney. Pfluegers Arch 343:189 – 220

Garcia-Perez A, Smith WL (1984) Apical-basolateral membrane asymmetry in canine cortical collecting tubule cells (bradykinin, arginine vasopressin, prostaglandin E2 interrelationships). J Clin Invest 74:63 – 74

Garg LC (1976) Failure of parathyroid hormone and cyclic AMP to inhibit renal carbonic anhydrase. Pfluegers Arch 367:103 – 104

Garg LC, Knepper MA, Burg MB (1981) Mineralocorticoid effects on Na – K-ATPase in individual nephron segments. Am J Physiol 240:F536 – F544

Garg LC, Narang N, Wingo CS (1985) Glucocorticoid effects on Na-K-ATPase in rabbit nephron segments. Am J Physiol 248:F487 – F491

Geck P, Heinz E (1986) The Na – K-2Cl cotransport system. J Membr Biol 91:97 – 105

Geering K, Girardet M, Bron C, Kraehenbühl J-P, Rossier BC (1982) Hormonal regulation of (Na^+, K^+)-ATPase biosynthesis in the toad bladder: effect of aldosterone and 3,5,3′-triiodo-L-thyronine. J Biol Chem 257:10338 – 10343

Gmaj P, Murer H (1986) Cellular mechanisms of inorganic phosphate transport in kidney. Physiol Rev 66:36 – 70

Gmaj P, Murer H, Kinne R (1979) Calcium ion transport across plasma membranes isolated from rat kidney cortex. Biochem J 178:549 – 557

Gögelein H, Greger R (1984) Single channel recordings from basolateral and apical membranes of renal proximal tubules. Pfluegers Arch 401:424 – 426

Gögelein H, Greger R (1986a) Na^+ selective channels in the apical membrane of rabbit late proximal tubules (pars recta). Pfluegers Arch 406:198 – 203

Gögelein H, Greger R (1986b) A voltage-dependent ionic channel in the basolateral membrane of late proximal tubules of the rabbit kidney. Pfluegers Arch 407:S142 – S148

Goligorsky MS, Loftus DJ, Hruska KA (1986) Cytoplasmic calcium in individual proximal tubular cells in culture. Am J Physiol 251:F938 – F944

Good DW (1985) Sodium-dependent bicarbonate absorption by cortical thick ascending limb of rat kidney. Am J Physiol 248:F821 – F829

Good DW, Knepper MA, Burg MB (1984) Ammonia and bicarbonate transport by thick ascending limb of rat kidney. Am J Physiol 247:F35 – F44

Gottschalk CW, Mylle M (1959) Micropuncture study of the mammalian urinary concentrating mechanism: evidence for the countercurrent hypothesis. Am J Physiol 196:927 – 936

Grantham JJ, Orloff J (1968) Effect of prostaglandin E1 on the permeability response of the isolated collecting tubule to vasopressin, adenosine 3′,5′-monophosphate, and theophylline. J Clin Invest 47:1154 – 1161

Greger R (1981a) Cation selectivity of the isolated perfused cortical thick ascending limb of Henle's loop of rabbit kidney. Pfluegers Arch 390:30 – 37

Greger R (1981b) Chloride reabsorption in the rabbit cortical thick ascending limb of the loop of Henle: a sodium dependent process. Pfluegers Arch 390:38–43
Greger R (1985) Ion transport mechanisms in thick ascending limb of Henle's loop of mammalian nephron. Physiol Rev 65:760–797
Greger R, Schlatter E (1981) Presence of luminal K^+, a prerequisite for active NaCl transport in the cortical thick ascending limb of Henle's loop of rabbit kidney. Pfluegers Arch 392:92–94
Greger R, Lang F, Oberleithner H (1978) Distal site of calcium reabsorption in the rat nephron. Pfluegers Arch 374:153–157
Greger R, Schlatter E, Lang F (1983a) Evidence for electroneutral sodium chloride cotransport in the cortical thick ascending limb of Henle's loop of rabbit kidney. Pfluegers Arch 396: 308–314
Greger R, Oberleithner H, Schlatter E, Cassola AC, Weidtke C (1983b) Chloride activity in cells of isolated perfused cortical thick ascending limbs of rabbit kidney. Pfluegers Arch 399:29–34
Greger R, Weidtke C, Schlatter E, Wittner M, Gebler B (1984) Potassium activity in cells of isolated perfused cortical thick ascending limbs of rabbit kidney. Pfluegers Arch 401:52–57
Grenier FC, Rollins TE, Smith WL (1981) Kinin-induced prostaglandin synthesis by renal papillary collecting tubule cells in culture. Am J Physiol 241:F94–F104
Greven J, Heidenreich O (1975) A micropuncture study of the effect of isoprenaline on renal tubular fluid and electrolyte transport in the rat. Naunyn-Schmiedebergs Arch Pharmacol 287: 117–128
Gross JB, Kokko JP (1977) Effects of aldosterone and potassium-sparing diuretics on electrical potential differences across the distal nephron. J Clin Invest 59:82–89
Gross JB, Imai M, Kokko JP (1975) A functional comparsion of the cortical collecting tubule and the distal convoluted tubule. J Clin Invest 55:1284–1294
Gross P, Minuth WW, Kriz W, Frömter E (1986) Electrical properties of renal collecting duct principal cell epithelium in tissue culture. Pfluegers Arch 406:380–386
Guder WG, Rupprecht A (1975) Antagonism between parathyroid hormone and norepinephrine on cyclic adenosine-3′:5′-monophosphate (cAMP) levels in isolated tubules from rat kidney cortex. Pfluegers Arch 354:177–186
Guggino WB (1986) Functional heterogeneity in the early distal tubule of the *Amphiuma* kidney: evidence for two modes of Cl^- and K^+ transport across the basolateral cell membrane. Am J Physiol 250:F430–F440
Guggino WB, Boulpaep EL, Giebisch G (1982a) Electrical properties of chloride transport across the *Necturus* proximal tubule. J Membr Biol 65:185–196
Guggino WB, Windhager EE Boulpaep EL, Giebisch G (1982b) Cellular and paracellular resistances of the *Necturus* proximal tubule. J Membr Biol 67:143–154
Guggino WB, London R, Boulpaep EL, Giebisch G (1983) Chloride transport across the basolateral cell membrane of the *Necturus* proximal tubule: dependence on bicarbonate and sodium. J Membr Biol 71:227–240
Gullner H-G, Campbell WB, Pettinger WA (1979) Role of substance P in water homeostasis. Life Sci 24:2351–2360
Gutsche H-U, Müller-Suur R, Hegel U, Hierholzer K (1980) Electrical conductivity of tubular fluid of the rat nephron: micropuncture study of the diluting segment in situ. Pfluegers Arch 383: 113–121
Häberle DA, Baeyer H v (1983) Characteristics of glomerulotubular balance. Am J Physiol 244:F355–F366
Hall DA, Varney DM (1980) Effect of vasopressin on electrical potential difference and chloride transport in mouse medullary thick ascending limb of Henle's loop. J Clin Invest 66:792–802
Hall JE, Guyton AC, Trippodo NC, Lohmeier TE, McCaa RE, Cowley AW, Jr (1977) Intrarenal control of electrolyte excretion by angiotensin II. Am J Physiol 232:F538–F544
Hammerman MR (1985) Interaction of insulin with the renal proximal tubular cell. Am J Physiol 249:F1–F11
Hammerman MR, Karl IE, Hruska KA (1980) Regulation of canine renal vesicle Pi transport by growth hormone and parathyroid hormone. Biochim Biophys Acta 603:322–335
Hammond TG, Yusufi ANK, Knox FG, Dousa TP (1985) Administration of atrial natriuretic factor inhibits sodium-coupled transport in proximal tubules. J Clin Invest 75:1983–1989
Handler JS, Orloff J (1981) Antidiuretic hormone. Annu Rev Physiol 43:611–624

Handler JS, Preston AS, Orloff J (1969) Effect of adrenal steroid hormones on the response of the toad's urinary bladder to vasopressin. J Clin Invest 48:823–833
Harris PJ (1979) Stimulation of proximal tubular sodium reabsorption by Ile5 angiotensin II in the rat kidney. Pfluegers Arch 381:83–85
Harris PJ, Navar LG (1985) Tubular transport responses to angiotensin. Am J Physiol 248: F621–F630
Harris PJ, Young JA (1977) Dose-dependent stimulation and inhibition of proximal tubular sodium reabsorption by angiotensin II in the rat kidney. Pfluegers Arch 367:295–297
Hebert SC, Andreoli TE (1984a) Effects of antidiuretic hormone on cellular conductive pathways in mouse medullary thick ascending limbs of Henle: II. Determinants of the ADH-mediated increases in transepithelial voltage and in net Cl^- absorption. J Membr Biol 80:221–233
Hebert SC, Andreoli TE (1984b) Control of NaCl transport in the thick ascending limb. Am J Physiol 246:F745–F756
Hebert SC, Culpepper RM, Andreoli TE (1981) NaCl transport in mouse medullary thick ascending limbs. Am J Physiol 241:F412–F442
Hebert SC, Friedman PA, Andreoli TE (1984) Effects of antidiuretic hormone on cellular conductive pathways in mouse medullary thick ascending limbs of Henle: I. ADH increases transcellular conductance pathways. J Membr Biol 80:201–219
Helman SI, Grantham JJ, Burg MB (1971) Effect of vasopressin on electrical resistance of renal cortical collecting tubules. Am J Physiol 220:1825–1832
Hierholzer K, Wiederholt M, Holzgreve H, Giebisch G, Klose RM, Windhager EE (1965) Micropuncture study of renal transtubular concentration gradients of sodium and potassium in adrenalectomized rats. Pfluegers Arch 285:193–210
Hierholzer K, Kawamura S, Seldin DW, Kokko JP, Jacobson HR (1980) Reflection coefficients of various substrates across superficial and juxtamedullary proximal convoluted segments of rabbit nephrons. Miner Electrolyte Metab 3:172–180
Higashihara E, Stokes JB, Kokko JP, Campbell WB, DuBose Jr TD (1979) Cortical and papillary micropuncture examination of chloride transport in segments of the rat kidney during inhibition of prostaglandin production. J Clin Invest 64:1277–1287
Hoffmann N, Thees M, Kinne R (1976) Phosphate transport by isolated renal brush border vesicles. Pfluegers Arch 362:147–156
Holt WF, Lechene C (1981) ADH-PGE2 interactions in cortical collecting tubule. I. Depression of sodium transport. Am J Physiol 241:F452–F460
Horisberger J-D, Diezi J (1984) Inhibition of aldosterone-induced antinatriuresis and kaliuresis by actinomycin D. Am J Physiol 246:F201–F204
Horster M, Schmid H, Schmid U (1980) Aldosterone in vitro restores nephron Na-K-ATPase of distal segments from adrenalectomized rabbits. Pfluegers Arch 384:203–206
Hoshi T, Sudo K, Suzuki Y (1976) Characteristics of changes in the intracellular potential associated with transport of neutral, dibasic and acidic amino acids in *Triturus* proximal tubule. Biochim Biophys Acta 448:492–504
Howlin KJ, Alpern RJ, Berry CA, Rector FC Jr (1986) Evidence for electroneutral sodium chloride transport in rat proximal convoluted tubule. Am J Physiol 250:F644–F648
Huang C-L, Lewicki J, Johnson LK, Cogan MG (1985) Renal mechanism of action of rat atrial natriuretic factor. J Clin Invest 75:769–773
Hunter M, Lopes AG, Boulpaep EL, Giebisch GH (1984) Single channel recordings of calcium-activated potassium channels in the apical membrane of rabbit cortical collecting tubules. Proc Natl Acad Sci USA 81:4237–4239
Iino Y, Imai M (1978) Effects of prostaglandins on Na transport in isolated collecting tubules. Pfluegers Arch 373:125–132
Iino Y, Troy JL, Brenner BM (1981) Effects of catecholamines on electrolyte transport in cortical collecting tubule. J Membr Biol 61:67–73
Imai M (1977) Function of the thin ascending limb of Henle of rats and hamsters perfused in vitro. Am J Physiol 232:F201–F209
Imai M (1979) The connecting tubule: a functional subdivision of the rabbit distal nephron segments. Kidney Int 15:346–356
Imai M (1981) Effects of parathyroid hormone and N_6, O_2-dibutyryl cyclic AMP on Ca^{2+} transport across the rabbit distal nephron segments perfused in vitro. Pfluegers Arch 390:145–151

Imai M (1984) Functional heterogeneity of the descending limbs of Henle's loop. II. Interspecies differences among rabbits, rats, and hamsters. Pfluegers Arch 402:393–401
Imai M, Kokko JP (1974) Sodium chloride, urea, and water transport in the thin ascending limb of Henle. J Clin Invest 53:393–402
Imai M, Nakamura R (1982) Function of distal convoluted and connecting tubules studied by isolated nephron fragments. Kidney Int 22:465–472
Imai M, Hayashi M, Araki M (1984) Functional heterogeneity of the descending limbs of Henle's loop. I. Internephron heterogeneity in the hamster kidney. Pfluegers Arch 402:385–392
Imbert M, Chabardès D, Montegut M, Clique A, Morel F (1975) Vasopressin dependent adenylate cyclase in single segments of rabbit kidney tubule. Pfluegers Arch 357:173–186
Jackson BA, Braun-Werness JL, Kusano E, Dousa TP (1983) Concentrating defect in the adrenalectomized rat. J Clin Invest 72:997–1004
Jacobson HR, Kokko JP, Seldin DW, Holmberg C (1982) Lack of solvent drag of NaCl and $NaHCO_3$ in rabbit proximal tubules. Am J Physiol 243:F342–F348
Jamison RL, Bennett CM, Berliner RW (1967) Countercurrent multiplication by the thin loops of Henle. Am J Physiol 212:357–366
Jamison RL, Buerkert J, Lacy F (1973) A micropuncture study of Henle's thin loop in Brattleboro rats. Am J Physiol 224:180–185
Jayakumar A, Cheng L, Liang CT, Sacktor B (1984) Sodium gradient-dependent calcium uptake in renal basolateral membrane vesicles: effect of parathyroid hormone. J Biol Chem 259: 10827–10833
Joergensen KE, Kragh-Hansen U, Roeigaard-Petersen H, Sheikh MI (1983) Citrate uptake by basolateral and luminal membrane vesicles from rabbit kidney cortex. Am J Physiol 244:F686–F695
Joergensen PL (1980) Sodium and potassium ion pump in kidney tubules. Physiol Rev 60:864–917
Johnson MD, Malvin RL (1977) Stimulation of renal sodium reabsorption by angiotensin II. Am J Physiol 232:F298–F306
Jose PA, Felder Ra, Holloway RR, Eisner GM (1986) Dopamine receptors modulate sodium excretion in denervated kidney. Am J Physiol 250:F1033–F1038
Kahn AM, Shelat H, Weinman EJ (1985) Urate and p-aminohippurate transport in rat renal basolateral vesicles. Am J Physiol 249:F654–F661
Kaissling B (1985) Cellular heterogeneity of the distal nephron and its relation to function. Klin Wochenschr 63:868–876
Kaneda Y, Bello-Reuss E (1983) Effect of dopamine on phosphate reabsorption in isolated perfused rabbit proximal tubules. Miner Electrolyte Metab 9:147–150
Karniski LP, Aronson PS (1985) Chloride/formate exchange with formic acid recycling: a mechanism of active chloride transport across epithelial membranes. Proc Natl Acad Sci USA 82:6362–6365
Katz AI, Doucet A, Morel F (1979) Na-K-ATPase activity along the rabbit, rat, and mouse nephron. Am J Physiol 237:F114–F120
Kauker ML, Crofton JT, Share L, Nasjletti A (1984) Role of vasopressin in regulation of renal kinin excretion in Long-Evans and diabetes insipidus rats. J Clin Invest 73:824-831
Kawamura S, Imai M, Seldin DW, Kokko JP (1975) Characteristics of salt and water transport in superficial and juxtamedullary straight segments of proximal tubules. J Clin Invest 55:1269–1277
Kimura G, Spring KR (1979) Luminal Na^+entry into *Necturus* proximal tubule cells. Am J Physiol 236:F295–F301
Kinne R (1976) Properties of the glucose transport system in the renal brush border membrane. Curr Top Membr Transp 8:209–267
Kinne R, Kinne-Safran E, Schölermann B, Schütz H (1986) The anion specificity of the sodium-potassium-chloride cotransporter in rabbit kidney outer medulla: studies on medullary plasma membranes. Pfluegers Arch 407:S168–S173
Kinsella J, Cujdik T, Sacktor B (1984a) Na^+-H^+ exchange in isolated renal brush-border membrane vesicles in response to metabolic acidosis: kinetic effects. J Biol Chem 259: 13224–13227
Kinsella J, Cujdik T, Sacktor B (1984b) Na^+-H^+ exchange activity in renal brush border membrane vesicles in response to metabolic acidosis: The role of glucocorticoids. Proc Natl Acad Sci US 81:630–634

Kinsella JL, Freiberg JM, Sacktor B (1985) Glucocorticoid activation of Na^+/H^+ exchange in renal brush border vesicles: kinetic effects. Am J Physiol 248:F233 – F239
Kinsella JL, Cujdik T, Sacktor B (1986) Kinetic studies on the stimulation of $Na^+ - H^+$exchange activity in renal brush border membranes isolated from thyroid hormone-treated rats. J Membr Biol 91:183 – 191
Kirschenbaum MA, Lowe AG, Trizna W, Fine LG (1982) Regulation of vasopressin action by prostaglandins. J Clin Invest 70:1193 – 1204
Knepper M, Burg M (1983) Organization of nephron function. Am J Physiol 244:F579 – F589
Knox FG, Haas JA (1982) Factors influencing renal sodium reabsorption in volume expansion. Rev Physiol Biochem Pharmacol 92:75 – 113
Koeppen BM (1985) Conductive properties of the rabbit outer medullary collecting duct: inner stripe. Am J Physiol 248:F500 – F506
Koeppen BM (1986) Conductive properties of the rabbit outer medullary collecting duct: outer stripe. Am J Physiol 250:F70 – F76
Koeppen BM, Giebisch G (1983) Electrophysiology of mammalian renal tubules: inferences from intracellular microelectrode studies. Annu Rev Physiol 45:497 – 517
Koeppen BM, Helman SI (1982) Acidification of luminal fluid by the rabbit cortical collecting tubule perfused in vitro. Am J Physiol 242:F521 – F531
Koeppen BM, Biagi BA, Giebisch GH (1983) Intracellular microelectrode characterization of the rabbit cortical collecting duct. Am J Physiol 244:F35 – F47
Kokko JP (1970) Sodium chloride and water transport in the descending limb of Henle. J Clin Invest 49:1838 – 1846
Kokko JP (1982) Transport characteristics of the thin limbs of Henle. Kidney Int 22:449 – 453
Kokko JP, Burg MB, Orloff J (1971) Characteristics of NaCl and water transport in the renal proximal tubule. J Clin Invest 50:69 – 76
Kondo Y, Imai M, Kangawa K, Matsuo H (1986) Lack of direct action of α-human atrial natriuretic polypeptide on the in vitro perfused segments of Henle's loop isolated from rabbit kidney. Pfluegers Arch 406:273 – 278
Koseki C, Hayashi Y, Torikai S, Furuya M, Ohnuma N, Imai M (1986) Localization of binding sites for α-rat atrial natriuretic polypeptide in rat kidney. Am J Physiol 250:F210 – F216
Kramer HJ (1981) Natriuretic hormone – a circulating inhibitor of sodium- and potassium-activated adenosine triphosphatase: its potential role in body fluid and blood pressure regulation. Klin Wochenschr 59:1225 – 1230
Kramer HJ, Klingmüller D, Flachskampf FA, Düsing R (1983) Substance P-induced changes in kidney function in the conscious rat: relation to the renal prostaglandin system. Renal Physiol 6:10 – 18
Landwehr DM, Klose RM, Giebisch G (1967) Renal tubular sodium and water reabsorption in the isotonic sodium chloride-loaded rat. Am J Physiol 212:1327 – 1333
Lang F, Quehenberger P, Greger R, Oberleithner H (1978) Effect of benzolamide on luminal pH in proximal convoluted tubules of the rat kidney. Pfluegers Arch 375:39 – 43
Lang F, Messner G, Wang W, Oberleithner H (1983) Interaction of intracellular electrolytes and tubular transport. Klin Wochenschr 61:1029 – 1037
Lang F, Messner G, Rehwald W (1986a) Electrophysiology of sodium-coupled transport in proximal renal tubules. Am J Physiol 250:F953 – F962
Lang F, Oberleithner H, Giebisch G (1986b) Electrophysiological heterogeneity of proximal convoluted tubules in *Amphiuma* kidney. Am J Physiol 251:F1063 – F1072
Lang F, Defregger M, Paulmichl M (1986c) Apparent chloride conductance of subconfluent Madin Darby canine kidney cells. Pfluegers Arch 407:158 – 162
Lang F, Oberleithner H, Kolb A, Paulmichl M, Völkl H, Wang W (1988) Interaction of intracellular pH and cell membrane potential. In: Häussinger (ed) pH homeostasis. Academic Press, London New York
Lapointe J-Y, Laprade R, Cardinal J (1984) Transepithelial and cell membrane electrical resistances of the rabbit proximal convoluted tubule. Am J Physiol 247:F637 – F649
Lapointe J-Y, Laprade R, Cardinal J (1986) Characterization of the apical membrane ionic permeability of the rabbit proximal convoluted tubule. Am J Physiol 250:F339 – F347
Lee CO, Taylor A, Windhager EE (1980) Cytosolic calcium ion activity in epithelial cells of *Necturus* kidney. Nature (London) 287:859 – 861

Lee S-MK, Chekal MA, Katz AI (1983) Corticosterone binding sites along the rat nephron. Am J Physiol 244:F504 – F509
Le Grimellec C (1975) Micropuncture study along the proximal convoluted tubule. Pfluegers Arch 354:133 – 150
Lifschitz MD, Schrier RW, Edelman IS (1973) Effect of actinomycin D on aldosterone-mediated changes in electrolyte excretion. Am J Physiol 224:376 – 380
Lipton P, Edelman IS (1971) Effects of aldosterone and vasopressin on electrolytes of toad bladder epithelial cells. Am J Physiol 221:733 – 741
Lo C-S, Lo TN (1979) Time course of the renal response to triiodothyronine in the rat. Am J Physiol 236:F9 – F13
Lorentz WB Jr (1974) The effect of cyclic AMP and dibutyryl cyclic AMP on the permeability characteristics of the renal tubule. J Clin Invest 53:1250 – 1257
Lorentz WB Jr (1976) Effect of parathyroid hormone on renal tubular permeability. Am J Physiol 231:1401 – 1407
Lucci MS, Warnock DG (1979) Effects of anion-transport inhibitors on NaCl reabsorption in the rat superficial proximal convoluted tubule. J Clin Invest 64:570 – 579
Madsen KM, Tisher CC (1986) Structural-functional relationships along the distal nephron. Am J Physiol 250:F1 – F15
Malnic G, Giebisch G (1972) Some electrical properties of distal tubular epithelium in the rat. Am J Physiol 223:797 – 808
Malnic G, de Mello Aires M, Giebisch G (1972) Micropuncture study of renal tubular hydrogen ion transport in the rat. Am J Physiol 222:147 – 158
Marchand GR, Ott CE, Cuche JL, Knox FG (1976) Comparison of proximal tubule fluid-to-plasma ultrafiltrate chloride ratio in rats and dogs. J Appl Physiol 40:1009 – 1011
Maren TH (1967) Carbonic anhydrase: chemistry, physiology, and inhibition. Physiol Rev 47:595 – 781
Marsh DJ (1970) Solute and water flows in the thin limbs of Henle's loop in the hamster kidney. Am J Physiol 218:824 – 831
Marver D (1984) Evidence of corticosteroid action along the nephron. Am J Physiol 246:F111 – F123
McKinney TD, Myers P (1980a) Bicarbonate transport by proximal tubules: effect of parathyroid hormone and dibutyryl cyclic AMP. Am J Physiol 238:F166 – F174
McKinney TD, Myers P (1980b) PTH inhibition of bicarbonate transport by proximal convoluted tubules. Am J Physiol 239:F127 – F134
Moore RD (1983) Effects of insulin upon ion transport. Biochim Biophys Acta 737:1 – 49
Morel F, Doucet A (1986) Hormonal control of kidney functions at the cell level. Physiol Rev 66:377 – 468
Morel F, Chabardès D, Imbert-Teboul M, Le Bouffant F, Hus-Citharel A, Montégut M (1982) Multiple hormonal control of adenylate cyclase in distal segments of the rat kidney. Kidney Int 21:S55 – S62
Morgan T, Berliner RW (1969) A study by continuous microperfusion of water and electrolyte movements in the loop of Henle and distal tubule of the rat. Nephron 6:388 – 405
Münich G, Dietl P, Oberleithner H (1986) Chloride transport in the diluting segment of the K^+ adapted frog kidney: effect of amiloride and acidosis. Pfluegers Arch. 407:S60 – S65
Mujais SK, Chekal MA, Hayslett JP, Katz AI (1986) Regulation of renal Na^+-K^+-ATPase in the rat: role of increased potassium transport. Am J Physiol 251:F199 – F207
Murer H, Burckhardt G (1983) Membrane transport of anions across epithelia of mammalian small intestine and kidney proximal tubule. Rev Physiol Biochem Pharmacol 96:1 – 51
Murer H, Hopfer U, Kinne R (1976) Sodium/proton antiport in brush-border-membrane vesicles isolated from rat small intestine and kidney. Biochem J 154:597 – 604
Nadler SP, Hebert SC, Brenner BM (1986) PGE_2, forskolin, and cholera toxin interactions in rabbit cortical collecting tubule. Am J Physiol 250:F127 – F135
Nakajima T, Naitoh F, Kuruma I (1977) Dopamine-sensitive adenylate cyclase in the rat kidney particulate preparation. Eur J Pharmacol 41:163 – 169
Nishimura H (1985) Endocrine control of renal handling of solutes and water in vertebrates. Renal Physiol 8:279 – 300
Nizet A, Lefebvre P, Crabbé J (1971) Control by insulin of sodium potassium and water excretion by the isolated dog kidney. Pfluegers Arch 323:11 – 20

Northrup TE, Krezowski PA, Palumbo PJ, Kim JK, Hui YSF, Dousa TP (1979) Insulin inhibition of hormone-stimulated protein kinase systems of rat renal cortex. Am J Physiol 236:E649 – E654
O'Brien KP, Williamson HE (1971) The natriuretic action of histamine. Eur J Pharmacol 16:385 – 390
Oberleithner H, Guggino W, Giebisch G (1982a) Mechanism of distal tubular chloride transport in *Amphiuma* kidney. Am J Physiol 242:F331 – F339
Oberleithner H, Giebisch G, Lang F, Wang W (1982b) Cellular mechanism of the furosemide sensitive transport system in the kidney. Klin Wochenschr 60:1173 – 1179
Oberleithner H, Lang F, Greger R, Wang W, Giebisch G (1983a) Effect of luminal potassium on cellular sodium activity in the early distal tubule of *Amphiuma* kidney. Pfluegers Arch 396: 34 – 40
Oberleithner H, Ritter M, Lang F, Guggino W (1983b) Antracene-9-carboxylic acid inhibits renal chloride reabsorption. Pfluegers Arch 398:172 – 174
Oberleithner H, Lang F, Wang W, Messner G, Deetjen P (1983c) Evidence for an amiloride sensitive Na^+ pathway in the amphibian diluting segment induced by K^+ adaption. Pfluegers Arch 399:166 – 172
Oberleithner H, Guggino W, Giebisch G (1985) Resistance properties of the diluting segment of *Amphiuma* kidney: influence of potassium adaption. J Membr Biol 88:139 – 147
O'Neil RG (1981) Potassium secretion by the cortical collecting tubule. Fed Proc 40:2403 – 2407
O'Neil RG (1983) Voltage-dependent interaction of barium and cesium with the potassium conductance of the cortical collecting duct apical cell membrane. J Membr Biol 74:165 – 173
O'Neil RG, Boulpaep EL (1979) Effect of amiloride on the apical cell membrane cation channels of a sodium-absorbing, potassium-secreting renal epithelium. J Membr Biol 50:365 – 387
O'Neil RG, Boulpaep EL (1982) Ionic conductive properties and electrophysiology of the rabbit cortical collecting tubule. Am J Physiol 243:F81 – F95
O'Neil RG, Helman SI (1977) Transport characteristics of renal collecting tubules: influences of DOCA and diet. Am J Physiol 233:F544 – F558
Palmer LG (1986) Patch-clamp technique in renal physiology. Am J Physiol 250:F379 – F385
Palmer LG, Frindt G (1986) Single channel in the apical membrane of the rabbit corical collecting tubule. Renal Physiol 9:50/3
Paulmichl M, Defregger M, Lang F (1986) Effects of epinephrine on electrical properties of Madin-Darby canine kidney cells. Pfluegers Arch 406:367 – 371
Paulmichl M, Friedrich F, Lang F (1987) Effects of bradykinin on electrical properties of Madin-Darby canine kidney epithelioid cells. Pfluegers Arch 408:408 – 413
Pennell JP, Lacy FB, Jamison RL (1974) An in vivo study of the concentrating process in the descending limb of Henle's loop. Kidney Int 5:337 – 347
Petty KJ, Kokko JP, Marver D (1981) Secondary effect of aldosterone on Na – K ATPase activity in the rabbit cortical collecting tubule. J Clin Invest 68:1514 – 1521
Pfaller W (1982) Structure function correlation on rat kidney. Adv Anat Embryol Cell Biol 70:1 – 106
Pollock DM, Arendshorst WJ (1986) Effect of atrial natriuretic factor on renal hemodynamics in the rat. Am J Physiol 251:F795 – F801
Pollock DM, Mullins MM, Banks RO (1983) Failure of atrial myocardial extract to inhibit renal Na^+, K^+-ATPase. Renal Physiol 6:295 – 299
Puschett JB, Zurbach P (1976) Acute effects of parathyroid hormone on proximal bicarbonate transport in the dog. Kidney Int 9:501 – 510
Radke KJ, Selkurt EE, Willis LR (1985) The role of histamine H1 and H2 receptors in the canine kidney. Renal Physiol 8:100 – 111
Rector FC Jr (1983) Sodium, bicarbonate, and chloride absorption by the proximal tubule. Am J Physiol 244:F461 – F471
Reif MC, Troutman SL, Schafer JA (1984) Sustained response to vasopressin in isolated rat cortical collecting tubule. Kidney Int 26:725 – 732
Rocha AS, Kokko JP (1973a) Membrane characteristics regulating potassium transport out of the isolated perfused descending limb of Henle. Kidney Int 4:326 – 330
Rocha AS, Kokko JP (1973b) Sodium chloride and water transport in the medullary thick ascending limb of Henle. J Clin Invest 52:612 – 623
Rocha AS, Kokko JP (1974) Permeability of medullary nephron segments to urea and water: effect of vasopressin. Kidney Int 6:379 – 387

Rocha AS, Kudo LH (1982) Water, urea, sodium, chloride, and potassium transport in the in vitro isolated perfused papillary collecting duct. Kidney Int. 22:485 – 491

Rossier BC, Geering K, Gäggeler HP, Claire M, Corvol P (1980) Testosterone: a specific competitive antagonist of aldosterone in the toad bladder. Am J Physiol 239:F433 – F439

Rossier BC, Geering K, Atkinson J, Roch-Ramel F (1985) Renal receptors. In: Seldin and Giebisch (eds) The kidney: physiology and pathophysiology, vol 1. Raven, New York, pp 775 – 806

Rouffignac C de, Elalouf JM (1983) Effects of calcitonin on the renal concentrating mechanism. Am J Physiol 245:F506 – F511

Rouffignac C de, Morel F (1969) Micropuncture study of water, electrolytes, and urea movements along the loops of Henle in psammomys. J Clin Invest 48:474 – 486

Rouffignac C de, Corman B, Roinel N (1983) Stimulation by antidiuretic hormone of electrolyte tubular reabsorption in rat kidney. Am J Physiol 244:F156 – F164

Roy DR, Jamison RL (1985) Countercurrent system and its regulation. In: Seldin, Giebisch (eds) The kidney: physiology and pathophysiology, vol 2. Raven, New York, pp 903 – 932

Sackin H (1986a) Electrophysiology of salamander proximal tubule. I. Effects of rapid cooling. Am J Physiol 251:F319 – F333

Sackin H (1986b) Electrophysiology of salamander proximal tubule. II. Interspace NaCl concentrations and solute-coupled water transport. Am J Physiol 251:F334 – F347

Sackin H, Boulpaep EL (1981) Isolated perfused salamander proximal tubule. II. Monovalent ion replacement and rheogenic transport. Am J Physiol 241:F540 – F555

Samarzija I, Frömter E (1982a) Electrophysiological analysis of rat renal sugar and amino acid transport. III. Neutral amino acids. Pfluegers Arch 393:199 – 209

Samarzija I, Frömter E (1982b) Electrophysiological analysis of rat renal sugar and amino acid transport. IV. Basic amino acids. Pfluegers Arch 393:210 – 214

Samarzija I, Frömter E (1982c) Electrophysiological analysis of rat renal sugar and amino acid transport. V. Acidic amino acids. Pfluegers Arch 393:215 – 221

Samarzija I, Hinton BT, Frömter E (1982) Electrophysiological analysis of rat renal sugar and amino acid transport. II. Dependence on various transport parameters and inhibitors. Pfluegers Arch 393:190 – 197

Sands JM, Ivy EJ, Beeuwkes R III (1985) Transmembrane potential of renal papillary epithelial cells: effect of urea and DDAVP. Am J Physiol 248:F762 – F766

Sands JM, Knepper MA, Spring KR (1986) Na – K – Cl cotransport in apical membrane of rabbit renal papillary surface epithelium. Am J Physiol 251:F475 – F484

Sansom SC, O'Neil RG (1986) Effects of mineralocorticoids on transport properties of cortical collecting duct basolateral membrane. Am J Physiol 251:F743 – F757

Sasaki S, Imai M (1980) Effects of vasopressin on water and NaCl transport across the in vitro perfused medullary thick ascending limb of Henle's loop of mouse, rat, and rabbit kidneys. Pfluegers Arch 383:215 – 221

Sato M, Dunn MJ (1986) Osmolality, vasopressin-stimulated cAMP, and PGE_2 synthesis in rat collecting tubule cells. Am J Physiol 250:F802 – F810

Schafer JA, Troutman SL, Watkins ML, Andreoli TE (1978) Volume absorption in the pars recta. I. "Simple" active Na^+ transport. Am J Physiol 234:F332 – F339

Schild L, Giebisch G, Karniski L, Aronson PS (1986) Chloride transport in the mammalian proximal tubule. Pfluegers Arch 407:S156 – S159

Schlatter E, Schafer JA (1987) Electrophysiological studies in principal cells of rat cortical collecting tubules. ADH increases the apical membrane Na^+ -conductance. Pfluegers Arch 409:81 – 92

Schlatter E, Greger R, Weidtke C (1983) Effect of "high ceiling" diuretics on active salt transport in the cortical thick ascending limb of Henle's loop of rabbit kidney: correlation of chemical structure and inhibitory potency. Pfluegers Arch 396:210 – 217

Schmidt-Nielsen B (1977) Excretion in mammals: role of the renal pelvis in the modification of the urinary concentration and composition. Fed Proc 36:2493 – 2503

Schütz W, Schnermann J (1972) Pelvic urine composition as a determinant of inner medullary solute concentration and urine osmolarity. Pfluegers Arch 334:154 – 166

Schuster VL, Kokko JP, Jacobson HR (1984a) Angiotensin II directly stimulates sodium transport in rabbit proximal convoluted tubules. J Clin Invest 73:507 – 515

Schuster VL, Kokko JP, Jacobson HR (1984b) Interactions of lysylbradykinin and antidiuretic hormone in the rabbit cortical collecting tubule. J Clin Invest 73:1659 – 1667

Schwartz GJ, Burg MB (1978) Mineralocorticoid effects on cation transport by cortical collecting tubules in vitro. Am J Physiol 235:F576 – F585

Scoble JE, Mills S, Hruska KA (1985) Calcium transport in canine renal basolateral membrane vesicles: effects of parathyroid hormone. J Clin Invest 75:1096 – 1105

Seely JF (1973) Variation in electrical resistance along length of rat proximal convoluted tubule. Am J Physiol 225:48 – 57

Silbernagl S (1979) Renal transport of amino acids. Klin Wochenschr 57:1009 – 1019

Shareghi GR, Agus ZS (1982) Magnesium transport in the cortical thick ascending limb of Henle's loop of the rabbit. J Clin Invest 69:759 – 769

Shareghi GR, Stoner LC (1978) Calcium transport across segments of the rabbit distal nephron in vitro. Am J Physiol 235:F367 – F375

Sonnenberg H, Cupples WA, De Bold AJ, Veress AT (1982) Intrarenal localization of the natriuretic effect of cardiac atrial extract. Can J Physiol Pharmacol 60:1149 – 1152

Sonnenberg H, Honrath U, Chong CK, Wilson DR (1986) Atrial natriuretic factor inhibits sodium transport in medullary collecting duct. Am J Physiol 250:F963 – F966

Sosa RE, Volpe M, Marion DN, Atlas SA, Laragh JH, Vaughan ED Jr, Maack T (1986) Relationship between renal hemodynamic and natriuretic effects of atrial natriuretic factor. Am J Physiol 250:F520 – F524

Steinhausen M, Weis S, Fleming J, Dussel R, Parekh N (1986) Responses of in vivo renal microvessels to dopamine. Kidney Int 30:361 – 370

Steinmetz PR (1986) Cellular organization of urinary acidification. Am J Physiol 251:F173 – F187

Steven K (1974) Effect of peritubular infusion of angiotensin II on rat proximal nephron function. Kidney Int. 6:73 – 80

Stokes JB (1981) Potassium secretion by cortical collecting tubule: relation to sodium absorption, luminal sodium concentration, and transepithelial voltage. Am J Physiol 241:F395 – F402

Stokes JB, Kokko JP (1977) Inhibition of sodium transport by prostaglandin E_2 across the isolated, perfused rabbit collecting tubule. J Clin Invest 59:1099 – 1104

Stokes JB, Ingram MJ, Williams AD, Ingram D (1981) Heterogeneity of the rabbit collecting tubule: localization of mineralocorticoid hormone action to the cortical portion. Kidney Int 20: 340 – 347

Stone DK, Seldin DW, Kokko JP, Jacobson HR (1983) Anion dependence of rabbit medullary collecting duct acidification. J Clin Invest 71:1505 – 1508

Stoner LC (1985) The movement of solutes and water across the vertebrate distal nephron. Renal Physiol 8:237 – 248

Suki WN, Rouse D (1981) Hormonal regulation of calcium transport in thick ascending limb renal tubules. Am J Physiol 241:F171 – F174

Suki WN, Rouse D, Ng RCK, Kokko JP (1980) Calcium transport in the thick ascending limb of Henle: Heterogeneity of function in the medullary and cortical segments. J Clin Invest 66:1004 – 1009

Szénási G, Bencsáth P, Takács L (1986) Supersensitivity of the renal tubule to catecholamines in the chronically denervated canine kidney. Pfluegers Arch 406:57 – 59

Tabei K, Imai M (1986) Permselectivity for cations over anions in the upper portion of descending limbs of Henle's loop of long-loop nephron isolated from hamsters. Pfluegers Arch 406:279 – 284

Tago K, Schuster VL, Stokes JB (1986a) Regulation of chloride self exchange by cAMP in cortical collecting tubule. Am J Physiol 251:F40 – F48

Tago K, Schuster VL, Stokes JB (1986b) Stimulation of chloride transport by $HCO_3 - CO_2$ in rabbit cortical collecting tubule. Am J Physiol 251:F49 – F56

Tago K, Warden DH, Schuster VL, Stokes JB (1986c) Effects of inhibitors of Cl conductance on Cl self-exchange in rabbit cortical collecting tubule. Am J Physiol 251:F1009 – F1017

Talor Z, Arruda JAL (1986) Na-Ca exchange in renal tubular basolateral membranes. Mineral Electrolyte Metab 12:239 – 245

Taylor A, Windhager EE (1979) Possible role of cytosolic calcium and Na-Ca exchange in regulation of transepithelial sodium transport. Am J Physiol 236:F505 – F512

Tomita K, Pisano JJ, Knepper MA (1985) Control of sodium and potassium transport in the cortical collecting duct of the rat: effects of bradykinin, vasopressin, and deoxycorticosterone. J Clin Invest 76:132 – 136

Torikai S (1987) Effects of phenylisopropyladenosine on vasopressin-dependent cyclic AMP generation in defined nephron segments from rat. Renal Physiol 10:33–39

Trachewsky D, Oakes MLI, Morris DJ (1985) Induction of flavokinase (EC 2.7.1.26) by aldosterone in the rat kidney. Endocrinol 116:879–888

Turner RJ (1984) Sodium-dependent sulfate transport in renal outer cortical brush border membrane vesicles. Am J Physiol 247:F793–F798

Ullrich KJ (1979) Sugar, amino acid, and Na^+ cotransport in the proximal tubule. Annu Rev Physiol 41:181–195

Ullrich KJ, Rumrich G, Fuchs G (1964) Wasserpermeabilität und transtubulärer Wasserfluß corticaler Nephronabschnitte bei verschiedenen Diuresezuständen. Pfluegers Arch 280:99–119

Ullrich KJ, Capasso G, Rumrich G, Papavassiliou F, Klöss S (1977) Coupling between proximal tubular transport processes: studies with ouabain, SITS and HCO_3^- -free solutions. Pfluegers Arch 368:245–252

Ullrich KJ, Rumrich G, Klöss S (1980) Active sulfate reabsorption in the proximal convolution of the rat kidney: specificity, Na^+ and HCO_3^- dependence. Pfluegers Arch 383:159–163

Ullrich KJ, Rumrich G, Klöss S (1982) Reabsorption of monocarboxylic acids in the proximal tubule of the rat kidney. I. Transport kinetics of D-lactate, Na^+-dependence, pH-dependence and effect of inhibitors. Pfluegers Arch 395:212–219. II. Specificity for aliphatic compounds. Pfluegers Arch 395:220–226. III. Specificity for aromatic compounds. Pfluegers Arch 395: 227–231

Ullrich KJ, Fasold H, Rumrich G, Klöss S (1984) Secretion and contraluminal uptake of dicarboxylic acids in the proximal convolution of rat kidney. Pfluegers Arch 400:241–249

Vari RC, Ott CE (1982) In vivo proximal tubular fluid-to-plasma chloride concentration gradient in the rabbit. Am J Physiol 242:F575–F579

Velazquez H, Wright FS (1986) Effects of diuretic drugs on Na, Cl, and K transport by rat renal distal tubule. Am J Physiol 250:F1013–F1023

Velazquez H, Wright FS, Good DW (1982) Luminal influences on potassium secretion: chloride replacement with sulfate. Am J Physiol 242:F46–F55

Völkl H, Geibel J, Greger R, Lang F (1986) Effects of ouabain and temperature on cell membrane potentials in isolated perfused straight proximal tubules of the mouse kidney. Pfluegers Arch 407:252–257

Wambach G, Higgins JR (1978) Antimineralocorticoid action of progesterone in the rat: correlation of the effect on electrolyte excretion and interaction with renal mineralocorticoid receptors. Endocrinology 102:1686–1693

Wambach G, Higgins JR, Kem DC, Kaufmann W (1979) Interaction of synthetic progestagens with renal mineralocorticoid receptors. Acta Endocrinol 92:560–567

Wang W, Messner G, Oberleithner H, Lang F, Deetjen P (1984) The effect of ouabain on intracellular activities of K^+, Na^+, Cl^-, H^+ and Ca^{2+} in proximal tubules of frog kidneys. Pfluegers Arch 401:6–13

Wang W, Dietl P, Oberleithner H (1987) Evidence for Na^+ dependent rheogenic HCO_3^- transport in fused cells of frog distal tubules. Pfluegers Arch 408:291–299

Wangemann P, Wittner M, Di Stefano A, Englert HC, Lang HJ, Schlatter E, Greger R (1986) Cl^--channel blockers in the thick ascending limb of the loop of Henle. Structure activity relationship. Pfluegers Arch. 407:S128–S141

Wardener HE de, Clarkson EM (1985) Concept of natriuretic hormone. Physiol Rev 65:658–759

Warnock DG, Eveloff J (1982) NaCl entry mechanisms in the luminal membrane of the renal tubule. Am J Physiol 242:F561–F574

Weigt M, Dietl P, Oberleithner H (1987) Activation of luminal Na^+/H^+ exchange in distal nephron of frog kidney: an early response to aldosterone. Pfluegers Arch 408:609–614

Weinman EJ, Sansom SC, Knight TF, Senekjian HO (1982) Alpha and beta adrenergic agonists stimulate water absorption in the rat proximal tubule. J Membr. Biol 69:107–111

Windhager EE, Giebisch G (1961) Micropuncture study of renal tubular transfer of sodium chloride in the rat. Am J Physiol 200:581–590

Windhager E, Frindt G, Yang JM, Lee CO (1986) Intracellular calcium ions as regulators of renal tubular sodium transport. Klin Wochenschr 64:847–852

Wingo CS, Kokko JP, Jacobson HR (1985) Effects of in vitro aldosterone on the rabbit cortical collecting tubule. Kidney Int 28:51–57

Wirthensohn G, Guder WG (1986) Renal substrate metabolism. Physiol Rev 66:469 – 497
Wong NLM, Reitzik M, Quamme GA (1986a) Micropuncture study of the superficial nephron of *Cercopithecus aethiops.* Renal Physiol 9:29 – 37
Wong NLM, Whiting SJ, Mizgala CL, Quamme GA (1986b) Electrolyte handling by the superficial nephron of the rabbit. Am J Physiol 250:F590 – F595
Wright SH, Krasne S, Kippen I, Wright EM (1981) Na^+-dependent transport of tricarboxylic acid cycle intermediates by renal brush border membranes: effects on fluorescence of a potential-sensitive cyanine dye. Biochim Biophys Acta 640:767 – 778
Wright SH, Kippen I, Wright EM (1982) Stoichiometry of Na^+-succinate cotransport in renal brush-border membranes. J Biol Chem 257:1773 – 1778
Wuthrich RP, Vallotton MB (1986) Prostaglandin E_2 and cyclic AMP response to vasopressin in renal medullary tubular cells. Am J Physiol 251:F499 – F505
Yoshitomi K, Frömter E (1984) Cell pH of rat renal proximal tubule in vivo and the conductive nature of peritubular HCO_3^- -(OH^-) exit. Pfluegers Arch 402:300 – 305
Yoshitomi K, Frömter E (1985) How big is the electrochemical potential difference of Na^+ across rat renal proximal tubular cell membranes in vivo? Pfluegers Arch 405:S121 – S126
Yoshitomi K, Burckhardt B-C, Frömter E (1985) Rheogenic sodium-bicarbonate cotransport in the peritubular cell membrane of rat renal proximal tubule. Pfluegers Arch 405:360 – 366
Yukimura T, Ito K, Takenaga T, Yamamoto K, Kangawa K, Matsuo H (1984) Renal effects of a synthetic α-human atrial natriuretic polypeptide (α-hANP) in anesthetized dogs. Eur J Pharmacol 103:363 – 366
Zeidel ML, Silva P, Seifter JL (1986) Intracellular pH regulation and proton transport by rabbit renal medullary collecting duct cells: role of plasma membrane proton adenosine triphosphatase. J Clin Invest 77:113 – 120

Chapter 5

NaCl Transport in Amphibian Skin

E. Hviid Larsen

Contents

1 Introduction

The skin of amphibians is an osmoregulatory organ. The members of this class of vertebrates inhabit freshwater, brackish water, and different terrestrial environments like open fields, tropical rain forests, and deserts. The transport systems of the epithelial membranes of the skin are common for species belonging to quite different habitats with the capacity for transporting water and ions depending on the animal's adaptation, e.g., the field conditions where the animal was caught or the laboratory conditions controlled by the investigator. Thus, from the point of view that the skin serves as an experimental model for the analysis of epithelial transport functions, it is more important to choose appropriate laboratory conditions for the animals than to select certain species for the studies. For the study of cellular regulation of ion fluxes advantage can be taken

Zoophysiological Laboratory A, The August Krogh Institute, University of Copenhagen, Universitetsparken 13, DK-2100 Copenhagen Ø, Denmark

Advances in Comparative and Environmental Physiology
Vol. 1. Edited by R. Greger

of the fact that some amphibians during their annual cycle adapt to very different external osmotic conditions imposing different demands on their osmoregulatory organs. The skin is also a respiratory organ and, probably, it serves pH regulation. But little is known of how the membranes are differentiated for these functions. This chapter deals with the transport of sodium and chloride ions and discusses how the epithelium and its cell types are organized for carrying out a net transport of NaCl from the outside to the serosal side of the epithelium. The major part of the chapter focuses on the discussion of the biophysics of the transport systems and their localization in the epithelial cells which defines the asymmetry of epithelia and the rectifying properties of transepithelial pathways. Thus, the literature to be reviewed is limited, mainly to studies of isolated skin. Recent reviews of in vivo studies are found in Koefoed-Johnsen (1979), Kirschner (1983), and Katz (1986).

The chapter has two major sections. In the first section a model of the functional organization of amphibian skin epithelium is presented. The framework is the classical two-membrane theory of epithelial NaCl transport. However, the updated transport model incorporates two different cellular units, principal cells and mitochondria-rich cells. They are highly specialized for transporting sodium and chloride, respectively. In the second section, their membrane transport systems are discussed. The majority of studies of the ionic pathways has been carried out with methods analyzing macroscopic properties of the membranes, and such studies form the basis of the second section. Fluctuation analyses have for some years been carried out for the study of passive cation pathways of frog skin. The field has recently been extensively reviewed (Lindemann 1984; van Driessche and Zeiske 1985) and the results are briefly discussed here. Reviews with references to the early literature were given by Lindemann and Voûte (1976) and Erlij and Ussing (1978). Kristensen and Ussing (1985) discussed models of the functional organization of frog skin and other epithelia.

2 Functional Organization of Amphibian Skin

2.1 The Transport Model

The cellular organization of the amphibian skin is illustrated in Fig. 1A, and the schematic diagram in Fig. 1B shows the transport model. The model depicts the epithelium as consisting of two types of transporting compartments. (1) The principal cells of the granulosum, spinosum, and germinativum layers form a functional syncytium, i.e., a single transporting unit. (2) The mitochondria-rich (m.r.) cells constitute minor transport compartments. They operate in parallel with one another and with the major transport compartment of principal cells. Transport of ions across the epithelium takes place through these two types of cellular compartments, and along the paracellular pathway. The conductance of the paracellular pathway is defined by the permeability of the junctional membranes which join the apical part of the outermost living cells. In the remainder of Section 2, experimental studies leading to this model are discussed in detail,

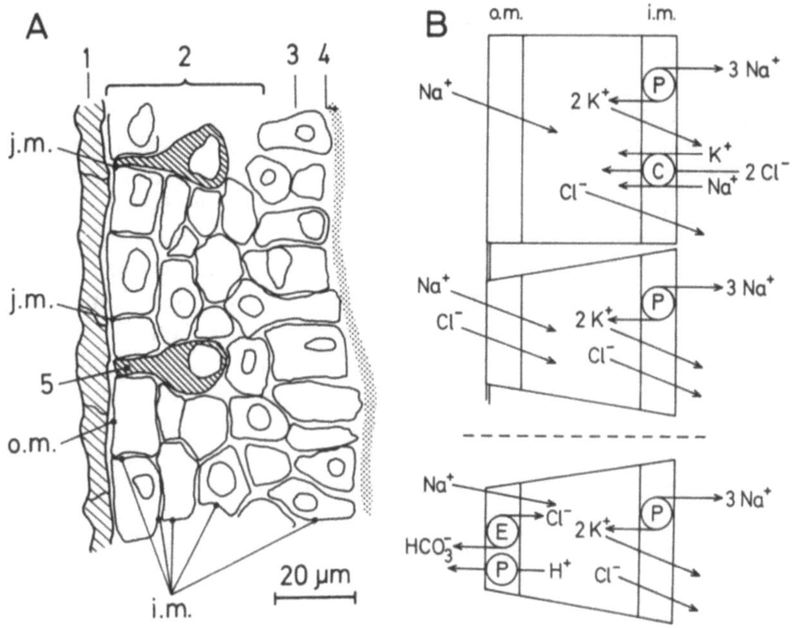

Fig. 1. A Amphibian skin epithelium showing: *1* layer of dead, highly permeable cornified cells; *2* layers of granulosum and spinosum cells; *3* germinativum cell layer; *4* basal lamina; *5* mitochondria-rich cell. *o.m.* Outward facing membrane; *i.m.* inward facing membrane, *i.e.*, the membranes lining the lateral intercellular spaces and the serosal facing membrane of germinativum cells; *j.m.* junctional membrane (*zonula occludentes*).
B Idealized representation of the functional organization of the epithelium with major transport pathways of the two types of transport compartments indicated. *Downhill arrow:* Passive pathway. *P* Primary active transport system; *C* cotransport system; *E* exchange system. In the upper mitochondria-rich cell major ionic pathways are shown with Ringer's on both sides and studied in the physiological range of potentials. Below the dashed line a model is shown of the mitochondria-rich cell bathed in a solution of low NaCl concentration on the outside

and results of other studies are discussed in the framework of the model. First, a few general remarks shall be given.

The model does not incorporate the glands. These are located in the corium with the ducts penetrating the epithelium and the openings facing the external environment at the level of the dead cornified cells. The glands secrete electrolytes (Koefoed-Johnsen et al. 1952b). The transport mechanisms were recently summarized by Mills (1985). In most studies of epithelial ion transport they are dormant. Because they do not seem to play a major role in the osmoregulatory function of the skin they shall not be considered here.

As indicated in Fig. 1B, the principal cells as well as the mitochondria-rich (m.r.) cells are furnished with a Na/K pump and a relatively large passive K permeability in the membranes facing the lateral intercellular spaces. Furthermore, the apical (outward facing) membrane is provided with a Na permeability which dominates its passive cation conductive properties. Basically, therefore, the cation pathways of both transport compartments are configured for active

Na absorption according to the classical two-membrane theory of epithelial NaCl uptake (Koefoed-Johnsen and Ussing 1958; Ussing and Windhager 1964). The new aspect extending the previous models is that the m.r. cells constitute a highly specialized pathway for transepithelial Cl ion flow (Voûte and Meier 1978). This cellular shunt is available for Cl transport in skins of room temperature-adapted animals. In the skins of frogs kept at 5 °C the much smaller fluxes of Cl seem to follow the paracellular pathway which is also taken by larger anions, like SO_4^{2-}, and a component of the passive transepithelial flux of Na (Ussing and Windhager 1964; Koefoed-Johnsen and Ussing 1974). The Na-dependent cotransport system in the inward facing membrane of principal cells is of significance for cell volume regulation (Sect. 2.4.2). Also included in Fig. 1B is a tentative model of the mitochondria-rich cell under conditions of ion uptake from very diluted solutions of NaCl (Erlij and Ussing 1978). The transport systems indicated shall be discussed in Section 3.2.4. Other transport systems are localized in the membranes, but they are not shown in Fig. 1B. These are, for example, a potassium channel in the outward facing membrane (Zeiske and van Driessche 1979), and a Na/H exchange system in the inward facing membrane (Ehrenfeld and Harvey 1987) of principal cells. They shall also be discussed in Section 3.

2.2 Ionic Components of Transepithelial Currents

Krogh's studies showed that the frog takes up Na and Cl through the skin and strongly indicated that the transport takes place at the expense of metabolic energy (Krogh 1939). The most important steps leading to the present concept of epithelial ion transport were taken about 10 years later by Ussing and his coworkers demonstrating that Na ions are actively transported, and that the transepithelial electrical potential difference is generated by an inwardly directed active Na flux shunted by a passive flow of Cl ions (Ussing 1949a, b; Ussing and Zerahn 1951; Koefoed-Johnsen et al. 1952a). These studies revealed the two

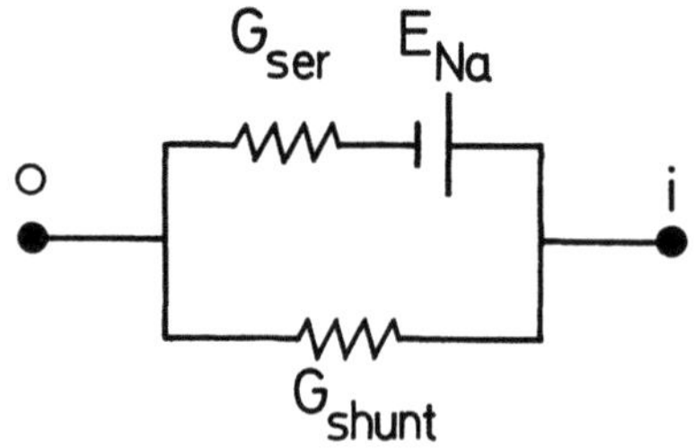

Fig. 2. The Ussing-Zerahn electrical circuit analog of amphibian skin epithelium. In this "black box treatment", the active Na pathway is represented by an electromotive force (E_{Na}), and a serial conductance (G_{ser}, which lumps the conductive elements carrying the active current; with a 3 Na/2 K pump in the inward facing membrane these are the Na conductance of the outward facing membrane, the K conductance of the inward facing membrane, and the pump). The parallel shunt conductance (G_{shunt}) was in the original treatment supposedly localized to the sodium-transporting compartment. Later, G_{shunt} was associated with paracellular ion flows. Recently, the passive Cl currents have been shown to flow through the mitochondria-rich cells. (After Ussing and Zerahn 1951)

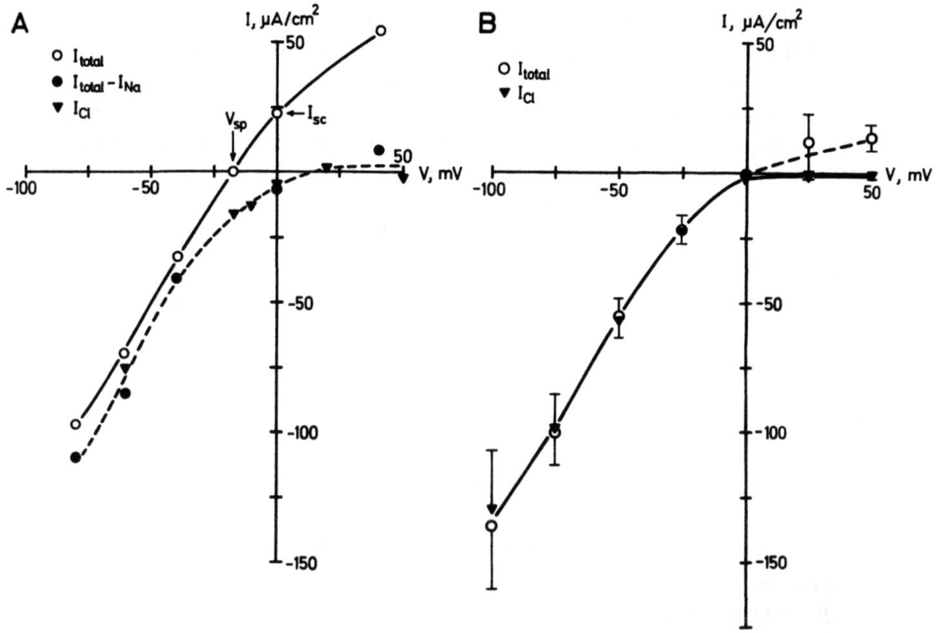

Fig. 3 A, B. Steady state current-voltage relationship of toad skin including the voltage dependence of the two major ionic components. *A* NaCl-Ringer's on both sides with a large active Na current under short-circuit conditions (the short-circuit current, I_{sc}, V = 0). V_{sp} Spontaneous skin potential (I_{total} = 0). *B* KCl-Ringer's on the outside, NaCl-Ringer's on the inside. Note that the large inward cation components of the total current have been reduced. Independent of the major cation of the external bath, the Cl pathway is strongly rectifying the current with a large conductance in the physiological range of potentials. *Sign conventions:* V is the potential of the outside bath with respect to that of the grounded serosal bath, inward currents are given a positive sign. (After Bruus et al. 1976; Willumsen and Larsen 1986)

major transepithelial ionic pathways of the skin, they provided methods for their experimental analysis, and an electric circuit analog of the skin was suggested which associates the electrical properties with the active and passive fluxes of ions (Fig. 2).

The contributions of the two ionic components, Na and Cl, to the steady state currents through the toad skin are illustrated by the current-voltage relationships depicted in Fig. 3 A. Under short-circuit conditions (bilateral NaCl Ringer's, V = 0) the active Na current accounts for almost all of the measured current. In the skin of European frogs (*Rana temporaria, R. esculenta*) the short-circuit current (I_{sc}) is carried by the active Na flux (Ussing and Zerahn 1951; Koefoed-Johnsen et al. 1952a; Kristensen 1983). In the skin of other species an active Cl flux is contained in I_{sc}. This was first observed for *Leptodactylus ocellatus* (Zadunaisky et al. 1963), and it has since been observed for other species, e.g., *B. bufo* (Bruus et al. 1976), *B. viridis* (Katz and Larsen 1984), and *R. pipiens* (Drewnowska and Biber 1985). Generally, the Cl currents under equilibrium conditions are of relatively small magnitude, and I_{sc} is a good measure of the active Na flux. In the hyperpolarizing region of the current-voltage curve

($V < V_{sp}$) the (outward) currents are almost entirely carried by the net inward flux of Cl ions. When the skin potential is reversed ($V > 0$ mV) the steady state Cl current virtually disappears and the (inward) currents are carried by Na. These relationships show that the active Na pathway as well as the Cl pathway are strongly rectifying: Large net fluxes of the two ions are maintained in an inward direction, but not in an outward direction. By comparing the results shown in Fig. 3 A, B, it can be seen that rectification of I_{Cl} occurs independently of whether Na or K is present as the major cation of the outside bathing solution (Larsen and Kristensen 1978). In the presence of amiloride in the external solution both frog and toad skin exhibit a similar strong Cl current rectification (Kristensen 1983; Dürr and Larsen 1986).

2.3 Energy Requirements of Transepithelial Currents

If the steady state ionic current is different from zero under conditions of equilibrium distribution of ions and no solvent flow, sufficient evidence is obtained for concluding that active transport is involved. If an electrochemical potential difference across the membrane prevails, the measurement of membrane currents does not, in general, provide the clue as to how the transport is energized. In this case, simultaneous measurements of unidirectional fluxes in both directions are required.

2.3.1 Flux-Ratio Analyses

Provided the ions do not interact with each other or with other moving molecules, and the transport is driven by the difference in electrochemical potential only, the ratio of steady state unidirectional fluxes is given by (Teorell 1949; Ussing 1949b):

$$J_j^{in}/J_j^{out} = [(c_j^o/c_j^i)\exp(-z_j FV/(RT))] \quad . \tag{1}$$

Here, c denotes concentrations (if the ionic strengths of the two bathing solutions are different, c_j^o and c_j^i should be multiplied by their respective activity coefficients), V is the potential difference across the barrier (the o-compartment with respect to i-compartment), subscripts o and i indicate compartments, j is the ion, and z_j, F, R, and T have their usual meanings. All quantities entering the analysis are easily measured, and Ussing (1949b) showed that Eq. (1) holds not only for homogeneous membranes but also for complex multimembrane systems like epithelia. With the equilibrium potential (V_j) given by the Nernst equation:

$$V_j = [RT/(z_j F)]\ln(c_j^o/c_j^i) \quad , \tag{2}$$

equation (1) can also be written as:

$$J_j^{in}/J_j^{out} = \exp[-z_j F(V-V_j)/(RT)] \quad , \tag{3}$$

where $-z_j F(V-V_j)$ is the work done per mole of j by the externally applied driving force. Recently, this tool was significantly refined as it could be proven that in a much more general form, the above expression is valid also under non-stationary conditions, i.e., during the period of time in which the isotope profile in the barrier is building up. Provided the time-dependent isotope fluxes on the left-hand side of the equation denote "appearing" fluxes of an ion taking a single pathway, the equation reads (Ussing 1978; Sten-Knudsen and Ussing 1981):

$$J_j^{in}(t)/J_j^{out}(t) = (c_j^o/c_j^i)\exp[N_A W/(RT)] \ , \qquad (4)$$

where N_A is Avogadro's number, and W is the work done on each isotope molecule in moving it through the barrier. This equation was shown to hold for any combination of transport mechanisms, that is, W may contain the work done by the electric field, an active transport system, or the drag exerted from solvent flow (Sten-Knudsen and Ussing 1981). For example, with no solvent flow and provided single filing does not take place, Eq. (4) takes the following form for the active sodium pathway:

$$J_j^{in}(t)/J_j^{out}(t) = \exp[-F(V - V_{Na} - E_{Na})/(RT)] \ , \qquad (5)$$

where V_{Na} is the transepithelial Na-equilibrium potential [Eq. (2)], and E_{Na} accounts for the coupling of the sodium flux to an exergonic process. The right-hand side of Eq. (5) is identical with the expression for stationary conditions (Ussing and Zerahn 1951).

2.3.2 Analysis of the Active Na Pathway

The Na ions cross the epithelium along different pathways with unidirectional fluxes of different ratios and rate constants. It follows from the above theory that a time-variant Na-flux ratio of the epithelium is composed of time-invariant flux ratios of individual pathways with the fastest (paracellular) route governing the ratio during the initial period of tracer appearance. Eskesen et al. (1985) showed that the presteady state flux ratio of sodium under open-circuit conditions with time increases from a value compatible with electrodiffusion (the paracellular pathway) to a steady state value that is two orders of magnitude larger. They developed a method for calculating the time-invariant flux ratio of the cellular pathway. The flux ratio of the cellular pathway was determined under experimental conditions of varying rates of active Na transport, and E_{Na} estimated from Eq. (5) (Eskesen and Ussing 1985). It was found (see Table 1) that E_{Na} depends on the experimental conditions, and that the largest values are obtained in skins exhibiting small active Na fluxes (low external Na concentrations). The maximum E_{Na} was estimated to be about 200 mV.

An entirely different approach for estimating E_{Na} takes its starting point in the circuit diagram of Fig. 2. The total skin conductance, G_t, which is easily measured, is given by:

$$G_t = G_{shunt} + G_{ser} \ . \qquad (6)$$

Furthermore, if the epithelium is continuously short-circuited, the following relation holds:

$$E_{Na} = I_{sc}/G_{ser} \ , \qquad (7)$$

Table 1. Estimated values of the electromotive force (E_{Na}) of the active Na pathway of amphibian skin (mean ± SE, n = number of experiments)

E_{Na} (mV)	Experimental conditions	Method	Ref.
146 ± 12 n = 12	Ringer's both sides, open circuit [a]	Presteady state flux-ratio analysis	Eskesen and Ussing 1985 (*R. temporaria*)
191 ± 5 n = 4	Diluted (1/10) Ringer's outside, open circuit [a]		
180 ± 17 n = 8	Diluted (1/10) Ringer's outside, open circuit [b]		
200 ± 2 n = 3	Diluted (1/20) Ringer's outside, open circuit [b]		
124 ± 2 n = 8	Sulfate Ringer's, short circuit [a]	Analog circuit analysis	Larsen 1973 (*B. bufo*)
205 ± 20 n = 8	1 μM amiloride, short circuit [a]		

[a] Whole skin.
[b] Isolated epithelium.

where I_{sc} is the steady state short-circuit current (the active Na current). These equations were used for calculating E_{Na} of toad skin exposed to sulfate Ringer's on both sides (Larsen 1973). The shunt conductance was calculated from the sum of the partial ionic conductances of the paracellular pathway using the equilibrium flux-conductance equation (Hodgkin 1951; Ussing and Zerahn 1951):

$$G_j = (z_j^2 F^2 J_j^{out})/(RT) \ , \tag{8}$$

where J_j^{out} is the steady state efflux of j (Na or SO_4). The crucial assumption is that the paracellular Na efflux is large compared to the backflux through the pump. This is probably a reasonable assumption for short-circuited toad skin as addition of amiloride to the external solution did not significantly reduce Na efflux. The results given in Table 1 show, as above, that E_{Na} depends on the rate of active Na flux, with a value of 200 mV when the active flux is significantly reduced (1 μM amiloride in the external solution).

With the stoichiometry of the Na pump being 3 Na/1 ATP, the maximum available energy (W_{Na}^{pump}) for active Na transport is:

$$W_{Na}^{pump} = -\Delta G_{ATP}/3F \ . \tag{9}$$

When the Gibb's free energy of the hydrolysis of ATP, ΔG_{ATP}, is about $-60000 \text{ J mol}^{-1}$ (Veech et al. 1979; Civan et al. 1983), W_{Na}^{pump} is close to 200 mV. The values found experimentally were equal to or less than 200 mV (Table 1). Being calculated from the flux-ratio equation [Eq. (5)], E_{Na} is expected to be less than 200 mV if the ratio of unidirectional passive K fluxes through the inward facing membrane is far from unity, i.e., if the potassium ion distribution is far from thermodynamic equilibrium. This may be the case when the active sodium flux is very large, as for the short-circuited preparation, and the cell depolarized (Nagel 1976; Harvey and Kernan 1984a; Willumsen and Larsen 1986).

However, when E_{Na} is being estimated by the analog circuit analysis, the energy dissipation associated with passive K transport through the serosal membrane K conductance is taken into account, and with this method a similar dependence of E_{Na} on active Na flux was observed (Table 1). The Na/ATP stoichiometry of the pump seems fixed under the above mentioned experimental conditions. Thus, the rate of oxygen uptake and the active sodium flux were found to be proportionally related with a stoichiometry of 18 mol Na mol^{-1} O_2 (Zerahn 1958). With six ATP formed per mole of O_2 consumed, it is indicated that three sodium ions are actively transported per ATP molecule hydrolzyed. This result was found irrespective of the magnitude of the Na flux, and the thermodynamic work done by moving the Na ions through the skin. From measurements of the lactate production in skins transporting Na under anaerobic conditions (Kristensen and Schousboe 1968), a ratio of 2.4 Na/1 ATP was estimated. However, this value was taken to be somewhat uncertain, as it was not possible to correct for the ATP consumption of other processes. Under these conditions as well, the free energy of ATP hydrolysis is sufficiently large for driving three Na ions per pump cycle (Kristensen and Schousboe 1969).

As an alternative hypothesis, it was suggested that with large cation fluxes through the pump, the ATP/ADP concentration ratio is reduced in the vicinity of the serosal membrane resulting in a reduction of ΔG_{ATP} near pump sites (Eskesen and Ussing 1985).

2.3.3 *Analysis of the Cl Fluxes*

Figure 4 collects results of steady state flux-ratio analyses of Cl transport across toad skin. The line indicates the theoretical relationship for simple passive transport [cf. Eq. (3)]. The measured flux ratios indicate electrodiffusion of Cl for $V < 0$ mV. Therefore, the large Cl currents observed at these potentials (Fig. 3)

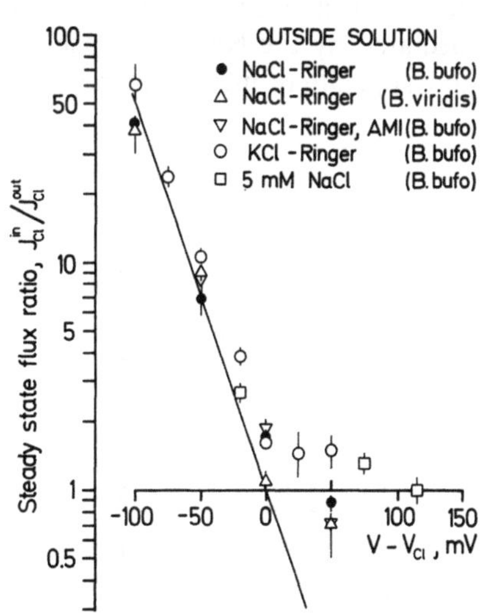

Fig. 4. Ratio of unidirectional steady state Cl fluxes as a function of externally applied driving force (toad skin). *Full line* indicates theoretical relationship for simple passive transport [independent ion movements, cf. Eq. (3)]. (Data from Bruus et al. 1976; Katz and Larsen 1984; Dürr and Larsen 1986; Willumsen and Larsen 1986)

are carried by passive pathways, which is also the case for frog skin (Koefoed-Johnsen et al. 1952a; Kristensen 1983).

For $V > 0$ mV, the flux ratio is close to unity, and independent of the driving force imposed on the ions (Fig. 4). It is conceivable then that self-exchange diffusion is the dominating mechanism of Cl transport in this range of potentials.

The ratios of unidirectional fluxes measured under equilibrium conditions (*B. bufo,* $V - V_{Cl} = 0$ mV) are significantly larger than unity, indicating inward active transport. Active Cl transport is observed in skins exhibiting active Na transport, and in skins with the transepithelial active Na flux eliminated with amiloride (Dürr and Larsen 1986) or by bathing the outside of the skin with a Na-free solution (Willumsen and Larsen 1986).

2.4 Principal Cell Compartment

In the absence of permeating anions the spontaneous transepithelial potential difference was found to respond with a change of 58 mV to a tenfold change of the Na concentration of the outside bathing solution. A similar change in V_{sp} occurred in response to a tenfold change of the K concentration of the serosal bath (Koefoed-Johnsen and Ussing 1958). It was also shown that the Na, K-ATPase inhibitor, ouabain, inhibits the short-circuit current when added to the inner bath (Koefoed-Johnsen 1957). From these observations the *two-membrane theory* was formulated stating that a K-permeable inward facing membrane contains Na/K pumps and that a Na-selective membrane is facing the apical solution (see Fig. 5). With such a configuration of cation pathways of the cell, the epithelium develops mucosa-negative potentials, transports NaCl in the inward direction under open circuit, generates a Na-carried steady state short-circuit current, and maintains ion gradients between intracellular water and extracellular fluid similar to other tissues. This powerful model of frog skin turned out to hold a general principle of the organization of absorbing epithelia, that is, transformation of chemical energy to transepithelial flows of solutes is mediated by the Na/K-ATPase which is localized in the serosal cell membrane and opposite to membrane pathways dissipating the established Na gradient.

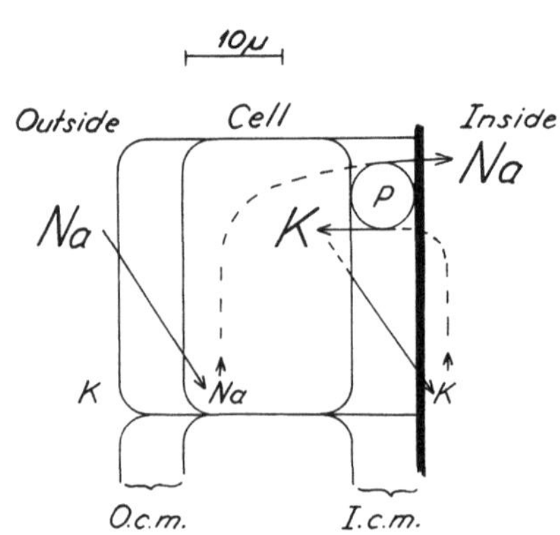

Fig. 5. The two-membrane theory of epithelial organization. Koefoed-Johnsen and Ussing (1958) suggested that the polarized nature of frog skin is due to asymmetrical distribution of the Na/K pump and the dissipative cation permeabilities in a specialized cell type of the epithelium. This principle is schematically depicted in their model shown here, and it is now recognized as a general principle of epithelial organization (Koefoed-Johnsen and Ussing 1958)

2.4.1 Localization of the Na-Selective Membrane. Intercellular Communication

It was speculated that the polarized cells are the germinativum cells. However, measurements of the intraepithelial electrical potential profile (Ussing and Windhager 1964) and fine structure analyses of the epithelium (Farquhar and Palade 1964) indicated that the cells are functionally coupled, and that the Na-selective barrier is the outward facing membrane of the outermost living cells with their junctional membranes constituting the major barrier for paracellular solute flows. In subsequent studies it was shown that the granulosum cells respond by swelling (the volume of more deeply located epithelial cells decreased) if the inward sodium flux is increased by voltage clamping, confirming that the outermost living cells participate in active transport of Na (Voûte and Ussing 1968). It was also found that when c^{o}_{Na} is stepped to a new value very fast, the transepithelial potential responds with such a small time delay that the diffusion distance of the Na-selective barrier from the external bulk solution (corrected for unstirred layer) can be no more than the thickness of the cornified cells (Fuchs et al. 1972).

That the cells of the epithelium communicate with one another through intercellular bridges of low resistance gained further support from two types of studies. Firstly, low-frequency impedance analysis indicated that frog skin can be represented by a simple two-condensor circuit diagram with a capacitance of the internal condensor being 65 times that of the external one. With the latter being about $1\ \mu F\ cm^{-2}$, and the total area of the membranes lining the lateral intercellular spaces being about 50 times that of the apical membranes, the large condensor can be identified with the laterobasal membranes of the epithelium if the cells form a syncytiumlike unit (Smith 1971). Secondly, very strong support was obtained by an electron microprobe analysis of intracellular electrolyte concentrations in frog skin (Rick et al. 1978). It was demonstrated that the respective concentrations of Na and K at steady states are similar in all epithelial layers, but often different from their concentrations in the mitochondria-rich cells. The Na concentration could be reduced in all cells by exposing the external border of the skin to the Na-channel blocker amiloride. Serosal application of ouabain resulted in an increase in cellular Na concentration and a decrease in cellular K concentration which could be prevented by exposure of the apical side of the epithelium to amiloride, or a Na-free Ringer's solution. Thus, the cells of all layers derive their Na from the external bath via the amiloride inhibitable permeability of the outward facing membrane. Furthermore, all cells extrude Na to the serosal solution via ouabain-sensitive pumps in the laterobasal membrane which exhibits virtually no passive sodium permeability. The electrolyte concentrations of the cornified cells were found to be equilibrated with those of the external bath, and it was concluded that the surface membrane of these cells is to be regarded as a nonselective, low resistance barrier for exchange of small ions.

The nonpolarized organization of cells of the deeper layers was also concluded from a radioautographic study of ^{3}H-ouabain distribution, which revealed ouabain-binding sites on all membranes lining intercellular spaces of the epithelium. Ouabain does not bind on outward facing membranes of the outermost living cells, nor on membranes of cornified cells (Mills et al. 1977).

2.4.2 Cell Water Volume

Cell volume response to change of composition and tonicity of the external bathing solutions has been studied with a method that uses thickness of the epithelium as a measure of epithelial volume (MacRobbie and Ussing 1961). It was shown that if the skin is bathed in SO_4-Ringer's the cells behave as an ideal osmometer. However, with Cl present in the internal bath a biphasic course of volume response was observed: Volume lost initially following an increase in tonicity of the internal bath is slowly regained. Likewise, volume expansion brought about by decreasing internal tonicity is followed by return of the volume toward the control value. These experiments indicated an intracellular Cl concentration of about 50 mM, and it was concluded that KCl could not be in Donnan equilibrium across the inward facing membrane. By assuming that it is the K permeability which is normally the rate-limiting pathway for net movement of KCl across the inward facing membrane, these results were not challenging the two-membrane model (Fig. 5) which predicts that it is the Na/K pump that keeps intracellular K concentration above its equilibrium value. Thus, the pump/leak theory of cell water volume control (Leaf 1956; Tosteson and Hoffman 1960), which is implicitly contained in the two-membrane model, seemed sufficient to explain the observed cell volume behavior if the additional assumption was made that the K permeability of the inward facing membrane is a volume-dependent variable.

Following serosal ouabain exposure a few skins only responded with a (very) small gain in volume. Somewhat puzzling in view of the pump/leak concept, the majority of skins responded with cell volume decrease (MacRobbie and Ussing 1961). A more recent study of the mechanisms of cell volume control provided evidence that the large intracellular Cl concentration is maintained by a Na-dependent, furosemide inhibitable Cl cotransport system in the inward facing membrane which operates in parallel with a passive Cl permeability (Ussing 1982a, b; Fig. 1B). A similar conclusion was reached by Ferreira and Ferreira (1981), who observed a decrease in intracellular Cl concentration and loss of intracellular water in response to removal of Na, or Cl, from the serosal bath.

According to Ussing's studies, the regulatory volume expansion following a stepwise increase in tonicity of the internal bath is brought about by stimulation of the cotransport system, whereby Na and Cl (together with K, see below) is taken up by the cell. In exchange for K, Na is returned to the serosal bath via the pump. The volume regulatory decrease is controlled by the passive Cl permeability which increases in response to volume expansion. Normally, however, the conductive Cl permeability must be small, since cell volume and intracellular K remained almost constant during a period of 1 h following exposure of the inside of the skin to furosemide, or a Na-free solution (Ussing 1982a). This notion implies that the rather fast exchange of ^{36}Cl between cells and serosal solution measured in skins bathed in Ringer's solutions (Ferreira and Ferreira 1981) is associated with an exchange diffusion system as was also concluded from a recent study by Stoddard et al. (1985) (see Sect. 3.5.2).

As to the stoichiometry of the cotransport system evidence has now been obtained that potassium ions are also involved and that 2 Cl is transported together with 1 Na and 1 K (Ussing 1985). Thus, it seems to be identical with the cotransport system in the surface membrane of nonpolarized cells (Geck and Heinz 1980). Recently, the dependence of intracellular Cl concentration on ion composition of the serosal bath was analyzed with the electron microprobe technique (Dörge et al. 1985). From this carefully planned study it was concluded that the Cl concentration is indeed regulated by a serosal cotransport system of the type proposed from the above mentioned studies of cell volume regulation. This study also demonstrated that bumetanide, besides furosemide, is a potent inhibitor of the cotransporter.

2.4.3 Membrane Potentials

When frog skin is bathed in Na_2SO_4-Ringer's on both sides, a large transepithelial potential difference of more than − 100 mV is often developed (outside solution negative). Recordings with a micropipette electrode advanced from the serosal solution showed that, on the average, − 64 mV is generated across the inward facing membrane, indicating a staircase-type transepithelial potential profile under these conditions. In several skins more than one plateau was encountered by further advancement of the electrode. By iontophoretical deposition of the dye, lithium carmine, it was verified that the first potential plateau was measured with the microelectrode tip in a germinativum cell (Ussing and Windhager 1964). Following short-circuiting the membrane potential changed to a value of about − 38 mV, i. e., the potential difference between external solution and cell reverses its polarity, while the serosal membrane is depolarized. A well-type potential profile of this magnitude measured under similar conditions was also obtained in a recent study where impalement artifacts were concluded to be of minor, or no significance (Schoen and Erlij 1985 a). Measurements by Nagel (1976) and Helman and Fisher (1977) have shown that also in skins exposed to NaCl-Ringer's on both sides the intracellular potential is negative with respect to the bathing solutions under short-circuit conditions. These studies showed that, on the average, the intracellular potential is about − 70 mV, but that a large variation of intracellular potentials is to be expected. Nagel (1976) reported values ranging from − 113 to − 32 mV in the short-circuited frog skin.

Nagel's studies provide three lines of evidence suggesting that variation of intracellular potentials among *short-circuited* skins is caused mainly by differences in their apical Na conductance. (1) Apical application of a large concentration of amiloride results in cellular hyperpolarization to near − 100 mV at the new steady state. Typically, this is associated with an increase, in the apical membrane's fractional resistance toward unity[1]. Thus, by eliminating the Na conduc-

[1] The fractional resistance of the outward facing membrane (fR_o) is defined by $fR_o = R_o/(R_o + R_i)$, where R_o and R_i are the resistances of the outward and inward facing membrane, respectively. fR_o is estimated from $\Delta V_o/\Delta V$, where ΔV_o and ΔV are the changes in apical membrane voltage and transepithelial voltage, respectively, in response to a brief transepithelial current pulse (Nagel 1976).

tance of the apical membrane the potential is driven toward the equilibrium potential of K, indicating that it is the *ratio* of the apical Na and serosal K conductances which directs the intracellular potential of the short-circuited preparation. (2) The intracellular potential varies among skins, but correlates with the short-circuit current with an intercept on the voltage axis ($I_{sc} = 0$) of about -100 mV, i.e., near the estimated K-equilibrium potential. This also indicates that skins exhibiting numerically small intracellular potentials are characterized by a relatively large apical Na conductance. (3) When the apical membrane's Na conductance is decreased by replacing external sodium with a non-penetrating cation, the cell hyperpolarizes toward the K-equilibrium potential (Nagel 1977). The finding (1), mentioned above, is now considered to be a good criterion for the success of impalement. The observation (2) was found also in subsequent studies by others (Harvey and Kernan 1984a; Garcia-Diaz et al. 1985; Willumsen and Larsen 1986), and it has been directly shown that skins generating large short-circuit currents indeed exhibit a large Na permeability of the apical membrane (Harvey and Kernan 1984b).

Since the Na/K pump is supposed to be rheogenic, it is expected that the pump current, as well, contributes to membrane potential (Thomas 1972; Lew et al. 1979). Experiments testing this notion were carried out with short-circuited frog skin (Nagel 1980). It was shown that sudden addition of a large amiloride concentration to the outside bath results in fast transient hyperpolarization of the cell to a value exceeding the estimated K-equilibrium potential (V_K^c). Such a hyperpolarization is not observed 3–5 min after addition of ouabain to the serosal bath. That the intracellular potential transiently exceeds V_K^c shortly after addition of amiloride was recently verified by Garzia-Diaz et al. (1985) who measured V_K^c as well. In the short-circuited skin with no current flow in the shunt, the potential (V_p) due to the pump current (I_p) is:

$$V_p = -I_p R_m \ , \qquad (10)$$

where R_m is the lumped resistance of outward and inward facing membranes. With a short-circuit current of 30 $\mu A\ cm^{-2}$, I_p is fairly large (10 $\mu A\ cm^2$), but V_p depends on the ionic conductances of the two membranes. Only if R_m is large is V_p expected to be of significant magnitude. This is why the rheogenic nature of the pump is revealed shortly after blockade of the Na conductance: R_m is now large and the pump current has not yet decreased. In the mitochondria-rich cell having a very large Cl conductance (see Sect. 2.5), V_p is expected to be small also after amiloride treatment (Larsen and Rasmussen 1985).

Under open circuit, the transcellular electrical potential profile depends also on the conductance in parallel with principal cells (Ussing and Windhager 1964; Schultz 1972; Lew et al. 1979). Junctional membranes, m. r. cells, and edge leaks contribute to this conductance (Figs. 1A and 2). For the limiting case of insignificant circular current flows, the apical membrane is expected to reverse its polarity, while the serosal membrane hyperpolarizes. Nagel (1976) and Harvey and Kernan (1984a) reported a "well-type" potential profile under open circuit. On the average, the skins generated transepithelial potentials of about -75 mV, indicating fairly large current flows through shunts. Tight skins developing transepithelial potentials of more than -110 mV exhibit a staircase potential profile (Ussing and Windhager 1964). Serosal membrane potentials of more than -115 mV recorded under open-circuit conditions (Nagel 1976) are incompatible with passive K net flux from cell to serosal bath. Rather than challenging the transport model, they indicate large microelectrode tip potentials or nonstationary conditions.

2.4.4 *Intracellular Sodium Concentration*

The first detailed study of the intracellular Na concentration (c^c_{Na}) of the short-circuited epithelium was carried out with the microprobe technique reporting values for c^c_{Na} of about 9 mmol kg^{-1} wet mass (Rick et al. 1978). With a mean dry mass of the tissue of 25 g/100 g (average for all tissues), the above concentration indicates a molar concentration of $c^c_{Na} = 12$ mM. A similar low intracellular Na concentration was obtained from the kinetics of tracer-Na exchange, 8 mM (Nielsen 1982) and 14 mM (Stoddard and Helman 1985). A low intracellular Na activity (a^c_{Na}) was measured with a single ion selective microelectrode and a subsequent impalement with the reference electrode, $a^c_{Na} = 14$ mM (Nagel et al. 1981) or with a doubled-barreled microelectrode, 11 mM (Harvey and Kernan 1984a). In all studies the reliability of measurements was verified by showing that the intracellular Na concentration (activity) estimated under control conditions was altered in a predictable manner either by interfering with the Na conductance of the outward facing membrane (amiloride, antidiuretic hormone, external Na replacement), or by inhibition of the Na/K pump with ouabain.

Low intracellular Na concentration together with a well-type potential profile of the short-circuited preparation clearly show that the condition for passive Na transport across the outward facing membrane is fulfilled. Harvey and Kernan (1984a) performed a detailed study of the response of intracellular potential and Na activity to a reduction of external Na activity from 79 mM (Ringer's strength) to 0.1 mM (dilution with distilled water) in frog skin under open-circuit conditions. While cellular Na activity dropped to 2 mM, apical membrane hyperpolarized to -101 mV (cell negative). From these values they calculated a driving force of:

$$\Delta \tilde{\mu}^{oc}_{Na}/F = (10^3 RT/F) \ln(0.1/2) + 101 = 25 \text{ mV} \tag{11}$$

in an inward direction. It was concluded that also in this case the condition for passive Na uptake across the outward facing membrane is fulfilled. It has been suggested that Na is actively transported across the outward facing membrane (Cuthbert 1972; Rotunno et al. 1966). The above results indicate, in agreement with a computer model analysis by Lew et al. (1979), that there is no need for such a transport mechanism in this membrane.

2.4.5 *Intracellular Potassium Concentration*

The K concentration measured with a microprobe technique, 118 mmol kg^{-1} wet mass (Rick et al. 1978), corresponds to 153 mM. The first study using an ion selective microelectrode technique for the measurements of intracellular K activity in short-circuited skin reported activities of about 132 mM (Nagel et al. 1981). This value was obtained by successive impalements with reference and ion selective microelectrodes. A subsequent study using simultaneous impalements with the two electrodes gave somewhat lower values, on the average about 92 mM with little variation among different skins (Garcia-Diaz et al. 1985). Still lower values, about 70 mM, were reported by Harvey and Kernan (1984a). It is not fully clarified why these measurements differ so much among different laboratories. Garcia-Diaz et al. (1985) discussed possible artifacts that may influence K-activity determinations, and they pointed out that small variations in tip potentials lead to large variations in calculated activities. This point is well illustrated if K-equilibrium potentials are compared, rather than the K activities (activity in parentheses, $a_K^i = 1.9$ mM), -108 mV (132 mM), -99 mV (92 mM), and -92 mV (70 mM), respectively. In all studies the K-equilibrium potential was numerically larger than the membrane potential, i.e., the condition for passive K ion flow from cell to (internal) bath is fulfilled.

Elimination of the apical membrane's Na conductance (amiloride, or Na-free external bath) results in cellular hyperpolarization with the potential transiently exceeding the K-equilibrium potential. By raising the serosal K concentration to 115 mM, the cell is rapidly depolarized (mean value of $V_c = -2 \pm 2$ V). These observations go along with the view that the pump is rheogenic (cf. Sects. 2.4.3 and 3.3.2), and that the serosal membrane is predominantly K permeable while the apical membrane has a vanishingly low K conductance (Garcia-Diaz et al. 1985).

2.4.6 *Intracellular Chloride Concentration*

With a microprobe technique the intracellular Cl concentration was found to be 36 mmol kg^{-1} wet mass, or, multiplied by 1.3 (see above), 47 mM (Rick et al. 1978). A similar value, 52 mM, was found in toad skin (Rick et al. 1980), and in a subsequent study of frog skin, 51 mM, with little variation among different cell layers (Dörge et al. 1985). They compare well with the above mentioned value of 50 mM calculated from the cell volume decrease associated with depletion of the intracellular Cl pool (MacRobbie and Ussing 1961). Similar concentrations of Cl were obtained from washout experiments, 47 – 60 mM (Ferreira and Ferreira 1981), 42 mM (Stoddard and Helman 1985), and chemical analysis, 54 mM (Nielsen 1985). For short-circuited preparations exposed to NaCl Ringer's on both sides the following mean values of Cl activities and intracellular potentials were reported, 18 mM, -78 mV (Nagel et al. 1981), 22 mM, -57 mV (Giraldez and Ferreira 1984), 15 mM, -90 mV (Biber et al. 1985) and 18 mM, -68 mV (Willumsen and Larsen 1986). Harvey and Kernan (1984a) reported a Cl activity of 20 mM under open-circuit conditions with an apical membrane potential of -38 mV (cell negative). Single- (Nagel et al. 1981; Biber et al. 1985) as well as a double-barreled microelectrode technique (Giraldez and Ferreira 1984; Harvey and Kernan 1984a; Willumsen and Larsen 1986) were employed on whole skins (Nagel et al. 1981; Biber et al. 1985), or isolated epithelia (Giraldez and Ferreira 1984; Harvey and Kernan 1984a; Willumsen and Larsen 1986). Thus, the activities reported are independent of the method. An intracellular activity of 20 mM indicates a concentration of 26 mM which is lower than the values obtained by all other methods. Willumsen and Larsen (1986) mounted the chamber with the isolated epithelium on the stage of a microscope so that they could verify that the electrode was advanced into a principal cell. It can be excluded, therefore, that the somewhat small activities obtained in microelectrode studies are caused by an occasional impalement of a mitochondria-rich cell of low Cl concentration (Sect. 2.5.1). It is possible that tip potentials of the ion selective and the reference electrode differ when the electrodes are located in a cell. Furthermore, determinations of cell water for expressing the measured amount of the element in terms of its concentration are also somewhat uncertain. Therefore, one cannot possibly expect excellent agreement between values of intracellular concentration of diffusible Cl estimated by these different methods. However, the discrepancies mentioned above are too large to be trivial, and they remain to be explained. For the present argumentation it is a fact that all estimates of the intracellular Cl activity give values significantly above the equilibrium value. Thus, in the short-circuited skin the intracellular electrochemical potential of Cl is at least 2800 J mol^{-1} above that of the bathing solutions, corresponding to an outward driving force of 30 mV or more.

With the double-barreled microelectrode technique Giraldez and Ferreira (1984) found that removal of Cl or Na from the serosal bath results in a significant decrease in intracellular Cl activity and they concluded, as Ferreira and Ferreira (1981), that a Na-dependent cotransport system in the inward facing membrane is involved in maintaining intracellular Cl above thermodynamic equilibrium. This is in agreement with the conclusion of Ussing (1982a, 1985) and Dörge et al. (1985) whose studies provided more detailed information on the cotransport system (Sect. 2.4.2). Following addition of amiloride, the cell hyperpolarized, but the Cl activity stayed constant during the first 30-min period of amiloride exposure (Nagel et al. 1981; Biber et al. 1985; Willumsen and Larsen 1986) or a^c_{Cl} slightly decreased (Harvey and Kernan 1984b). However, in all studies the activity remained significantly above its equilibrium value despite the outward driving force acting on passive Cl ion flows being almost doubled. This indicates a small passive Cl permeability. It should be noted, however, that the response of the cellular Cl activity to the addition of amiloride is not easily predicted. This is because the decrease in a^c_{Na} results in an increased driving force for the Cl uptake via the 1 Na/1 K/2 Cl cotransporter. For the limiting case with no Cl leaks and a constant intracellular K concentration, a doubling of a^c_{Cl} should occur if a^c_{Na} is reduced to one-fourth of its control value. The time it takes the cell to reach a new steady state depends generally on the Na, K pump flux and the turnover of the cotransporter. If the latter process is slow, as indicated in some studies (Ussing 1982a), it may take hours before an increase in a^c_{Cl} can be detected. That such an effect may be observed if one waits long enough was indicated in the study of Willumsen and Larsen (1986), who measured Cl activities of 40 mM (impalement from the outside), and 31 mM (impalement from the serosal side) in toad skin epithelia equilibrated for several hours with KCl-Ringer's on the outside.

Removal of Cl from the external bath had no effect on intracellular Cl concentration of short-circuited frog skin (Ferreira and Ferreira 1981; Biber et al.

1985; Dörge et al. 1985), despite the cell potential staying at about -80 mV during the observation period of 40 min (Biber et al. 1985). It was concluded that the electrodiffusive Cl permeability of the outward facing membrane is practically zero. Studies of exchange of ^{36}Cl between tracer-loaded, short-circuited frog skin and external compartments revealed small (Ferreira and Ferreira 1981) or immeasurable efflux (Stoddard et al. 1985) of Cl to the apical bath. At transepithelial potentials near 0 the Cl conductance is small but in the negative region of the current-voltage curve large passive Cl currents flow through the skin (Figs. 3 and 4). For investigating the Cl permeability of principal cells under these conditions Willumsen and Larsen (1986) followed a protocol by which the transepithelial potential was shifted from 30 mV (deactivated transepithelial Cl conductance) to -100 mV (fully activated passive Cl conductance, cf. Fig. 3). This maneuver changed the driving force imposed on the Cl flow across the apical membrane by about 100 mV, from 52 mV in an outward direction to 47 mV in an inward direction. Also, the driving force across the inward facing membrane changed, from 12 mV to 53 mV (both in cell to serosal bath direction). However, no change in the intracellular Cl activity was observed despite the transepithelial Cl conductance increased as much as 1.2 mS cm^{-2}. This result was obtained whether the epithelium was impaled from the cornified or the serosal side. They concluded that the large passive Cl currents do not pass through principal cells, implying that the conductive Cl permeability of their outward facing membrane is virtually zero.

There is an interesting observation which is difficult to reconcile with the model of Fig. 1B. Harvey and Kernan (1984a) observed a significant decrease in intracellular Cl activity in response to piretanide and SITS added to the external solution. The basis for these effects is not yet clear.

2.5 Mitochondria-Rich Cells

The flask-shaped mitochondria-rich (m.r.) cells constitute a minority cell type of anuran skin epithelium. The body is located in the second and third layer of principal cells. The neck is in the outermost granulosum layer and the apical membrane faces the subcorneal space. Thus, m.r. cells are in contact with the external solution from which NaCl is taken up by the skin (Fig. 1A; Budtz and L.O. Larsen 1973, 1975; Whitear 1975). The density is in the order of 10^5 cells cm^{-2}. The outward facing membrane area of each cell is about 10 μm^2. Usually, their total apical membranes occupy a few percent of the total area of the epithelial surface. However, m.r. cell density exhibits a considerable variation among different species, among different regions of the skin, and among individuals kept under similar laboratory conditions (Ehrenfeld et al. 1976; Ilic and Brown 1980; Brown et al. 1981; Willumsen and Larsen 1986). The density decreases if the animal is kept in solutions of high NaCl concentration and increases if the animal is kept in distilled water for about 1 week (Ilic and Brown 1980; Katz and Larsen 1984; Willumsen and Larsen 1986). Unlike principal cells, m.r. cells exhibit a significant carbonic anhydrase activity (Rosen and Friedley 1973) which increases in the presence of aldosterone (Voûte et al.

1975) and is located preferentially in the neck region (Zaccone et al. 1986). M.r. cells secrete methylene blue added to the serosal bath. The rate of secretion is significantly reduced if external NaCl concentration is decreased, or amiloride is added to the external, or ouabain to the serosal bath (Ehrenfeld et al. 1976). M.r. cells stain selectively with silver ions in the external solution (Rudneff 1865; Whitear 1975) like chloride cells of fish gill (Philpott 1965). Silver staining of the skin provides an easy and fast method for m.r. cell density measurements (Willumsen and Larsen 1986). A similar cell type is found in amphibian urinary bladder (Choi 1963), and in kidney collecting tubules of amphibians (Bargmann and Welsh 1972) and mammals (Rhodin 1958; Madsen and Tisher 1986).

Their role in transepithelial Cl transport was suggested from the finding that the passive Cl conductance of frog skin is positively correlated to the number of m.r. cells, and that m.r. cells, in contrast to principal cells, swell when exposed to K-gluconate Ringer's on the inside, and KCl Ringer's on the outside (Voûte and Meier 1978). The Cl pathways of principal cells are not configured for inward passive Cl transport (Fig. 1B; Sect. 2.4). Nevertheless, the Cl-equilibrium fluxes decrease or increase if apical Na conductance is inhibited or stimulated, respectively (Macey and Meyers 1963; Candia 1978; Kristensen 1978; Ques-von Petery et al. 1978; Kristensen 1982). On the sum of evidence Kristensen (1981) suggested that Cl ions flow through another cellular pathway, i.e., m.r. cells, which also transport Na.

2.5.1 Intracellular Ion Concentrations

The relative concentrations of Na and K of m.r. cells follow the well-known pattern of other cells, $c^c_{Na} = 11$ mmol kg^{-1} wet mass, and $c^c_K = 111$ mmol kg^{-1} wet mass. However, the Cl concentration is significantly smaller than the Cl concentration of principal cells, 12 mmol kg^{-1} wet mass (*R. temporaria, R. esculenta,* Rick et al. 1984) or 20 mmol kg^{-1} wet mass (*B. viridis,* Rick et al. 1980). From the decrease in m.r. cell volume following exposure of the epithelium to Cl-free Ringer's, and the estimated osmotically inactive volume of the cell, an intracellular Cl concentration of 20 mM was calculated (*B. marinus,* Larsen et al. 1987).

Addition of ouabain to the serosal bath resulted in an increase in c^c_{Na}, and a decrease in c^c_K, but these effects were not as pronounced as in principal cells. When the outer bathing medium was made Na-free, the effect of ouabain was reduced (Rick et al. 1978). Results of microprobe analysis indicate, therefore, that ouabain-sensitive Na/K pumps are localized in the inward facing membrane, and that Na can be taken up from the external solution. In these experiments, however, amiloride had little, if any, effect on ouabain-induced cation redistributions (Rick et al. 1978; see also Sect. 2.5.3).

2.5.2 Cl Conductance of Mitochondria-Rich Cells

The current density profile above the surface of the voltage-clamped epithelium has been studied by a vibrating probe voltage scanning technique (Foskett and

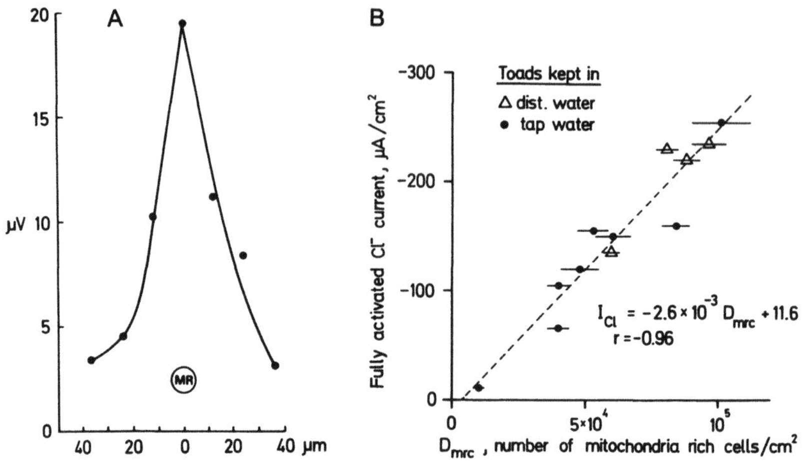

Fig. 6. A Current density profile measured by a vibrating probe moved from left to right at a constant height above frog skin epithelium clamped at −100 mV. As the electrode passes a mitochondria-rich cell (*MR*) the voltage output of the probe increases, indicating the presence of a current peak (Foskett and Ussing 1986). *B* Dependence of the Cl currents measured at −100 mV on the density of m.r. cells in toad skin. (After Willumsen and Larsen 1985)

Ussing 1986; Katz and Scheffey 1986). It was found that the large outward currents at −100 mV (cf. Fig. 3) are not uniformly distributed over the entire epithelial surface, but confined to areas above m.r. cells (see Fig. 6A). The peaks disappear if Cl of the external bath is replaced by gluconate. For obtaining good spatial resolution the animals were salt-adapted which reduces m.r. cell density by about 50%. Not all m.r. cells revealed large conductance, and some gland ducts were found to be extremely conductive. In another series of experiments with tap water-adapted frogs, the volume of single m.r. cells was measured with the video microscope technique of Spring and Hope (1978). It was found that voltage clamping the tissue to −150 mV induces a significant swelling of m.r. cells. Swelling was prevented by exposing the external side, or both sides, to Cl-free Ringer's. Also in these experiments only about 50% of the cells responded to voltage clamping (Foskett and Ussing 1986). Despite uncertainty as to whether all m.r. cells possess a Cl permeability, these studies present the proof that m.r. cells are sites of a large Cl ion-specific conductance.

From the peak voltage response (Fig. 6A) measured by the vibrating probe over 15 m.r. cells, Foskett and Ussing (1986) estimated a mean current of −1.2 nA per cell at −100 mV. A similar value, −2.6 nA per cell at −100 mV, was estimated from the relationship between total Cl currents and m.r. cell densities using the silver-staining technique (Fig. 6B, Willumsen and Larsen 1985). Considering that only about every second cell is conductive when measured with the vibrating probe technique, the fair agreement between the two estimates may seem surprising. It is very likely, though, that silver ions are precipitated only on the outside of those cells that are Cl conductive. Both estimates indicate a very large Cl conductance per m. r. cell, about 200 mS cm^{-2} apical surface area (at -100 mV).

At spontaneous potential m. r. cell volume decreased when external Cl was replaced by a nonpenetrating anion indicating a significant apical Cl permeability (Spring and Ussing 1986). Under short-circuit conditions, m. r. cells of toad skin epithelium (*B. marinus*) lost about 14% of their volume in response to a bilateral removal of Cl. Ouabain induced a significant gain in volume which was prevented by replacing the Cl ions of the bathing solutions by gluconate. The ouabain-induced volume increase was not prevented by unilateral Cl/gluconate substitutions. Thus, the apical as well as the serosal membrane is permeable to Cl at $V = 0$ mV (Larsen et al. 1987).

It can be concluded that the m. r. cells fulfill the condition for being a pathway for transepithelial Cl fluxes for $V \leq 0$ mV.

2.5.3 Active Na Transport by Mitochondria-Rich Cells

Under short-circuit conditions m. r. cell volume was found to decrease by 15% in the presence of amiloride (Fig. 7A). Five mM acetate was added to both bathing

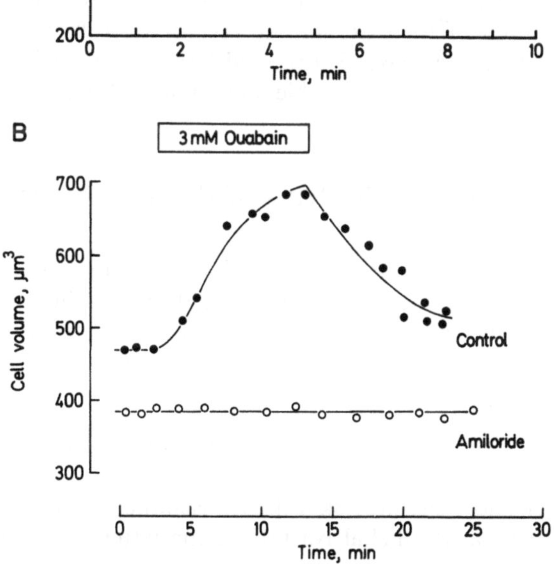

Fig. 7. A Volume response of a single m.r. cell of short-circuited toad skin epithelium to the addition of 50 µM amiloride in the external bath. *B* Addition of ouabain to the serosal solution results in volume increase of the m.r. cell. The presence of 50 µM amiloride in the apical solution prevents ouabain-induced cell swelling (Larsen et al. 1987, unpubl.)

solutions for providing the cells with a metabolizable substrate. A similar volume loss (11%) was seen following replacement of external Na ions with K or N-methyl-D-glucosamine. The ouabain-induced volume increase was prevented if the epithelium was exposed to amiloride (see Fig. 7B), so it can concluded that the apical membrane of m.r. cells contains a Na-specific permeability which is blocked by amiloride. From initial rates of volume gain following addition of ouabain to the serosal bath of short-circuited preparations it was estimated that the m.r. cell, on the average, generates an active Na current of 10 pA. With a density of 60000 m.r. cells cm^{-2}, the Na current carried by these cells is 0.6 $\mu A\ cm^{-2}$ epithelial surface. Thus, less than 3% of the short-circuit current is carried by m.r. cells (Larsen et al. 1987).

2.6 Principal Cell Compartment and Mitochondria-Rich Cells Behave as Noncoupled Units

Several lines of evidence strongly indicate that principal cells and mitochondria-rich cells do not communicate with one another through intercellular junctions. (1) Their steady-state Cl concentrations are different whether measured with an electron microprobe technique (Rick et al. 1984), or with quantitative light microscopy of cell volume response to Cl-free solutions (MacRobbie and Ussing 1961; Larsen et al. 1987). (2) During transient states large differences in their cation concentrations can be observed, while the concentrations of Na and K, respectively, show little variation among different layers of principal cells (Rick et al. 1978). (3) Serosal application of antidiuretic hormone results in an increase in Na concentration of principal cells only (Rick et al. 1984). (4) Principal cells shrink (MacRobbie and Ussing 1961), while m.r. cells swell (Larsen et al. 1987) in response to ouabain. (5) Under open-circuit conditions, exposure of the skin to Cl-free external solution results in m.r. cell volume decrease (Spring and Ussing 1986), whereas the principal cell volume remains constant (Ussing 1982a). (6) Dilution of the external bathing solution with distilled water results in m.r. cell volume increase (Spring and Ussing 1986) with virtually no effect on principal cell volume (MacRobbie and Ussing 1961).

3 Transport Systems of the Membranes

Much information on individual transport systems of anuran skin epithelium has been obtained during the last decade. With the introduction of ion selective microelectrode, patch-clamp, and fluctuation analysis techniques the field is entering a period of rapid development.

3.1 Sodium Permeability of Apical Membrane

In short-circuited skin the transepithelial Na efflux amounts to a few percent of the transepithelial Na influx, the latter being almost identical with the rate of transcellular active Na transport (Ussing and Zerahn 1951). The low intracellular Na concentration of about 10 mM (Sect. 2.4.4) indicates that the Na/K pump in the inward facing membrane is far from saturation. The rate-limiting process for the transcellular active Na flux, therefore, is the uptake of Na from the outer solution into the cells.

3.1.1 Macroscopic Properties of Sodium Permeability

The complexity of the apical pathway was indicated in studies showing that the unidirectional Na influx across the apical membrane saturates at large external Na concentrations (Biber and Curran 1970; Rotunno et al. 1970) as was found also for the unidirectional Na influx across the whole skin (Ussing 1949a; Kirschner 1955). The tracer-flux studies of the apical pathway showed that the Na influx is stimulated by antidiuretic hormone (Biber and Cruz 1973) and by short-circuiting the preparation (Biber and Sanders 1973), confirming that the saturating influx is associated with the transcellular active Na flux which responds in a similar way. It was speculated that the transport mechanism is that of a saturating carrier, rather than simple electrodiffusion. As mentioned in Section 2.4.4, active transport was also considered. It is noteworthy, however, that in response to sudden stepwise increase in c^{o}_{Na}, the Na current (I_{Na}) rapidly increases to a peak value and then "reclines" to its new steady state. It is the steady state, I_{Na}, and not the peak current, which saturates with increasing c^{o}_{Na}. Furthermore, external application of, for example, BIG (benzimidazolylguanidinum) prevents "recline" of I_{Na}, so that the resultant steady state currents do not exhibit saturation kinetics (Zeiske and Lindemann 1974).

Evaluation of the transport mechanism requires information on how the Na flux across the apical membrane depends on the transmembrane electrochemical potential difference. In the tracer kinetic studies the flux was estimated, but the driving force was unknown. In an attempt to control this variable Fuchs et al. (1977) applied the voltage-clamp technique to preparations bathed in K-Ringer's on the inside with a nonpermeating anion in both bathing solutions (sulfate). Due to the K-selective properties of the inward facing membrane, it was assumed that this treatment results in the elimination of its resistance and membrane potential. If these ideal conditions are established, the apical membrane is clamped at the command voltage, and the transcellular current is governed by the conductance of this membrane only. From measurements of ("near") instantaneous current voltage curves of the preparation before and after amiloride exposure, the voltage dependence of the amiloride blockable current was obtained.[2] The reversal potential, V_r, of these currents was independent of total transepithelial

[2] For successful application of the method the following requirements have to be met for amphibian skin. (1) The cells should be fully depleted of diffusible anions. The preparations were washed in K-sulfate Ringer's for several hours before being mounted in the chamber. (2) Exposure of the outer border to solutions of large Na activities has to be made so fast that current-voltage curves are obtained before the intracellular Na activity starts changing. This condition was fulfilled with a "fast-flow" chamber permitting a 95% exchange of solution on the outer side of the preparation within 25 ms (Lindemann et al. 1972).

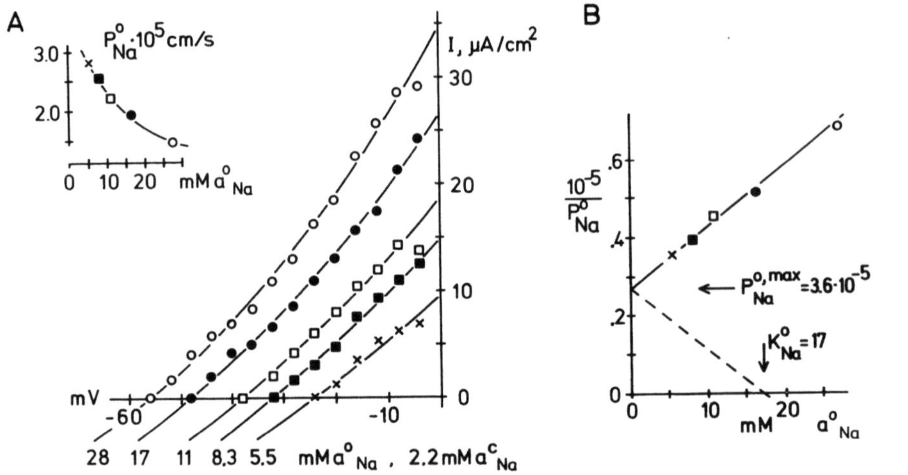

Fig. 8. A Current-voltage curves of apical Na conductance of K-depolarized frog skin. *Full lines* are best fit of constant field equation to experimental points. The five graphs are from a single preparation equilibrated with 2.8 mM external Na activity, but measured 10 s after exposure to the test activity shown on each graph. For small a^o_{Na}, near a^c_{Na}, the conductance approaches simple ohmic behaviour. Rectification is seen when a^o_{Na} is increased, i.e., as the difference between external and cell Na activity is increased (*graphs right to left*). This illustrates the pertinent qualitative property of a *symmetric, constant fieldlike* pathway. *Insets* show relationship between apical Na permeability and a^o_{Na} at constant a^c_{Na}, which is linearized when plotting $1/P^o_{Na}$ versus a^o_{Na} (Fuchs et al. 1977)

conductance, and V_r increased 58 mV per tenfold change of the apical Na activity (a^o_{Na}). These observations permit the conclusion that the amiloride-sensitive current is a Na current. At constant a^o_{Na}, the voltage dependence of inward Na currents ($V_r \leq V_o \leq 0$) could be fitted with the constant field equation for electrodiffusion (Goldman 1943; Hodgkin and Katz 1949). However, the calculated steady state Na permeability, P^o_{Na}, was found to decrease with increasing a^o_{Na} (see Fig. 8). The change in P^o_{Na} following an increase in a^o_{Na} paralleled the I_{Na} recline, and at constant a^c_{Na}, the steady state $1/P^o_{Na}$ was found to be a linear function of a^o_{Na}. Thus, the $I_{Na}-V_o$ relationship is described by the following expression for electrodiffusion:

$$I_{Na} = \frac{P^{o,max}_{Na} K^o_{Na}}{a^o_{Na} + K^o_{Na}} \cdot \frac{F^2 V_o/RT}{1-\exp(-FV_o/RT)} [a^o_{Na} - a^c_{Na}\exp(-FV_o/RT)] \ , \qquad (12)$$

where $P^{o,max}_{Na}$ is the permeability for $a^o_{Na} = 0$, K^o_{Na} an empirical constant, and V_o the membrane potential ($= \psi_o - \psi_c$, with ψ_o and ψ_c being the outer and cellular potentials). Equation (12) predicts saturation of the unidirectional Na influx at large a^o_{Na} with true "Michaelis-Menten" kinetics, if V_o and a^c_{Na} are kept constant. It was concluded that a mechanism of electrodiffusion explains the results well if the additional assumption is made that a slow regulatory process causes Na channels to close as a^o_{Na} is increased (Fuchs et al. 1977). The hypothesis of free diffusion has been tested by flux-ratio analysis of unidirectional, amiloride blockable Na fluxes through the apical membrane in frog skin. Over a wide range of electrochemical potential gradients good agreement was found between measured flux ratio and theoretical flux ratio of independent ion movements [cf. Eq. (1)] (Benos et al. 1983). Recently, the current-voltage curve of apical Na channels of skins bathed in NaCl-Ringer's on the inside was measured with a microelectrode technique (Schoen and Erlij 1985b). Their analysis covered apical membrane potentials over the range of ± 100 mV, and it was concluded that also under these conditions the Na-ion translocation through amiloride-sensitive channels is well described by the constant field equation.

The apical membrane's Na permeability is highly Na-specific, as no other cations, except Li (Zerahn 1955; Morel and Leblanc 1975; Li and Lindemann 1982), is known to be translocated through this pathway.

Addition of ouabain to the serosal solution results in inhibition of the unidirectional Na influx through the outward facing membrane (Biber 1971; Erlij and Smith 1973) and a decrease of its Na conductance (Larsen 1973). Ouabain-induced inhibition is not seen in skins preincubated in Na-free medium (Erlij and Smith 1973). It is indicated then that an increase in a_{Na}^{c} is required for inhibition of P_{Na}^{o}. A similar "feedback" mechanism has been demonstrated in mammalian colon (Frizzell and Schultz 1978; Schultz 1981). It may not be caused by a direct inhibitory effect of Na_c on P_{Na}^{o}. Cytosolic Ca concentration increases with a_{Na}^{c} due to decrease, or reversal, of Na – Ca exchange across the serosal membrane, and it has been suggested that Ca-controlled P_{Na}^{o} inhibition contributes (Grinstein and Erlij 1978; Taylor and Windhager 1979; Garty and Lindemann 1984). It is also possible that cytosolic pH decreases as a_{Na}^{c} increases by slowing down Na – H exchange across the serosol membrane. Intracellular acidification has been suggested to inhibit apical Na conductance (Ehrenfeld and Harvey 1987). Therefore, intracellular Na, Ca, and protons are all candidates as mediators of the feedback effect.

Depolarization of the apical membrane results in a slow (seconds) decrease of the Na conductance which was estimated to be more pronounced than that expected from a constant field effect caused by the membrane depolarization and a decrease of a_{Na}^{c} (Nagel and Essig 1982). Thus, V_o may control P_{Na}^{o}, but with such a long time constant that the process is not revealed by $I_{Na} - V_o$ relationships obtained within 200 ms [Eq. (12)].

The anion composition of the outside bathing solution also influences I_{sc}. The ranking order for stimulation of Na transport is $NO_3 > I > Br > Cl > SO_4$ (Huf 1972; Kristensen and Larsen 1978; Biber and Mullen 1980). With the apical membrane being impermeable to small anions (Sect. 2.4.6) the effect is probably caused by P_{Na}^{o} stimulation via anion interactions with the external aspect of the membrane, but the mechanism is unknown. Two inhibitors of the Cl conductance (in the m.r. cells, Sect. 2.5.2), furosemide (Kristensen and Larsen 1978) and MK-196 (Nagel et al. 1985) were found to stimulate the Na conductance.

3.1.2 Microscopic Properties of Sodium Permeability

The time-dependent P_{Na}^{o} inhibition in response to a stepwise increase in a_{Na}^{o} raises the question as to the underlying mechanism. The macroscopic Na permeability is the product of open-channel density and single-channel permeability. One might think of a process then by which binding of a Na ion to a modifier site causes the adjacent channel to close, leaving the permeability of remaining open channels unchanged (Fuchs et al. 1977).

The above hypothesis was tested by current fluctuation analysis of K-depolarized frog skin (Lindemann and van Driessche 1977; van Driessche and Lindemann 1979). As expected from the slowly relaxing P_{Na}^{o}, current fluctuations associated with binding/unbinding of Na ions to channels were found to occur at low frequencies (<1 Hz). A pure frequency spectrum of this process could not be obtained due to contamination by other low-frequency current components. Interaction of amiloride with the channel was found to be governed by larger on-off rate constants. The associated high-frequency fluctuations obtained with small concentrations of amiloride, could be analyzed in terms of a "pseudo" three-state channel model (Na_o-closed channel/open channel/amiloride-blocked channel). With this method the single-channel currents and channel densities could be estimated at varying a_{Na}^{o}. It was calculated that with $a_{Na}^{o} = 60$ mM and V_o near 0 mV, a single conducting unit translocates more than 10^6 ions s^{-1}. This was concluded in agreement with a porelike structure, as known mobile carriers exhibit orders of magnitude lower translocation rates (Armstrong 1975). At 0 membrane voltage, the single-channel current was found to increase linearly with a_{Na}^{o} ($a_{Na}^{o} \leq 60$ mM). The density of amiloride accessible channels was lower at higher a_{Na}^{o}. These findings meet the requirements of the "self-inhibition" process being associated with Na_o-dependent closure of Na channels. It follows that saturation of the macroscopic Na current at large a_{Na}^{o} is caused by channel closure and not saturation of single conducting units. The mechanism of Na-current stimulation by BIG and PCMBS (parachloromercuriphenyl sulfonate) was also analyzed by the method of amiloride-induced current fluctuations. The total population of electrically active Na channels, and the single-channel conductance, were affected little, or not at all, by these chemicals. It was concluded that the stimulation of the macroscopic current by these compounds, and the elimination of the a_{Na}^{o}-dependent saturation of I_{sc}, are associated with release of channels being closed by Na (Li and Lindemann 1983).

Hormonal stimulation of the Na current seems to be caused by another type of membrane effect. The addition of aldosterone to the serosal bathing solution results in an approximately 30-min delayed stimulation of the short-circuit current (Crabbé 1963; Sharp and Leaf 1964), which is caused by an increase in the Na conductance of the apical membrane (Nagel and Crabbé 1980). At the new steady state, the dependence of the macroscopic Na current on a_{Na}^{o} is not changed (Palmer et al. 1982). Amiloride-induced fluctuations of these currents in toad urinary bladder revealed that a major effect of aldosterone is an increase in the population of electrically active Na channels. The single-channel current was unaffected by aldosterone. It was suggested that the major fraction of the "dormant" pool from which electrically active channels are recruited already exists before the aldosterone treatment, and that the dependence on oxidative metabolism of the aldosterone effect is associated with activation of channel recruitment from this pool. It was not excluded that this steroid induces synthesis of new channels as well (Palmer et al. 1982). Aldosterone initiates molting in amphibian skin (Nielsen 1969; Larsen 1971a, 1972) and studies of the mechanisms of aldosterone actions, therefore, have to be carried out with other tight epithelia. The field has been reviewed by Garty (1986). Antidiuretic hormone, which stimulates active Na flux in tight epithelia (Fuhrman and Ussing 1951; Nagel 1978), was found to mediate its stimulatory effect via a similar, but much faster, recruitment of preexisting Na channels from an electrically silent pool (Li et al. 1982; Helman et al. 1983).

The mechanism of "feedback" inhibition of the macroscopic Na permeability was studied by current fluctuation analysis of K-depolarized frog skin (van Driessche and Erlij 1983). The cellular Na-activity was increased by pretreatment with ouabain, and it was found that the major effect on P^{o}_{Na} is a reduced density of amiloride-blockable Na channels. The current through remaining conducting units was not significantly changed. An ouabain-induced decrease in the number of amiloride-binding sites at the apical border of frog skin was previously reported by Cuthbert and Shum (1977). Abramcheck et al. (1985) studied systematically the effect of reducing the Na entry into frog skin on steady state Na-channel density. The Na flux was reduced with increasing concentrations of amiloride or the amiloridelike Na-channel inhibitor, CGS 4270. Independent of inhibitor, and with otherwise constant composition of the external bathing solution, significant increase in the number of electrically detectable channels was observed with reduced Na entry. They speculated that an intracellular mechanism of "autoregulation" adjusts the Na permeability of the apical membrane by recruiting conducting channels.

Since fluctuation analysis of the Na current in frog skin and other tight epithelia is based on binding/unbinding of amiloride to the apical membrane, detailed information on the molecular pharmacology of this diuretic has been obtained. Such a description has also led to the beginning of modeling the apical aspect of the Na channel. References to this literature are found in the review by Sariban-Sohraby and Benos (1986), and in recent papers by Warncke and Lindemann (1985), Hoshiko and van Driessche (1986), and Li et al. (1985, 1987).

3.2 Chloride Permeability of Mitochondria-Rich Cells

Analysis of the ionic components of the current-voltage curve showed that the Cl conductance behaves as an ideal current rectifier (Fig. 3). The proof that the Cl current flow through m.r. cells was presented in Section 2.5.2. First, a plausible mechanism of the voltage dependence of currents through these cells shall be discussed.

3.2.1 Voltage and Time Dependence of Cl Conductance

The Cl conductance is a continuous, saturating function of V over the physiological range of potentials, and is about half of its maximum value at $V = -50$ mV (see Fig. 9A). The initial experiments (Larsen and Kristensen 1978) were carried out with the skin of *B. bufo*, but a similar potential dependence characterizes the

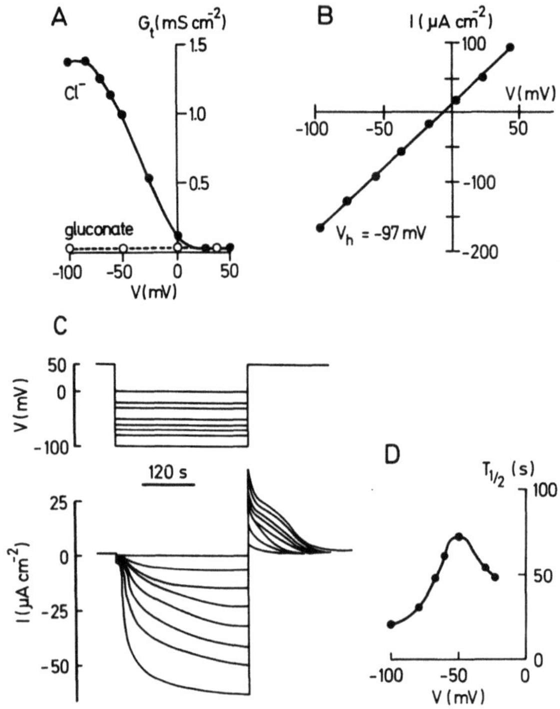

Fig. 9. A Steady state conductance-voltage (G_t-V) curve of toad skin exposed to NaCl-Ringer's on the inside, and KCl-Ringer's, or K gluconate-Ringer's on the outside, as indicated on the graphs. G_t calculated from the current response (ΔI) to a voltage pulse (ΔV) of 20 mV and 500 ms, $G_t = \Delta I/\Delta V$. In the absence of apical Na conductance, the skin conductance (m.r. cell conductance) is governed by a voltage-dependent apical Cl conductance. *B* "Near"-instantaneous current-voltage curve of toad skin of fully activated Cl conductance. NaCl-Ringer's on both sides, 60 μM amiloride in outer solution. From the holding potential of −97 mV a staircase-shaped voltage ramp of 20 mV steps, each with 50-ms duration, was applied. Notice that the "instantaneous" I-V curve is linear in contrast to the steady state I-V curve which exhibits strong outward rectification (cf. Fig. 3B). *C* Family of I-time curves in response to the V program on top. KCl-Ringer's on the outside. *D* Bell-shaped V dependence of the halftime of current activations. $T_{1/2}$ was read from the currents shown in *C* (Larsen and Rasmussen 1982)

Cl conductance of the skin of other anurans (*R. esculenta*, Kristensen 1983; *B. viridis*, Katz and Larsen 1984; *B. marinus*, Larsen et al. 1987). There is a notable variation among skins. Especially in skins of *R. esculenta* and *B. marinus* the conductance-voltage curve is displaced to the right along the V-axis resulting in large Cl conductance at 0 mV (with eliminated apical Na conductance). There is evidence that an intracellular cAMP-dependent pathway activates the Cl conductance (Cuthbert and Painter 1968; Mandel 1975) and that this metabolic pathway controls a rightward displacement of the G_{Cl}− V curve along the V-axis (Kristensen 1983; Katz and Larsen 1984). With apical Na conductance eliminated, a fast staircase-shaped voltage pulse applied across the skin of fully activated Cl conductance results in currents which depict a linear function of V (Fig. 9B). This important observation shows that the apical Cl conductance does not exhibit "instantaneous" rectification (Larsen and Kristensen 1978). Any model of the Cl conductance has to account for this observed difference in its steady state current-voltage curve (which is strongly nonlinear, Fig. 3), and its instantaneous current-voltage curve (which is linear, Fig. 9B). The model recently discussed by Biber et al. (1986) with an asymmetric distribution of fixed surface charges on the rate-limiting Cl-conducting membrane may well reproduce the observed steady-state rectification. It was not discussed, however, that this class of models essentially produces instantaneous rectification which is not met by results obtained with toad and frog skin (Larsen and Kristensen 1978; Larsen and Rasmussen 1982; Kristensen 1983).

A clue to the nature of the Cl-current rectification was obtained by studying the time course of current response to a shift in transepithelial potential. In Fig. 9C a family of current time curves of a skin exposed to a Na-free apical solution is shown. The skin was held at 50 mV and subsequently pulsed for 300 s to the series of eight potentials indicated at the top. It can be seen that current activations proceed with a sigmoidal course, and that the maximum rate of current activation depends on clamping potential. The halftime of current activation depicts a bell-shaped function of V (Fig. 9D). The V-induced conductance activation is reversible, as the return of V from pulse to holding potential results in slow current relaxations to the initial small steady state value (see Fig. 9C). Similar V-induced conductance activations are observed in skins exposed to NaCl, NaCl + amiloride, or choline-chloride in the external solution, but they are completely eliminated if the external Cl is replaced by gluconate or sulfate, the residual conductance being simple ohmic (Fig. 9A). Thus, the voltage-activated currents are controlled by a Cl ion-specific conductance. There is a notable variation among preparations. Especially the skin of *R. esculenta* exhibits a current activation course lasting for more than 10 min (Kristensen 1983). The example in Fig. 10 shows that, otherwise, these slowly developing currents are of the type generated by toad skin (cf. Fig. 9C). Likewise, the resulting steady state I – V relation of frog skin (Fig. 10) is similar to that of toad skin (Fig. 3). The basis for slow versus fast current activations are discussed in Section 3.2.2. Figure 11 shows the response of m.r. cell volume to activation of the Cl current in the absence of apical Na conductance. It can be seen that the cell undergoes reversible swelling with time courses similar to activation/deactivation of transepithelial conductance. Furthermore, cell swelling and current activation are both eliminated by replacing external Cl with gluconate (Fig. 11, right-hand panel). Accordingly, Cl ions are accumulated in the m.r. cell (together with K derived from the serosal bath) as its Cl permeability becomes activated. This finding provides strong evidence that the apical membrane Cl conductance slowly increases in response to a hyperpolarizing voltage step and, eventually, exceeds that of the serosal membrane. It was further shown that apical Cl-conductance activation is not triggered by the m.r. cell volume increase. Thus, osmotically swelling of m.r. cells in preparations clamped at 30 mV (deactivated Cl conductance) does not activate the Cl permeability. Nor is the Cl permeability deactivated by osmotically shrinking m.r. cells clamped at $V = -100$ mV (Larsen et al. 1987).

The above observations indicate that the Cl conductance is controlled by a voltage-gated permeability of a kinetic scheme similar to Hodgkin-Huxley gate-

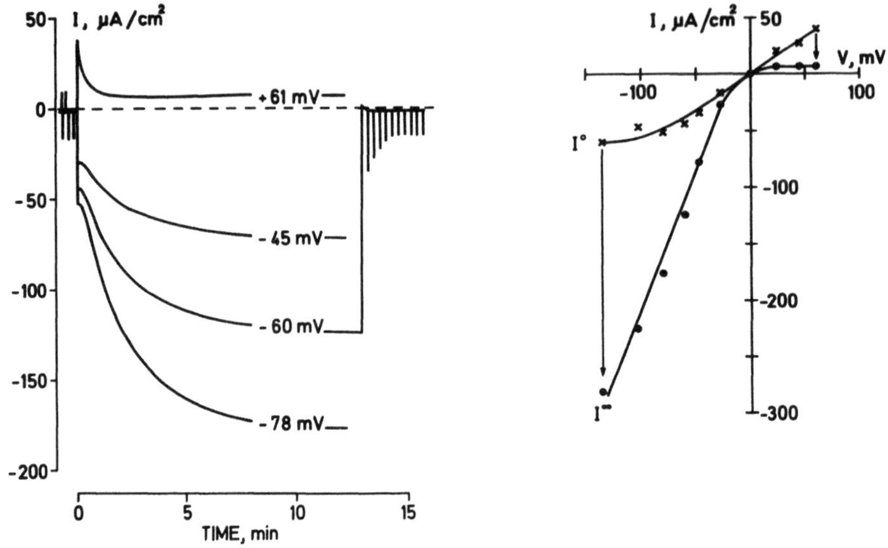

Fig. 10. Voltage-clamp currents of *R. esculenta* skin. Bilateral NaCl-Ringer's with 60 μM amiloride in apical bath. Current records in *left-hand panel* were obtained by stepping the transepithelial potential from 0 mV to the clamp voltage indicated on each record. Notice their long time constant. It is also seen that a significant conductance prevails at 0 mV, which is removed by stepping V to 61 mV. *Right hand panel*: x symbols are "zero-time" currents in response to shifting V from 0 mV holding potential; ● symbols are resultant steady state currents read 10–15 min after the V shift from 0 mV

type models. For clarifying this point, let us consider a membrane reaction of the type (Hodgkin and Huxley 1952):

$$\underset{(1-s)}{\text{channels closed}} \underset{\beta}{\overset{\alpha}{\rightleftharpoons}} \underset{(s)}{\text{channels open}} , \tag{13}$$

where α and β are the forward and backward rate coefficients which are instantaneous functions of the membrane potential, V:

$$\alpha = f(V) \; ; \tag{14}$$

$$\beta = h(V) \; ; \tag{15}$$

and s is the gating variable ($0 \leq s \leq 1$) with the Cl permeability (P_{Cl}) at any given membrane potential and time defined by:

$$P_{Cl} = P_{Cl}^{max} s^n \; ; \tag{16}$$

$$ds/dt = \alpha(1-s) - \beta s \; ; \tag{17}$$

$$s_{\infty} = \alpha/(\alpha+\beta) \; ; \tag{18}$$

P_{Cl}^{max} is the maximum Cl permeability and n is a positive integer. Equation (17) states that the transition of the gating variable from one steady state to another,

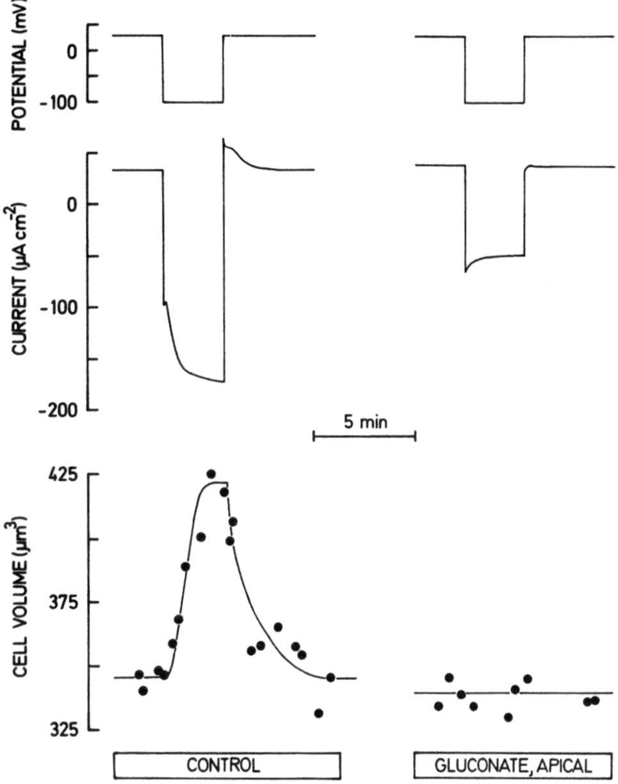

Fig. 11. Current and m.r. cell volume responses to shift in transepithelial potential (V) from 30 mV to −100 mV and back. Na-Ringer's on both sides and apical Na-conductance blocked by 50 μM amiloride in the external bath. *Left-hand panel*: Cl as major external anion. Cell volume increases reversibly with reversible activation of the Cl permeability. *Right-hand panel*: Following replacement of external Cl by gluconate, current activation and cell volume increase are eliminated (Larsen et al. 1987, unpubl.)

at constant V, is governed by first-order reaction kinetics with the time constant, $\tau = 1/(\alpha+\beta)$. Equation (18) provides the steady state value of s_∞ at a given V. For $n>1$ [Eq. (16)], the set of equations fits sigmoidal current activations rather than a monoexponential course. For the special case of a single energy barrier, Eqs. (14) and (15) take the form of Arrhenius expressions:

$$\alpha = \alpha_0 \exp(z\delta(V - V_{1/2})/(RT)) , \qquad (14A)$$

$$\beta = \beta_0 \exp(-z(1-\delta)(V - V_{1/2})/(RT)) , \qquad (15A)$$

where $V_{1/2}$ is the membrane potential for $s = 0.5$ with the associated rate coefficients, α_0 and β_0 ($\alpha_0 = \beta_0$). z is the charge of the hypothetical mobile membrane particles responding to a change in V, and δ ($0 \leq \delta \leq 1$) indicates the asymmetry of the system. As $\tau\ (=1/(\alpha+\beta)) \rightarrow 0$ for $V \rightarrow \pm\infty$, and with the maximum value of $\tau = 1/2\alpha_0$ at $V = V_{1/2}$, it is seen that τ is a bell-shaped function of V with its peak value where the permeability exhibits its steepest dependence on V. It was concluded that a mathematical model defined by the above set of equations promised to describe the experimental results well (Larsen 1982). In casu, the apical Cl permeability should be activated by membrane depolarization.

3.2.2 Quantitative Description of Transepithelial Currents

From the above analysis, a conceptually clear approach for testing the hypothesis of a voltage-gated membrane Cl permeability was indicated. However, the experiments needed for obtaining its voltage- and time-dependent states are met with difficulties. The apical membrane has to be voltage clamped for obtaining unambiguous proof that the rate coefficients are governed by the type of expressions indicated in Eqs. (14) and (15). Moreover, since the cell is small and the time-dependent conductance activation slow, the nonsteady state current flowing through the cell is expected to change its electrolyte and water content during the transition of the Cl permeability from one steady state to another. This implies that even with successfully applied one-membrane voltage clamp, the driving force of the membrane Cl current will be time-dependent due to the change in intracellular Cl concentration. As cellular Cl accumulation/depletion is associated with accumulation/depletion of cations, their membrane conductances may be expected to influence the kinetics of the nonsteady state currents as well. For more rigorous analysis of such a system, a computer model of the epithelium was constructed, and computed currents resulting from simple kinetic schemes analyzed and compared with experimental observations (Larsen and Rasmussen 1982, 1983, 1985). Examples of computed steady states and transient states of the epithelium shall be presented below. Patch-clamp analysis of the apical membrane of another NaCl-absorbing epithelium, A6 cells of the amphibian kidney, revealed voltage-dependent Cl-channel activity of a polarity similar to the proposed polarity of the apical Cl permeability of the m.r. cell (Nelson et al. 1984). This finding further substantiates the pertinence of quantitative analysis.

First, a brief, general description of the model is given. Further details, computing strategies, and procedures for finding membrane permeabilities and pump constants are found in the original papers. The recent version (Larsen and Rasmussen 1985) is based upon the model of Fig. 1 B (upper part) which assumes transepithelial ion flows through (1) a major Na-transporting compartment representing the functional syncytium of principal cells, (2) m.r. cells of specified density and apical area, (3) a single barrier representing the junctional membranes. Passive conductors are mathematically described by constant field equations, which for the ionic current, I_j, with membrane permeability, P_j, reads:

$$I_j = P_j \frac{z_j^2 F^2(\psi' - \psi'')/(RT)}{1 - \exp[z_j F(\psi' - \psi'')/(RT)]} (c_j' - c_j'' \exp[z_j F(\psi' - \psi'')/(RT)]) \; . \tag{19}$$

With the membrane potential defined as the potential of the left-side compartment (ψ') minus that of the right-side compartment (ψ'') (Fig. 1 B), inward currents become positive. The Na/K pump fluxes are represented by:

$$J_{Na}^{P} = J_{Na}^{P,max}(c_{Na}^{c}/(c_{Na}^{c} + K_{Na}^{P}))^3 \; , \tag{20}$$

$$J_K^P = -2J_{Na}^P/3 \ , \tag{21}$$

which saturate for $c_{Na}^c \gg K_{Na}^P$. The fluxes carried by the cotransport system of the principal cells are calculated by:

$$J_j^C = iK^C(c_{Na}^c x c_K^c x (c_{Cl}^c)^2 - c_{Na}^i x c_K^i x (c_{Cl}^i)^2) \ , \tag{22}$$

where $i = 1$ for j = Na and K, and $i = 2$ for j = Cl. The equations of the rate coefficients are constructed to fulfill the requirements of channel opening in the physiological range of potentials, saturation for $V < -80$ mV, and time constants of permeability activation and deactivation in the order of $10-100$ s:

$$\alpha = 0.0007 \frac{(V_o^{mrc} - 85)}{\exp((V_o^{mrc} - 85)/12) - 1} \ , \tag{14B}$$

$$\beta = 0.03 \frac{1}{\exp(-(V_o^{mrc} - 47)/6) + 1} \ , \tag{15B}$$

where the apical membrane potential of the m.r. cell, $V_o^{mrc} = \psi^o - \psi^c$. Cell volume ($v$) adjustments are allowed for by assuming the existence of fixed amounts, Q_x^c, of impermeable intracellular anions:

$$v = Q_x^c / c_x^c \tag{23}$$

with a concentration, c_x^c, governed by the assumption of isosmolarity in the serosal and cellular compartments. With requirements of mass conservation and electroneutrality in cellular compartments, the equations form a closed system.

The independent variables of the model were selected to reproduce the m.r. and principal cell components of the short-circuit current, and experimentally estimated intracellular concentrations of the two cell types (cf. Sects. 2.4 and 2.5). Their computed steady state values together with computed intracellular potentials are collected in Table 2. Steady state current-voltage curves are depicted in Fig. 12A. The active inward Cl flux is not included in the model, otherwise it produces the experimental $I-V$ relations well. In Fig. 12B the relationship between computed efflux of Cl (J_{Cl}^{out}) and transepithelial potential is shown. Notice the stimulation of Cl efflux when stepping its driving force down

Table 2. Model values of intracellular concentrations (c_j), intracellular potentials (V_c), volumes (vol), and active Na currents (I_{sc}) at V = 0 mV (open-circuit potential, $V_{sp} = -24.3$ mV)

		Mitochondria-rich cell compartments[a]	Principal cell compartment
c_{Na}	mM	12.2	11.9
c_K	mM	134.9	129.9
c_{Cl}	mM	19.5	39.6
V_c	mV	−44.7	−84.4
Vol	nl cm^{-2}	24.8	3401.0
I_{sc}	μA cm^{-2}	0.59	29.8

[a] 60,000 m.r. cells cm^{-2}; vol = 413 μm^3 per cell. Apical membrane area = 12 μm^2 per m.r. cell. Fractional membrane area of m.r. cells is 0.0072.

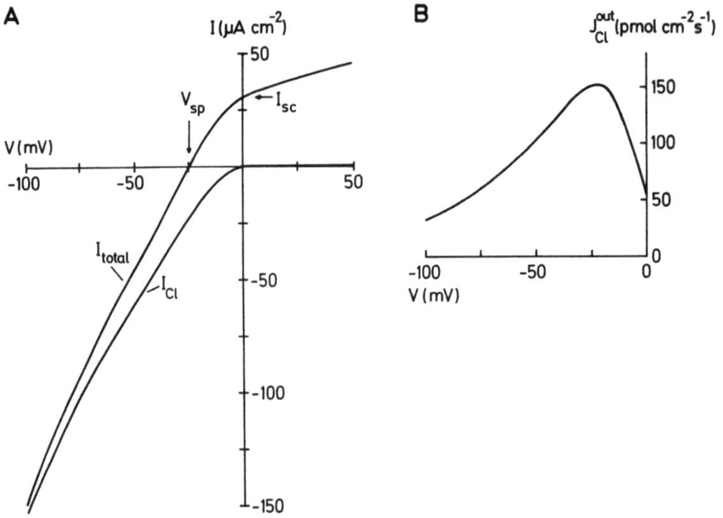

Fig. 12. *A* Computed steady state current-voltage curves of simplified model of the epithelium. I_{total} and I_{Cl} are total and Cl currents generated by the model as a function of the transepithelial potential (V). Rectification of I_{Cl}-V curves of the skin (Fig. 3) are produced by assuming that apical membrane potential of m.r. cells controls its Cl permeability [Eqs. (13 – 18)]. *B* Relationship between computed efflux of Cl (J_{Cl}^{out}) and transepithelial potential difference

from 0 to, e.g., – 50 mV. This paradoxical response has indeed been found for the in vitro preparation (Bruus et al. 1976; Willumsen and Larsen 1985; Dürr and Larsen 1986; Harck and Larsen 1986). In the model it is caused by opening of voltage-gated Cl channels in the apical membrane of the m.r. cells.

The significant apical Na permeability in m.r. cells implies that the intracellular potential under short-circuit conditions is depolarized (numerically less than the K-equilibrium potential of the serosal membrane) and the apical Cl permeability activated. In the example shown, the apical Cl permeability is about 12% of its fully activated value. Thus, conductive equilibrium fluxes of Cl flow through the m.r. cells at V = 0 mV. Elimination of the Na permeability results in elimination of I_{sc} (Fig. 13 A) and fast hyperpolarization of the principal cells, as was found in experiments with frog skin (Sect. 2.4.3). Model calculations predict, however, that the initial response of the intracellular potential of the m.r. cell is small (Fig. 13 B). This is due to the large membrane Cl conductance which "clamp" the potential to a value near the Cl-equilibrium potential. As the Na pump depletes the intracellular Na pool, the Cl ion pool is also depleted (Fig. 13 C). As a result, the cell hyperpolarizes, and the apical Cl permeability is slowly deactivated (Fig. 13 D). Eventually, at the new steady state the intracellular potential is close to the K-equilibrium potential. It follows that the transepithelial conductance of the model exhibits an initial, fast decrease caused by the instantaneous closure of apical Na channels, and a subsequent, slow decrease associated with time-dependent deactivation of the Cl permeability (see Fig. 13 C). Over the past years several of these predictions have been reported in the literature:

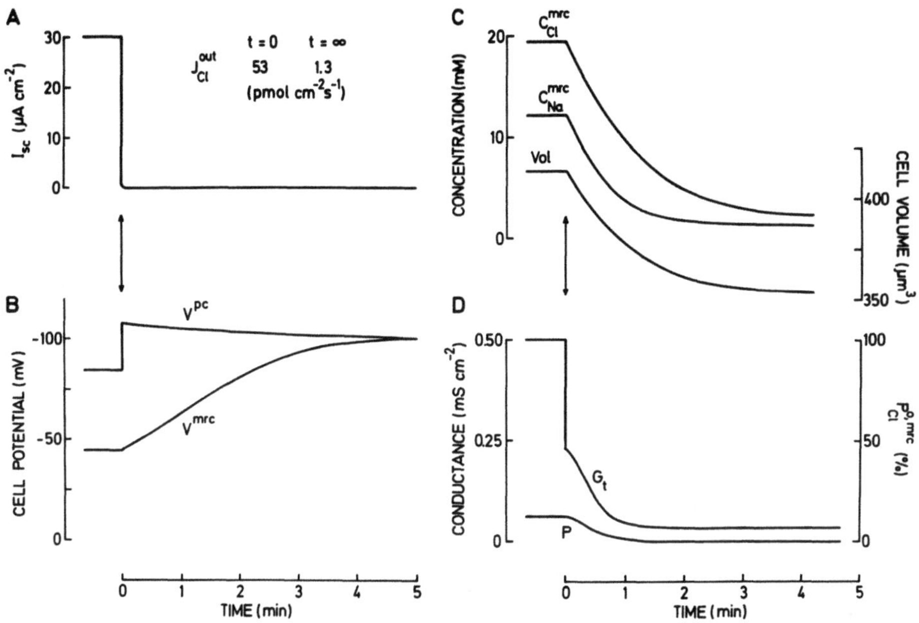

Fig. 13A–D. Computer simulation of amiloride effects on short-circuited preparation (bilateral NaCl-Ringer's, V = 0 mV). At zero time (*arrows*) apical Na permeability was set to a value near zero (1×10^{-10} cm s^{-1}). *A* The short-circuit current instantaneously drops to near zero. In the *inset* the computed Cl efflux (J_{Cl}^{out}) at steady states is shown before (t = 0) and after (t = ∞) "amiloride exposure". *B* Intracellular potential of principal cell compartment ($V^{p.c.}$) hyperpolarizes to a value above the potassium equilibrium potential, while that of m.r. cells ($V^{m.r.c}$) is "clamped" by the large Cl conductance to the value prevailing before amiloride exposure (the Cl-equilibrium potential, $V_{Cl}^{m.r.c} = -44.7$ mV at t = 0). *C* Elimination of Na influx through the apical membrane results in depletion of m.r. cell Na pool by the pump. The graphs show the associated decreases of intracellular Na and Cl concentrations and m.r. cell volume. *D* The slow decrease in $c_{Cl}^{m.r.c}$ results in slow hyperpolarization of m.r. cells which, in turn, leads to deactivation of the apical Cl permeability (P_{Cl}^{o}). Thus, the transepithelial conductance (G_t) responds instantaneously to Na-channel closure, and slowly to Cl-permeability deactivation

1. At the new steady state computed Cl equilibrium fluxes are significantly reduced (cf. Fig. 13A). Thus, the model predicts the well-established effect of eliminating the apical sodium permeability on the steady state Cl fluxes through short-circuited skin (Macey and Meyers 1963; Candida 1978; Kristensen 1978, 1982; Ques-von Petery et al. 1978).
2. The above "dual" effects of a Na-channel blockade on the transepithelial conductance have been observed in experiments with frog skin following apical exposure to amiloride (Kristensen 1983; Nagel et al. 1983; Erlij and Schoen 1983).
3. Associated with the depletion of intracellular Na and Cl ion pools, the m.r. cell volume decreases (Fig. 13C). The time course and magnitude of this volume response fit experimental observations well (Fig. 7A; Larsen et al. 1987).

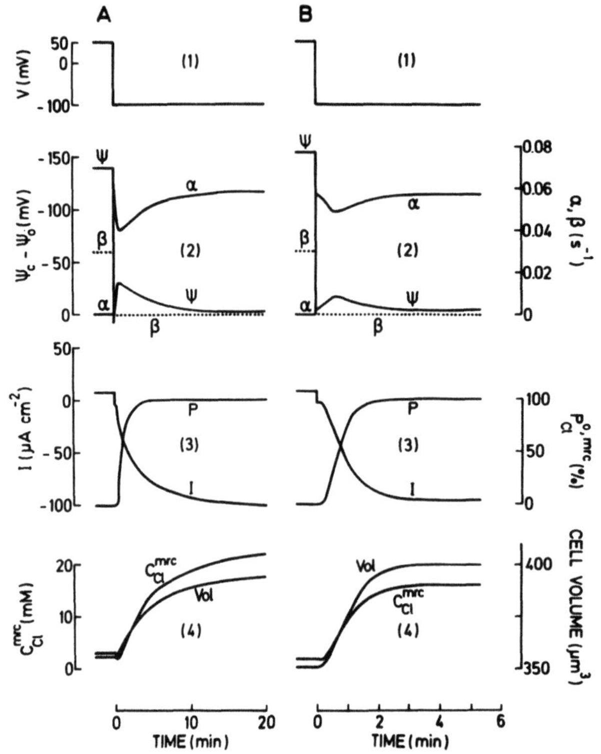

Fig. 14A, B. Computed transepithelial currents generated by the voltage-clamp program indicated in (*1*). $c^o_{Na} = 0$ mM. The *A* panel is computed for the K permeability of the serosal membrane of m.r. cell = 1.3×10^{-4} cm s^{-1}. In the *B* panel the K permeability was set to a tenfold larger value. (*2*) Time course of apical membrane potential and rate coefficients following a step of V from 50 to −100 mV. (*3*) Time course of apical Cl-permeability activation and transepithelial current. (*4*) Time course of increase in Cl concentration of the m.r. cells and the associated cell volume response

It can be concluded that, in principle, we understand the effects of amiloride on transepithelial fluxes and conductances without having to assume that this diuretic inhibits Cl channels directly, nor do we need to consider amiloride-induced ion redistributions between discrete lateral interspace permeability barriers as hypothesized by Essig (1982).

Figure 14 shows examples of computed voltage clamp-induced transient states with a Na-free apical compartment (K substitution). As compared with previous model analyses (Larsen and Rasmussen 1982, 1983) the new aspect of these computations is that the model now incorporates a realistic ratio of the inner membrane Cl permeability and the fully activated apical Cl permeability as revealed by the close resemblance between computed (Fig. 14) and observed (Fig. 11) volume response to voltage clamping. Thus, these computations cover what can be assumed to be realistic cellular KCl accumulation during activation of the apical Cl conductance. The two examples are computed with small and high K permeability of the serosal membrane of the m.r. cells. For both cases at the holding potential of 50 mV, the apical membrane potential is large, and α zero, i.e. the apical Cl permeability is deactivated. Following shift of V to −100 mV [Fig. 14(1)], α and β instantaneously take up their new values according to the new apical potential [near zero, see Fig. 14(2)]. In turn, this results in delayed P^o_{Cl} activation [Fig. 14(3)]. As the apical conductance is increased the voltage divider ratio decreases and the apical membrane hyperpolarizes. This effect is more pronounced when the serosal membrane K conductance is small. Thereby, α decreases, but only transiently; as P_{Cl}^{o} is further activated and KCl accumulates in the cell, the apical potential depolarizes and α increases accordingly. Thus, for both cases the Cl permeability activation is governed by a *time-variant* rate coefficient. Following activation of the apical Cl permeability, Cl ions enter the m.r. cells from the apical compartment together with K derived from the serosal compartment. Thereby, m.r. cell volume increases [Fig. 14(4)]. Independent of serosal K conductance, current activation and volume increase proceed with a similar time course, as was found in experiments with toad skin (Larsen et al. 1987). However, only if the serosal membrane K conductance is not rate-limiting for cellular accumulation of KCl, then the time course of the current transient is similar to that of Cl-permeability activation. This result offers a simple explanation for the slow versus fast current activations observed among different preparations exhibiting similar steady state current-voltage relations (cf. Figs. 9C and 10; Larsen et al. 1987).

The general conclusion to be made from the examples shown in Figs. 12–14 is that the model, supposedly oversimplified in many respects, reproduces so many experimental observations that it may be regarded as a useful framework for extension, rather than being rejected. It is of interest then that a more detailed analysis has shown that the assumption of rate coefficients being independent of the prehistory of the membrane [Eqs. (14B), (15B)] is probably wrong. Thus, the epithelium exhibits a significant current delay when activated from a potential at which the Cl conductance is already partially activated. It is an essential property of the HH equations that they predict the disappearance of the delay under such conditions (Larsen and Rasmussen 1982). It is not very interesting to construct better fitting kinetic schemes as long as the time-dependent conductance changes of the apical and serosal membrane of the m.r. cell are unknown. The degrees of freedom are too large, and experimental testing of alternative descriptions is, as yet, not possible. It is emphasized that a well-fitting mathematical model does not imply any particular physical mechanism of the membrane process.

3.2.3 Concentration Dependence and Selectivity

The early studies by Linderholm (1954) indicated that the conductance of frog skin exhibits a stronger dependence on external Cl concentration than on external Na concentration. Koefoed-Johnsen and Ussing (1958) took advantage of this property of the "shunt" in their study of the K permselectivity of the inward facing membrane. The in vivo studies with frogs by Kirschner (1970) showed that also the Nernstian c_{Na}^{o} dependence of the skin potential is obtained with outside NaCl concentrations below 10 mM, but not at larger Na concentrations unless the external Cl is replaced by SO_4. He forwarded the hypothesis that a passive Cl permeability of the skin is activated by raising the external Cl concentration. Subsequent in vitro studies of Cl fluxes in the short-circuited skin confirmed that a conductive Cl permeability is activated by raising c_{Cl}^{o} (Mandel and Curran 1972; Ques-von Petery et al. 1978; Kristensen 1978, 1983). Kristensen (1982) showed that bromide as well, but not iodide and SCN, can activate the Cl conductance, whereas all of these ions pass the Cl_o-activated conductance according to SCN:Br:Cl:I = 1.7:1.3:1:0.8, He concluded that regulation of the Cl permeability occurs at a Cl (and Br) ion-specific site remote from a poor anion-selective translocation site of the channel.

In the short-circuited toad skin (and in skin of *R. pipiens*, Biber et al. 1980; Drewnowska and Biber 1985) the Cl influx contains large Cl self-exchange components and a minor conductive component through the cellular conductance. There is evidence that the paracellular Cl flux is very small and masked by these two cellular components (Bruus et al. 1976; Kristensen 1978; Harck and Larsen 1986; Sect. 3.7). Therefore, investigation of the c^o_{Cl} dependence of the conductive Cl influx through m.r. cells can be carried out with the hyperpolarized skin in which exchange fluxes are absent and the Cl conductance is brought to full activation (Figs. 4 and 9A). In toad skin clamped at −80 mV, the increase in c^o_{Cl} produces a sigmoidal increase in the conductive Cl influx which tends to saturate at c^o_{Cl}>100 mM (see Fig. 15A) (Harck and Larsen 1986). Iodide influx was

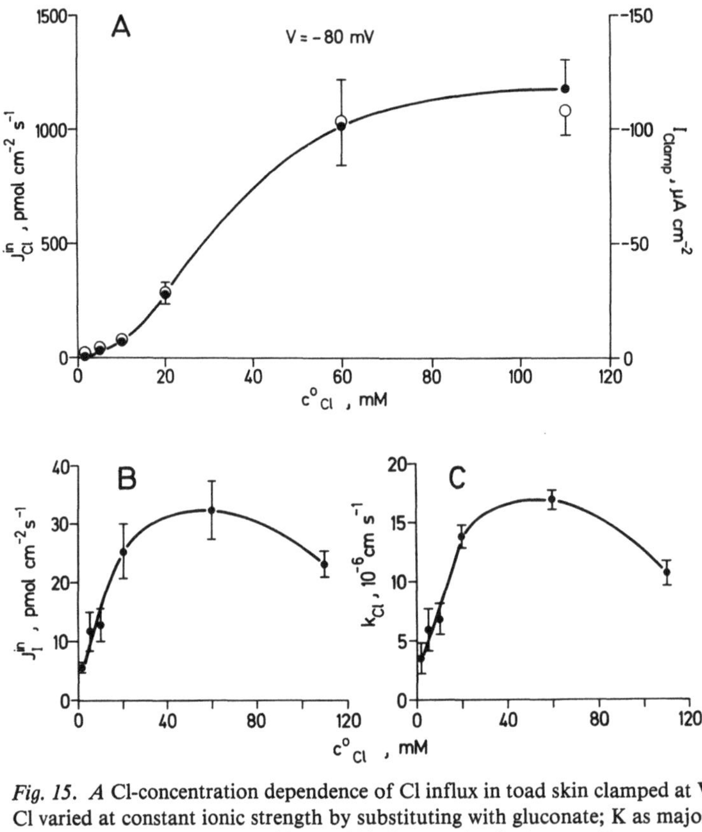

Fig. 15. A Cl-concentration dependence of Cl influx in toad skin clamped at V = −80 mV. External Cl varied at constant ionic strength by substituting with gluconate; K as major cation in the external bath; NaCl-Ringer's as serosal bath. Cl influx carries the (outward) current and both are stimulated along a sigmoidal curve by raising c^o_{Cl}. *B* Simultaneously measured iodide influx reveals activation of an iodide pathway as c^o_{Cl} is increased. *C* Rate coefficient of Cl influx as a function of external Cl concentration. Note similar dependence of k_{Cl} and iodide influx on c^o_{Cl} (Harck and Larsen 1986)

measured as well, but the external concentration of this ion was kept constant at 3 mM. The results showed that the pathway taken by this anion is activated with increasing c^o_{Cl}, and that the $J^{in}_I - c^o_{Cl}$ relationship is similar to the c^o_{Cl} dependence of the rate coefficient for Cl influx (Fig. 15B, C). It was further shown that the iodide flux ratio exhibits a V dependence similar to the Cl flux ratio (cf. Fig. 4), and that both iodide and chloride influxes are inhibited by furosemide added to the apical solution. Thus, these ions pass through the same V-dependent, furosemide inhibitable anion conductance which, at constant V, is activated by increasing the external Cl concentration. The selectivity of the pathway was found to be independent of the mole fraction ($0.48 \leq c^o_{Cl}/c^o_I \leq 37$) with a halide sequence of conductive ion translocations, Br : Cl : I = 1.3 : 1 : 0.7, that is, identical to the conductive anion selectivity of short-circuited frog skin.

External bromide activates, whereas external iodide inhibits the anion conductance, indicating an inhibitory modifier site with high iodide affinity. It was speculated that self-inhibition causes the Cl influx to increase asymptotically to a maximum value at large external Cl concentrations (Fig. 15A), but channel saturation could not be excluded (Harck and Larsen 1986).

3.2.4 Cl : Cl Exchange Diffusion and Active Transport

Flux ratio analysis of steady-state unidirectional Cl fluxes revealed three mechanisms of Cl transport through the skin: electrodiffusion, self-exchange diffusion, and active transport (Fig. 4). A useful framework for interpretation of their physiological significance was suggested from the following three types of findings (Kristensen and Larsen 1978). Firstly, the electrodiffusive Cl flux and Cl-self exchange flux (characterizing the conductive and nonconductive state, respectively) are positively correlated: Skins exhibiting large conductive Cl fluxes in the hyperpolarizing region of the current-voltage curve also show large self-exchange fluxes under short-circuit conditions, and vice versa. Secondly, inhibitors (furosemide, phloretin, diamox, SCN) of the conductive Cl flux were found also to inhibit the Cl self-exchange flux. Thirdly, of particular interest, the carbonic anhydrase inhibitor, diamox, inhibits the exchange and active flux of Cl. As a working hypothesis it was suggested that the different mechanisms of Cl transport represent two states of the same transport system (Kristensen and Larsen 1978; Harck and Larsen 1986):

1. In one configuration, the system facilitates Cl transport driven by the electrical potential difference. In toad skin this configuration is expressed in the physiological region of potentials. With the driving force of the Cl flux being in the inward direction, passive Cl uptake is insured.
2. In the other configuration, electroneutral transport takes place. This configuration is expressed for $V \geq 0$ mV, that is, the passive Cl net flux is zero when its driving force is outwardly directed. The exchange partners may be a halide : halide ion pair, or they may be a heterogeneous pair of the type, halide : HCO_3, whereby active inward transport of the halide ion takes place.

The chloride transport system in red cell membrane exhibits nonconductive/conductive transport modes as well as self-exchange/heterogeneous exchange modes (Tosteson 1959; Vestergaard-Bogind and Lassen 1974; Knauf et al. 1977; Wieth 1979; Wieth and Brahm 1985). Moreover, self-inhibition by binding of anions to an inhibitory modifier site with high iodide affinity was described for this system (Dalmark 1976; Wieth and Bjerrum 1982). It was suggested, therefore, that the anion transport system of amphibian skin is related to the anion exchanger in red cell (Kristensen and Larsen 1978; Harck and Larsen 1986). DIDS which is a potent inhibitor of Cl transport in red cell also inhibits the Cl conductance of short-circuited toad skin (*B. marinus*, Beauwens et al. 1986).

When frogs and toads are exposed to a diluted external NaCl concentration, Cl is taken up by active transport (Jørgensen et al. 1954). Bicarbonate excretion associated with this uptake from a Na-free external solution was first noted by Krogh (1937). In a careful study of net fluxes it was indicated that Cl and HCO_3 are exchanged stoichiometrically when the frog is in a Na-free solution of low Cl concentration (Garcia-Romeu et al. 1969). It was speculated that bicarbonate is derived from blood. But as we now know that a significant carbonic anhydrase activity is localized in m.r. cells (Rosen and Friedley 1973; Voûte et al. 1975), it is conceivable that these cells contribute to the production of bicarbonate ions exchanging for Cl (Erlij and Ussing 1978). The localization of the carbonic anhydrase in the "neck" of the m.r. cell (Zaccone et al. 1986) may also point to a role, primarily, for apical membrane transport (see Fig. 1 B, lower panel).

There is no experimental evidence for apical active Na transport from solutions of low NaCl concentration (Harvey and Kernan 1984a). The amiloride blockable Na channel facilitates electrodiffusion of Na (Fuchs et al. 1977; Benos et al. 1983; Schoen and Erlij 1985b) and with an electromotive force of 200 mV, the Na/K ATPase in the inward facing membrane has the capacity for energizing a transcellular uptake of Na from a 0.01 mM apical solution (Eskesen and Ussing 1985).

However, we are left with the question of how active Cl uptake is energized. Since the ouside solution was slightly acidified when Na was taken up from low concentrations of its salt, proton excretion, active Na uptake, and Cl/HCO_3 exchange by frog skin was discussed by Ussing (1949a) as possible mechanisms in vivo. Further evidence for H transport by the skin in vivo was published by Garcia-Romeu et al. (1969) and Kirschner et al. (1973) who found linear relationships between H efflux and net Na influx with a high correlation coefficient. However, these fluxes do not seem to be associated with a forced exchange system, as they can be dissociated from one another. For example, Diamox abolishes H excretion, while active Na uptake is unaffected (Emilio et al. 1970; Ehrenfeld and Garcia-Romeu 1977), and short-circuiting the skin stimulates Na flux and reduces proton secretion (Ehrenfeld and Garcia-Romeu 1977). Erlij and Ussing (1978) summarized the findings by suggesting that a proton pump is localized in the apical membrane of m.r. cells (Fig. 1 B, lower panel). According to this configuration of apical transport systems, it is the proton ATPase which drives the active uptake of Cl by raising the intracellular chemical potential of HCO_3. As proton excretion is postulated to be associated with active uptake of Cl, the model predicts cellular alkalization in response to stimulation of active Cl transport. Apical alkalization in response to Cl uptake from a Na-free solution under open-circuit conditions is also expexted and observed (Garcia-Romeu et al. 1969), as the protons under these conditions have to leave the cells via the serosal membrane in order to maintain electroneutral, transepithelial ion transport. As yet, there is no direct evidence for a proton ATPase in the m.r. cells of the skin. However, proton pumps are present in m.r. cells of urinary bladder (Gluck et al. 1982; Steinmetz and Andersen 1982), contributing to urine acidification (Ludens and Fanestil 1972; Frazier 1974). A proton ATPase has been localized in a similar cell type of kidney collecting tubule (Gluck and Al-Awqati 1984; Schwartz and Al-Awqati 1985; Madsen and Tisher 1986).

3.3 The Sodium/Potassium Pump

The Na/K pump in the serosal membrane constitutes the primary active transport system for establishing transmembrane ionic gradients which, in turn, drive ions through passive, co- and countertransport systems. Thereby, the pump participates in the uptake of Na and Cl (Koefoed-Johnsen and Ussing 1958), in regulation of cell volume (Ussing 1982a) and maintenance of intracellular concentrations of Ca (Taylor and Windhager 1979) and protons (Cragoe et al. 1987; Ehrenfeld and Harvey 1987).

3.3.1 Na/ATP Stoichiometry of the Pump

It is generally assumed that 1 mol ATP is used when 3 mol Na is pumped. Zerahn's (1958) studies led to this hypothesis, and experiments with toad urinary bladder suggested a similar ratio (for discussion of this literature, see Canessa et al. 1978). The implication of 3 Na ions transported per cycle was tested in experiments with K-depolarized frog skin (Larsen et al. 1979). Intracellular Na activity was varied by exposing the external side of the skin to Na activities between 2.5 and 60 mM and Na entry was stimulated by 1 mM BIG in apical solution. Still, the apical membrane behaves as an ideal sodium electrode (Fig. 16B) permitting calculation of a^{c}_{Na} from the reversal potential of amiloride blockable currents (cf. Fig. 8). Steady-state short-circuit current was taken as a measure of Na-pump flux. A distinctive S-shaped relationship between a^{c}_{Na} and Na-pump flux was found (Fig. 16C, D). Following Garay and Garrahan (1973), if cellular Na is in equilibrium with n Na-pump sites of identical Na affinity, and assuming that a pump cycle is initiated only if all n sites are occupied by Na, the Na flux carried by the pump (J^{P}_{Na}) obeys an equation of the form:

$$J^{P}_{Na} = J^{P,max}_{Na}/[1 + K^{P}_{Na}/a^{c}_{Na}]^{n} \ , \tag{24}$$

where $J^{P,max}_{Na}$ is the flux at nonlimiting intracellular Na activity, and K^{P}_{Na} is the apparent dissociation constant of the Na-site complex. It can be seen that Eq. (24) provides an excellent fit to experimental points for n = 3 (Fig. 16C, D).

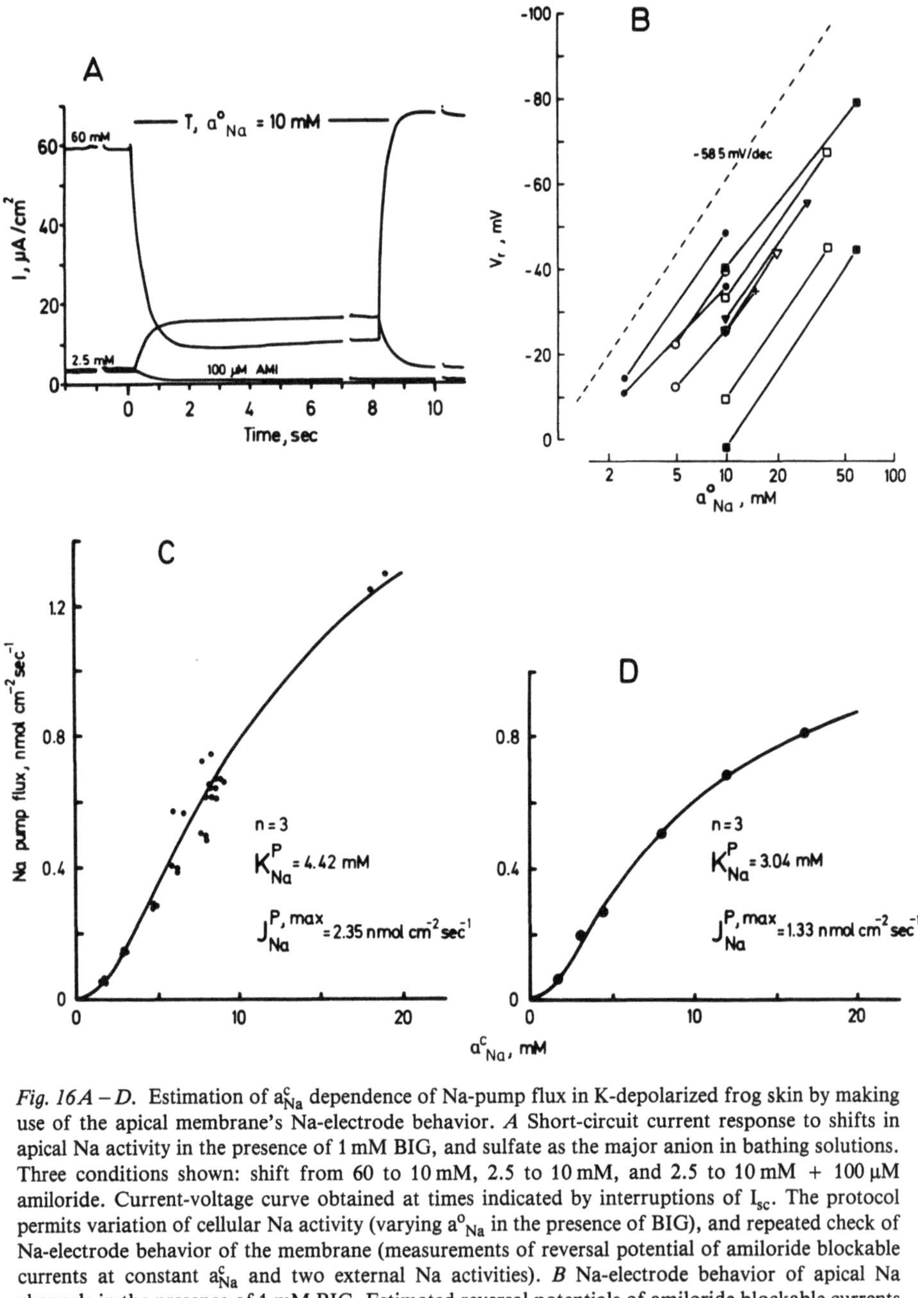

Fig. 16A – D. Estimation of a^c_{Na} dependence of Na-pump flux in K-depolarized frog skin by making use of the apical membrane's Na-electrode behavior. *A* Short-circuit current response to shifts in apical Na activity in the presence of 1 mM BIG, and sulfate as the major anion in bathing solutions. Three conditions shown: shift from 60 to 10 mM, 2.5 to 10 mM, and 2.5 to 10 mM + 100 μM amiloride. Current-voltage curve obtained at times indicated by interruptions of I_{sc}. The protocol permits variation of cellular Na activity (varying a^o_{Na} in the presence of BIG), and repeated check of Na-electrode behavior of the membrane (measurements of reversal potential of amiloride blockable currents at constant a^c_{Na} and two external Na activities). *B* Na-electrode behavior of apical Na channels in the presence of 1 mM BIG. Estimated reversal potentials of amiloride blockable currents obtained at two external Na activities, however, for each preparation, at constant a^c_{Na}. Nernst slope of ideal electrode indicated by *dashed line. C, D* Representative examples of relationship between cellular Na activity and Na-pump flux (calculated from steady-state I_{sc} in pre-equilibration solution). Full graphs are best fit of Eq. (24) with n = 3 to experimental points. For eight preparations, the maximum Na-pump flux was $J^{P,max}_{Na} = 1.95 \pm 0.37$ nmol cm^{-2} s^{-1}, and $K^P_{Na} = 3.4 \pm 0.3$ mM (Larsen et al. 1979, unpubl.)

It can be concluded, in agreement with red cell (Garay and Garrahan 1973), that the Na pump in frog skin behaves as a unit with three noninteracting binding sites of identical affinity which all have to bind Na for ion translocation to occur. Nielsen (1982) measured the sodium transport pool from tracer-Na uptake kinetics and found a quantitatively similar sigmoidal relationship [Eq. (24), n = 3] between steady-state a^c_{Na} and short-circuit current when he varied transport rate with amiloride, antidiuretic hormone, and theophylline. The microprobe technique (Rick et al. 1984) and ion selective microelectrode technique (Garcia-Diaz et al. 1986) for measurement of intracellular Na revealed a positive correlation between active sodium flux and a^c_{Na} of individual preparations. However, for each preparation there were too few data pairs, or the a^c_{Na} range was too narrow, for evaluating quantitative relationships.

3.3.2 Na/K Stoichiometry of the Pump

The Na pump in frog skin operates in a rheogenic fashion (Nielsen 1979a; Nagel 1980; Garcia-Diaz et al. 1985; Sect. 2.4.3), but attempts to establish its Na/K stoichiometry have led to different conclusions. In isolated epithelium serosal K activates tracer-Na extrusion from the cell along a sigmoidal relationship. Na extrusion was ouabain inhibitable showing its association with the pump, and the sigmoidal relationship was well fitted by a kinetic model assuming two external K-binding sites on the Na pump (Aceves 1977). Garcia-Diaz et al. (1985) estimated the passive K current (I_K) across the serosal membrane in short-circuited frog skin from:

$$I_K = G_K(V^c_K - V^c) \; . \qquad (25)$$

The K-equilibrium potential (V^c_K) and intracellular potential (V^c) were measured with microelectrodes. The K conductance (G_K) of the serosal membrane was tentatively assumed equal to serosal membrane conductance. For the range of short-circuit currents (I_{sc}) between 5 and 70 $\mu A \; cm^{-2}$, ($I_{sc} - I_K$) was linearly dependent on I_K with a slope of 0.31 (r = 0.96), compatible with the notion that three Na ions are exchanged for two K ions per pump cycle. Nielsen (1979b) arrived at a similar conclusion by showing that the short-circuit current transiently decreased to about 33% of its steady state value following the blocking of the K conductance of the serosal membrane by Ba. As discussed by Nielsen (1985), this type of experiment is expected to provide a correct estimate of the Na/K ratio of the pump, if the Cl permeability of the serosal membrane is low. This condition is very likely fulfilled in short-circuited skins before, and shortly after, serosal K-channel blockade (Sect. 3.5). The above results are consistent with a Na/K stoichiometry of 3/2, which, furthermore, is independent of pump activity (Garcia-Diaz et al. 1985). Cox and Helman (1986a) measured serosal ^{42}K influx into isolated epithelium of frog skin. With Na_2SO_4-Ringer's as bathing solutions, 94% of the K influx was inhibited by ouabain, indicating insignificant components through pathways in parallel with the pumps. In preparations with short-circuit currents about 10 $\mu A \; cm^{-2}$, a Na/K coupling ratio near 3/2 was obtained. In epithelia generating larger Na currents, the estimated ratio was larger and in the most active preparation ($I_{Na} = 70\ \mu A\ cm^{-2}$) an apparent pump stoichiometry of Na/K ≈ 6 was calculated. If it is postulated that the Na/K stoichiometry is fixed and equal to 3/2, it follows that serosal ^{42}K does not have access to all pumps in very active skins. As yet, there is no direct proof of this notion.

3.4 Passive Potassium Permeabilities

The major passive K permeability resides in the inward facing membrane shunting inwardly pumped K back into serosal solution. Tracer-kinetic measurements of this K permeability are impeded by the fact that the inward facing membrane lines maze of lateral intercellular spaces of low K concentration (Fig. 1 A). Thus, tracer K leaving the cellular compartment is taken up by cells again. Kinetics of tracer fluxes between multicellular structures and bathing solution are discussed by Harris and Burn (1949).

3.4.1 Potassium Permeability of the Serosal Membrane

Some qualitative properties of serosal K pathways have been elucidate from ^{42}K efflux studies of isolated epithelia. The efflux occurs via a pathway which can be blocked by Ba (Nielsen 1985). Ba ions also block a major conductance of the serosal membrane (Nagel 1979). The K efflux is reduced in the presence of amiloride in the apical solution, probably caused by membrane hyperpolarization (Ferreira 1979). The addition of ouabain results in stimulation of K efflux and as this is associated with membrane depolarization, the ouabain-induced effect may be caused by this increase in driving force of the K flux through the conductive pathway (Cox and Helman 1986b). Reduced recycling following pump inhibition has also been discussed to cause ouabain-induced ^{42}K-efflux stimulation (Ferreira 1979).

With Cl-Ringer's containing ouabain on the serosal side, K efflux was reduced by furosemide added to the inner bathing solution. Also, basolateral Cl efflux was inhibited by furosemide in such preparations. Under these conditions the ions seem translocated via an electroneutral mechanism (Cox and Helman 1986b) which also carries Na (Cox and Helman 1983, 1986c). This nonconductive K pathway is most likely the Na/2Cl/K cotransporter (Ussing 1985; Dörge et al. 1985) discussed in Section 2.4.2.

With SO_4-Ringer's bathing the serosal side of ouabain-treated preparations K-flux ratio analysis revealed a flux-ratio exponent of 2.9. It was concluded that the electrodiffusive K-flux components are mediated via single-file channels (Cox and Helman 1986b). Eskesen and Ussing (1986) arrived at a similar conclusion from a study of presteady state K-fluxes across whole epithelia of significant apical K permeability and eliminated pump activity (Na-free apical bath). This method permits separation of a cellular pathway with a time-independent flux ratio from the paracellular leak (Sect. 2.3.1). It was found that the transcellular pathway obeys the flux-ratio equation with a flux-ratio exponent of 2–3. This result was obtained irrespective of anion composition and K concentration of bathing solutions. Potassium net flux was always in the direction of decreasing electrochemical potential, supporting the assumption of electrodiffusive K transport in the preparations.

3.4.2 Potassium Permeability of the Apical Membrane

Generally, the K permeability of the outward facing membrane is vanishingly low. However, at low K concentrations of both bathing solutions inward active K transport was found under open-circuit conditions (Levi and Ussing 1949). Recently, a saturating K conductance of the apical membrane has been disclosed in *R. temporaria* which in the presence of high external K concentration may carry a significant inward short-circuit current (Zeiske and van Driessche 1979). In such skins the apical membrane is depolarized by raising the external K concentration (Nagel and Hirschmann 1980), and at low external K concentration I_{sc} contains a significant outward K current (Nielsen 1984). The apical K conductance is stimulated significantly if the frog is kept in 15 mM-K solutions for a few days (van Driessche 1984), and frogs loaded with K by injection excrete K via the skin (Frazier and Vanatta 1981). These findings point to a role of amphibian skin in maintaining K balance of the animal. It is not known whether one and the same K pathway carries the above K fluxes, or whether more than one pathway participates in the exchange of this ion across the apical border of the skin. A barium and cesium blockable apical K conductance has been studied in details by current fluctuation analysis. This channel is also permeable to Rb, NH_4, and Tl. The cation selectivity varies with the external mole fraction of the permeating ions, permitting the conclusion that the channel holds more than one ion at a time (Zeiske and van Driessche 1983). It is conceivable then that it is identical with the single-file K channel in the apical membrane described by Eskesen and Ussing (1986). In the review by van Driessche and Zeiske (1985) K-conductive pathways in frog skin are compared with those of other epithelia.

3.5 Chloride Permeability of Principal Cells

Absence of Cl permeability of the apical membrane of principal cells excludes this compartment as a pathway for transepithelial Cl fluxes (Sect. 2.4.6). Besides the Na-dependent cotransport system, two other mechanisms, both located in the serosal membrane, have been studied.

3.5.1 Chloride Conductance of the Serosal Membrane

With a Cl concentration in principal cells between 25 and 50 mM and a serosal Cl concentration of 115 mM, the potential at which the Cl ions are distributed at thermodynamic equilibrium across the inward facing membrane is -21 mM $\geq V^c \geq -39$ mV. In the short-circuited skin the intracellular potential is about -70 mV, and if apical Na conductance is reduced, V^c is instantaneously driven toward the K-equilibrium potential near -100 mV, while a_{Cl}^{c} remains constant over a period of 30 min (Sects. 2.4.3 and 2.4.6). It follows that the Cl conductance of the serosal membrane contributes very little to the total conductance of the membrane. The insignificant K loss and volume decrease in response to exposure of the serosal side to Na-free arginine-Ringer's, or furosemide, also indicate that the inner membrane normally has a small leakage permeability to Cl (Ussing 1982a). However, the Cl conductance may be activated by osmotically swelling the cells, whereby Cl (and K) are lost, and cell volume returns toward the control value. Ussing (1986) showed that the Cl conductance is also activated by raising the serosal K concentration, or by inhibiting the K conductance by Ba. These results strongly indicate that membrane depolarization activates the Cl conductance. In its activated state the Cl pathway is also permeable to other small anions, like SCN, Br, I, and NO_3, and it can be blocked by MK-196. With the patch-clamp technique, poor anion-selective, DIDS inhibitable Cl channels, which are activated by membrane depolarization, were found in the serosal membrane of another tight NaCl absorbing epithelium (rabbit urinary bladder, Hanrahan et al. 1985). Thus, the Cl conductance of the serosal membrane of tight epithelia has many properties in common with the apical Cl conductance of the m.r. cell (cf. Sect. 3.2).

3.5.2 Electroneutral Chloride Transports Across the Serosal Membrane

Stoddard et al. (1985) studied tracer-Cl fluxes across the serosal membrane in frog skin under conditions where the conductive Cl transport, according to Ussings investigations, should be low. It was concluded that the membrane mediates large, but electrically silent Cl fluxes via a SITS inhibitable exchange system. In studies of intracellular pH regulation it was also suggested that Cl is transported across the serosal membrane by an electroneutral mechanism. Cell pH (pH_c) was measured by serosal impalements of isolated epithelia with a liquid-resin pH-sensitive and reference microelectrode (Duffey et al. 1986; Duranti et al. 1986) or by ^{31}P-nuclear magnetic resonance (NMR) spectrometry of whole skins (Civan and Peterson-Yantorno 1986). Independent of method, principal cells were alkalized by removal of Cl from the external bathing solutions with pH_c being $0.1-0.2$ units higher in preparations exposed to Cl-free serosal solution (gluconate substitution, microelectrode measurements) or to bilateral Cl-free solutions (sulfate substitution, NMR spectroscopy). The effect was abolished by addition of SITS, suggesting Cl/HCO_3 transport in the serosal membrane. The addition of SITS alone had variable effects among the preparations, with pH_c either slightly increasing or slightly decreasing, which was assumed to reflect uneven trapping of metabolically produced protons and bicarbonate ions by the cells (Civan and Peterson-Yantorno 1986). According to these authors, cellular alkalization in Cl-free media, therefore, is not caused by inhibition of the exchanger but is due to reversal of the net HCO_3 flux driven by the outwardly directed Cl gradient. Duranti et al. (1986) showed that acidification of the mucosal Cl-free bath is correlated with alkalization of the serosal bath under control conditions. This was taken to indicate that the alkalization of the serosal bath can be monitored by following pH changes of the mucosal bath. With this method they found that serosal application of DIDS or meclofenamate inhibits proton excretion, whereas bumetanide and depolarization of the basolateral membrane with Ba have no effect on mucosal acidification. Taken together with the demonstrated cellular alkalization in response to Cl-free serosal solution, they suggested that an electroneutral $Cl:HCO_3$ exchange transport system is localized in the basolateral membrane. It is conceivable that it is a common, reversible, SITS- (and DIDS-)sensitive anion exchange system which cycles in heterogeneous ($Cl:HCO_3$) and homogeneous (Cl: Cl) modes.

3.6 Proton Pathways in Principal Cells

Intracellular pH regulation is also associated with proton transport across the apical and serosal membranes. Net proton excretion by frog skin exposed on the outside to 1 mM-Na_2SO_4 under open-circuit conditions is 88 ± 23 pmol $cm^{-2} s^{-1}$ (Ehrenfeld et al. 1985). Total oxygen consumption of frog skin under similar conditions is in the order of 100 pmol $cm^{-2} s^{-1}$ (Zerahn 1958), indicating the maximum flux of metabolically produced protons. It is strongly indicated then that the principal cell compartment contributes significantly to the above high rate of proton excretion. With an apical membrane potential of, on the average, -47 mV, and a mucosal and cell pH of 7.34 and 7.2, respectively, it was proven that exit of protons across the apical membrane is due to active transport against an electrochemical potential difference of 39 mV (Ehrenfeld et al. 1985). Proton excretion could be dissociated from Na uptake and was found to be reduced by mucosal acidification, deoxygenation, or apical hyperpolarization with a reversal potential of about -130 mV (cell negative). It was concluded that a rheogenic proton pump contributing to the membrane potential is localized in the apical membrane. Ehrenfeld et al. (1985) and Harvey and Ehrenfeld (1986) speculated that membrane hyperpolarization caused by proton-pump activity contributes to driving Na across the apical membrane from diluted external solutions. However, at steady state, the serosal membrane potential has to be numerically less than V_K^s to insure passive K net flux from the cell into the intercellular space [Eq. (25)]. Thus, at known polarity of frog skin, it is impossible for the cell to make use of a V_H of -130 mV, which at a steady state and $V = 0$ mV, requires $c_K^c \geq 2.4 \exp(-F(-0.13)/(8.31 \cdot 295)) = 400$ mM. Alternatively, and this is not definitively excluded, the transepithelial potential difference should develop mucosa-positive polarity. It should not be overlooked then that such a regulation of apical Na uptake imposes a larger load on the m.r. cell pathway for active transepithelial Cl uptake.

In two preliminary reports it was concluded that metabolically produced protons also leave the cell across the serosal membrane via an electroneutral amiloride-sensitive Na/H exchange system (Cragoe et al. 1987; Ehrenfeld and Harvey 1987).

3.7 Junctional Membrane Permeabilities

In the first study of the "shunt" in frog skin it was found that its resistance varies enormously among preparations (Koefoed-Johnsen et al. 1952a). We now know that the passive Cl conductance is localized in m.r. cells whose density varies just as much (Fig. 6B). Thus, to find the range of the junctional Cl permeability (P_{Cl}^{jm}), we may take the upper value of the Cl resistance (R_{Cl}). Assuming constant field conditions of the junctional path we have (Finkelstein and Mauro 1963; Sten-Knudsen 1978):

$$P_{Cl}^{jm} = \frac{(RT)^2[(1-\Phi)\ln(c_{Cl}^{i}\cdot \Phi/c_{Cl}^{o})]}{-R_{Cl}F^3V(c_{Cl}^{i}\cdot \Phi - c_{Cl}^{o})} \ . \qquad (26)$$

With $\Phi = \exp(FV/RT)$, the external bath diluted to one-tenth of the Ringer's of the internal bath, $V = -60$ mV, and an upper value of R_{Cl} of 74,000 $\Omega \cdot cm^2$ (Koefoed-Johnsen et al. 1952a), we calculate from Eq. (26), $P_{Cl}^{jm} \leq 5.8 \cdot 10^{-8}$ cm s^{-1}. Recent estimates of junctional membrane Cl permeabilities of amphibian skin indeed converge toward values in the order of 10^{-8} cm s^{-1}.

In short-circuited toad skin exposed to NaCl-Ringer's outside the Cl efflux is 64 ± 18 pmol cm^{-2} s^{-1} ($n = 16$), but no more than 1.1 ± 0.2 pmol cm^{-2} s^{-1} ($n = 8$) when apical Cl is replaced by gluconate (Bruus et al. 1976). Frog skin studied under similar experimental conditions exhibits Cl effluxes which are also reduced with the external Cl concentration (Mandel and Curran 1972). Effluxes in amphibian skin clamped at $V = 0$ mV, and with external Cl replaced by a non-permeating anion are collected in Table 3 together with their associated rate coefficients defined by:

$$k_{Cl} = J_{Cl}^{out}/c_{Cl}^{i} \ . \qquad (27)$$

The studies of the two European species were both carried out which the classical Ussing-designed chamber, so contributions from edge-damaged areas are presumably insignificant (cf. Harck and Larsen 1986). As discussed in Section 3.2.3, removal of external Cl results in closure of the apical Cl channels of m.r. cells. It is possible then that the Cl effluxes of Table 3 are governed by the native junctional membrane permeability of the respective preparations. However, this is not easily tested. Kristensen (1972) showed that the efflux of Cl towards 1 mM

Table 3. Estimates of upper limiting value of junctional membrane Cl permeability of amphibian skin epithelium[a]

J_{Cl}^{out} (pmol cm^{-2} s^{-1})	k_{Cl} (cm s^{-1})	Species	References
1.1 ± 0.2 (8)	0.97×10^{-8}	*B. bufo*	Bruus et al. (1976)
25 ± 5 (6)	22×10^{-8}	*R. temporaria*	Kristensen (1978)
28 ± 5 (16)	24×10^{-8}	*L. ocellatus*	Ques-von Petery et al. (1978)
6 ± 3 (8)	5×10^{-8}	*R. pipiens*	Biber et al. (1980)

[a] Cl efflux (J_{Cl}^{out}) and rate coefficient [k_{Cl}, Eq. (27)] for short-circuited amphibian skin exposed on the outside to Cl-free Ringer's. Mean ± SE with number of preparations in parentheses.

Table 4. Estimates of upper limiting value of junctional membrane Na permeability of amphibian skin epithelium[a]

J_{Na}^{out} (pmol cm^{-2} s^{-1})	k_{Na} (cm s^{-1})	V (mV)	Species	Reference[b]
2.3 ± 0.3 (24)	2.0×10^{-8}	0	*B. bufo*	[1]
3.7 ± 0.6 (9)	3.5×10^{-8}	0	*R. pipiens*	[2]
12.2 ± 2.1 (9)	11×10^{-8}	0	*B. bufo*	[3a]
1.9 ± 0.4 (8)	1.7×10^{-8}	0	*B. bufo*	[3b]
8.2 ± 0.8 (36)	8.2×10^{-8}	0	*R. pipiens*	[4]
2.9 ± 0.7 (8)	2.6×10^{-8}	0	*B. bufo*	[5]
3.4 ± 1.5 (7)	0.8×10^{-8}	−107 ± 7	*R. temporaria*	[6a]
0.7 ± 0.5 (4)	0.15×10^{-8}	−108 ± 6	*R. temporaria*	[6b]
6.9 ± 0.7 (7)	0.45×10^{-8}	−104 ± 8	*R. temporaria*	[6c]

[a] Mean ± SE (number of preparations) for Na efflux (J_{Na}^{out}) and its rate coefficient [k_{Na}, cf. Eqs. (27, 28) at transepithelial potential (V, serosal reference) indicated.

[b] [1] Larsen 1973 (bilateral Na_2SO_4-Ringer's); [2] Helman and Miller 1974 (bilateral NaCl-Ringer's); [3] Bruus et al. 1976 (a: bilateral NaCl-Ringer's; b: Na gluconate-Ringer's outside); [4] Biber and Mullen 1977 (bilateral NaCl-Ringer's); [5] Harck and Larsen 1986 (bilateral NaCl-Ringer's); [6] Eskesen and Ussing 1985 (a: bilateral NaCl-Ringer's; b: 1/10 NaCl-Ringer's outside; c: 1/10 NaCl-Ringer's outside with amiloride).

external Cl exhibits a dependence on the transepithelial potential which is predicted by constant-field theory applied to a single-barrier transepithelial pathway. The same theory was tested for the efflux in toad skin bathed on the outside with sulfate-Ringer's, but with less success (Bruus et al. 1976). The rate coefficient of sulfate efflux in toad skin is $0.16(\pm 0.03) \times 10^{-8}$ cm s^{-1} (Harck and Larsen 1986). Thus, on the sum of evidence, the junctional anion permeability is no more than 10^{-8} cm s^{-1} for toad skin; for isolated frog skin the anion permeability of junctional membranes may be as small, but the data indicate that it may also be larger.

Two methods can be applied for estimating the Na permeability of the junctional membranes. The first relies on the assumption that the transcellular Na efflux is impeded by the serosal membrane and that this, supposedly small, component can be estimated by blocking the apical Na permeability with amiloride. The second method is based on separation of cellular and paracellular fluxes by presteady state flux-ratio analysis.

Measured Na effluxes and calculated rate coefficients are collected in Table 4. Some studies (Larsen 1973; Kirschner 1973) reported lack of amiloride inhibition of Na efflux, whereas others noted statistically significant effects: a small stimulation at large external Na concentration, and a small decrease in skins bathed with 6 mM NaCl-Ringer's on the outside (Biber and Mullen 1977). The value from their study (Table 4) is for skins exposed to NaCl-Ringer's. Not all studies analyzed for transcellular components. Therefore, the conclusion is nothing more than an indication of a junctional membrane Na permeability of about 2×10^{-8} cm s^{-1} for short-circuited toad skin exposed to solutions of Ringer's strength on the outside, and a value for frog skin which may be somewhat larger under similar conditions. Presteady state flux-ratio analysis under open-circuit conditions was carried out with skins of cold-adapted frogs in order to obtain as high transepithelial potentials as possible (Eskesen and Ussing 1985). The rate coefficients (permeability coefficients) given in Table 4 have been calculated from the estimated Na efflux (J_{Na}^{jm}) through the paracellular pathway;

$$k_{Na} = \frac{J_{Na}^{jm} \cdot (1 - \exp(F \cdot V/(R \cdot T)))}{-c_{Na}^{i} \cdot F \cdot V/(R \cdot T)} , \tag{28}$$

where V is the transepithelial potential difference (serosa reference). In this study, selecting for high-resistance preparations, the paracellular efflux was also estimated as the total efflux following addition of amiloride to the external solution, but with the skin clamped at the spontaneous potential recorded under control conditions. Independent of method these values (Table 4) are significantly lower than those obtained under short-circuit conditions. The general conclusion to be made from the values given in Tables 3 and 4 is that the junctional Cl and Na permeability is in the order of 10^{-8} cm s^{-1}, but that they may depend on experimental conditions and perhaps also on animal species. It is evident, however, that the junctional membranes do not constitute a significant pathway for transepithelial flow of small ions. If, as an example representative for short-circuited toad skin, we assume values of $P_{Cl}^{jm} = 1 \times 10^{-8}$ cm s^{-1} and $P_{Na}^{jm} = 2 \times 10^{-8}$ cm s^{-1}, respectively, the junctional membrane conductance (G^{jm}) calculated by:

$$G^{jm} = (P_{Cl}^{jm} + P_{Na}^{jm}) \cdot c_{NaCl} \cdot F^2/(R \cdot T) \tag{29}$$

turns out to be no more than 13 μS cm^{-2} (c_{NaCl} = 115 mM). This is less than 2% of the total skin conductance measured under similar conditions.

When values larger than the above estimate are measured they are supposedly not indicating edge leaks, but rather variations among preparations. Such a variation may be correlated with the molting cycle which has a period of 4–6 days for room temperature adapted animals (L.O. Larsen 1976). The passive leak increases dramatically during natural molting (Jørgensen 1949) and aldosterone-induced shedding of the cornified cell layer (Nielsen 1969; Larsen 1971 a, b). It has been suggested that the "new" outermost living cell layer has not developed fully differentiated tight junctions at the time it starts taking over apical transport functions. A nonspecific increase in skin permeability is also seen when the skin is bathed with a hypertonic Ringer's solution on the outside (Ussing and Andersen 1956), or following addition of large concentrations of silver (Curran 1972) or copper ions (Ferreira 1978) to the external bath. (This effect on skin permeability is different from the specific Cl-conductance inhibition observed at small concentrations of heavy metal ions, Koefoed-Johnsen and Ussing 1958, 1974.) The osmotically induced permeability increase is reversible. Sulfate and Na efflux may increase by an order of magnitude (Ussing and Windhager 1964) and lanthanum, which normally cannot penetrate tight junctions, readily passes from the outer bath into the lateral intercellular spaces (Erlij and Matinez-Palomo 1972). The functional significance of these effects in amphibian skin is unknown, but it has been suggested that the osmotic sensitivity of tight junctions of other high resistance epithelia participates in regulating transepithelial flows of solutes (DiBona 1985).

4 Conclusions

The amphibian skin is differentiated for performing steady state transport of NaCl. Strongly rectifying cellular pathways for both ions, and a vanishingly small dissipative permeability of junctional membranes are developed. The early analyses of transport pathways provided fundamental methods for the study of membrane transport. The first model of frog skin constituted the theoretical framework for the study of transport in other epithelia. These studies of epithelial transport were conducted when the biophysical description of excitable cells were also at the beginning. At that time the two areas of membrane biology had overlapping problems, that is, distinguishing between active transport and diffusion, proving coupling of ion transport and metabolic pathways, separating ionic components of membrane currents and measuring their driving forces, and establishing the theory of ion-specific membrane permeabilities. Because understanding the functional organization of the excitable cell required detailed study of time-dependent states of the membrane, and further investigation of epithelial organization and transport focused on time-independent states, the literature diverged. With the development of techniques for the study of single-channel events, and the recognition that also epithelial membranes contain dynamic permeabilities the two areas of membrane transport meet again. With methods of fluctuation analysis, patch clamp, and vesicle studies it is now possible to investigate ionic pathways and their regulation at the molecular level.

Another line of questioning has developed with the recognition that the skin is a heterocellular epithelium with a minority cell type, mitochondria-rich cells, which constitute the pathway for transport of chloride. The functional organization of these cells is different from that of principal cells. Studies of their differentiation, regulation of their density, and interdependence of fluxes carried by different cell types now come into focus. Similar questions are raised in studies of other heterocellular epithelia, for example, operculum epithelium of euryhaline teleosts, amphibian urinary bladder, and collecting duct of mammalian nephron. Despite the diversity of osmoregulatory organs, at the level of their cellular organization common questions are still being raised.

References

Abramcheck FJ, Driessche W van, Helman SI (1985) Autoregulation of apical membrane Na^+ permeability of tight epithelia. *Noise analysis with amiloride and CGS 4270.* J Gen Physiol 85:555 – 582

Aceves J (1977) Pumped movements of sodium and potassium ions in the isolated epithelium of the frog skin. Pfluegers Arch 371:201 – 209

Armstrong CM (1975) Ionic pores, gates, and gating currents. Q Rev Biophys 7:179 – 209

Bargmann W, Welsh U (1972) Über Kanalschenzellen und dunkle Zellen in Nephron von Anuren. Z Zellforsch 134:193 – 204

Beauwens R, Beaujean MZ, Zizi M, Rentmeesters M, Crabbé J (1986) Increased chloride permeability of amphibian epithelia treated with aldosterone. Pfluegers Arch 407:620 – 624

Benos DJ, Hyde BA, Latorre R (1983) Sodium flux ratio through the amiloride-sensitive entry pathway in frog skin. J Gen Physiol 81:667 – 685

Biber TUL (1971) Effect of changes in transepithelial transport on the uptake of sodium across the outer surface of the frog skin. J Gen Physiol 58:131 – 144
Biber TUL, Cruz LJ (1973) Effect of antidiuretic hormone on sodium uptake across outer surface of frog skin. Am J Physiol 225:912 – 917
Biber TUL, Curran PF (1970) Direct measurement of uptake of sodium at the outer surface of the frog skin. J Gen Physiol 56:83 – 99
Biber TUL, Mullen TL (1977) Effect of inhibitors on transepithelial efflux of Na and nonelectrolytes in frog skin. Am J Physiol 232:C67 – C75
Biber TUL, Mullen TL (1980) Effect of external cation and anion substitutions on sodium transport in isolated frog skin. J Membr Biol 52:121 – 132
Biber TUL, Sanders ML (1973) Influence of transepithelial potential difference on the sodium uptake at the outer surface of the isolated frog skin. J Gen Physiol 61:529 – 551
Biber TUL, Walker TC, Mullen TL (1980) Influence of extracellular Cl concentration on Cl transport across isolated skin of *Rana pipiens*. J Membr Biol 56:81 – 92
Biber TUL, Drewnowska K, Baumgarten CM, Fisher RS (1985) Intracellular Cl activity changes of frog skin. Am J Physiol 249:F432 – F438
Biber TUL, DeSimone JA, Drewnowska K (1986) Potential dependence of unidirectional chloride fluxes across isolated frog skin. Biochim Biophys Acta 862:27 – 38
Brown D, Grosso A, De Sousa RC (1981) The amphibian epidermis: distribution of mitochondria-rich cells and the effect of oxytocin. J Cell Sci 52:197 – 213
Bruus K, Kristensen P, Larsen EH (1976) Pathways for chloride and sodium transport across toad skin. Acta Physiol Scand 97:31 – 47
Budtz P, Larsen LO (1973) Structure of the toad epidermis during the moulting cycle. I. Light microscopic observations on *Bufo bufo* (L.). Z Zellforsch 144:353 – 368
Budtz P, Larsen LO (1975) Structure of the toad epidermis during the moulting cycle. II. Electron microscopic observations on *Bufo bufo* (L.). Cell Tissue Res 159:459 – 483
Candida O (1978) Reduction of chloride fluxes by amiloride across the short circuited frog skin. Am J Physiol 234:F437 – F445
Canessa M, Labarca DR, DiBona DR, Leaf A (1978) Energetics of sodium transport in toad urinary bladder. Proc Natl Acad Sci USA 75:4591 – 4595
Choi JK (1963) The fine structure of the urinary bladder of the toad, *Bufo marinus*. J Cell Biol 16:53 – 72
Civan MM, Peterson-Yantorno K (1986) Intracellular pH regulation in frog skin: a ^{31}P-nuclear magnetic resonance study. Am J Physiol 251:F831 – F838
Civan MM, Peterson-Yantorno K, DiBona DR, Wilson DF, Erecinska M (1983) Bioenergetics of Na^+ transport across frog skin: chemical and electrical measurements. Am J Physiol 245:F621 – F700
Cox TC, Helman SI (1983) Effects of ouabain and furosemide on basolateral membrane Na efflux of frog skin. Am J Physiol 245:F312 – F321
Cox TC, Helman SI (1986a) Na^+ and K^+ transport at basolateral membranes of epithelial cells. I. *Stoichiometry of the Na, K-ATPase*. J Gen Physiol 87:567 – 583
Cox TC, Helman SI (1986b) Na^+ and K^+ transport at basolateral membranes of epithelial cells. II. *K^+-efflux and stoichiometry of the Na, K-ATPase*. J Gen Physiol 87:485 – 502
Cox TC, Helman SI (1986c) Na^+ and K^+ transport at basolateral membranes of epithelial cells. III. *Voltage independence of basolateral Na^+ efflux*. J Gen Physiol 87:503 – 509
Crabbé J (1963) The sodium retaining action of aldosterone. Presses Acad Eur SC, Bruxelles, pp 1 – 119
Cragoe JE, Ehrenfeld J, Harvey BJ (1987) Evidence for sodium-hydrogen ion exchange in the basolateral membrane of the isolated frog skin epithelium: kinetic studies. J Physiol (London) 382:143P (C. 113)
Curran PF (1972) Effect of silver ion on permeability properties of frog skin. Biochim Biophys Acta 288:90 – 97
Cuthbert AW (1972) A double series pump model for transporting epithelia. J Theor Biol 36:555 – 580
Cuthbert AW, Painter E (1968) The effect of theophylline on chloride permeability and active chloride transport in various epithelia. J Pharm Pharmacol 20:492 – 495
Cuthbert AW, Shum WK (1977) Does intracellular sodium modify membrane permeability to sodium ions? Nature (London) 266:468 – 469

Dalmark M (1976) Effects of halides and bicarbonate on chloride transport in human red blood cells. J Gen Physiol 67:223–234
DiBona DR (1985) Functional analysis of tight junction organization. Pfluegers Arch 405 (Suppl 1): S59–S66
Drewnowska K, Biber TUL (1985) Active transport and exchange diffusion of Cl across the isolated skin of *Rana pipiens*. Am Physiol 249:F424–F431
Dörge A, Rick R, Beck F, Thurau K (1985) Cl transport across the basolateral membrane in frog skin epithelium. Pfluegers Arch 405 (Suppl 1):S8–S11
Driessche W van (1984) Physiological role of apical potassium channels in frog skin. J Physiol (London) 356:79–95
Driessche W van, Erlij D (1983) Noise analysis of inward and outward Na^+ currents across the apical border of ouabain-treated frog skin. Pfluegers Arch 398:179–188
Driessche W van, Lindemann B (1979) Concentration-dependence of currents through single sodium-selective pores in frog skin. Nature (London) 282:519
Driessche W van, Zeiske W (1985) Ionic channels in epithelial cell membranes. Physiol Rev 65:833–903
Duffey ME, Kelepouris E, Peterson-Yantorno K, Civan MM (1986) Microelectrode study of intracellular pH in frog skin: dependence on serosal chloride. Am J Physiol 251:F468–F474
Duranti E, Ehrenfeld J, Harvey BJ (1986) Acid secretion through the *Rana esculenta* skin: involvement of an anion-exchange mechanism at the basolateral membrane. J Physiol (London) 378:195–211
Dürr JE, Larsen EH (1986) Indacrinone (MK-196) – a specific inhibitor of the voltage-dependent Cl^- permeability in toad skin. Acta Physiol Scand 127:145–153
Ehrenfeld J, Garcia-Romeu F (1977) Active hydrogen excretion and sodium absorption through isolated frog skin. Am J Physiol 233:F46–F54
Ehrenfeld J, Harvey BJ (1987) Evidence for basolateral sodium/hydrogen ion exchange in frog-skin epithelium and its role in sodium transport and intracellular pH regulation. J Physiol (London) 382:144P (C. 114)
Ehrenfeld J, Masoni A, Garcia-Romeu F (1976) Mitochondria-rich cells of frog skin in transport mechanisms: morphological and kinetic studies on transepithelial excretion of methylene blue. Am J Physiol 231:120–126
Ehrenfeld J, Garcia-Romeu F, Harvey BJ (1985) Electrogenic active proton pump in *Rana esculenta* skin and its role in sodium ion transport. J Physiol (London) 359:331–355
Emilio MG, Machado MM, Menano HP (1970) The production of a hydrogen ion gradient across the isolated frog skin. Quantitative aspects and effect of acetazolamide. Biochim Biophys Acta 203:394–409
Erlij D, Martinez-Palomo A (1972) Opening of tight junctions in frog skin by hypertonic urea solutions. J Membr Biol 9:229–240
Erlij D, Schoen H (1983) Barriers for Na and Cl movements across tight epithelia. Proc Int Un Physiol Sci XV:457
Erlij D, Smith MW (1973) Sodium uptake by frog skin and its modification by inhibitors of trans-epithelial sodium transport. J Physiol (London) 228:221–239
Erlij D, Ussing HH (1978) Transport across amphibian skin. In: Giebiesch G, Tosteson DC, Ussing HH (eds) Membrane transport in biology, vol 3. Transport across multimembrane systems. Springer, Berlin Heidelberg New York, pp 175–208
Eskesen K, Ussing HH (1985) Determination of the electromotive force of active sodium transport in frog skin epithelium (*Rana temporaria*) from presteady-state flux ratio experiments. J Membr Biol 86:105–111
Eskesen K, Ussing HH (1986) Single-file diffusion through K^+ channels in frog skin epithelium. J Membr Biol 91:245–250
Eskesen K, Lim JJ, Ussing HH (1985) Evaluation of transport pathways for Na^+ across frog skin epithelium by means of presteady-state flux ratio. J Membr Biol 86:99–104
Essig A (1982) Influence of cellular and paracellular conductance patterns on epithelial transport and metabolism. Biophys J 38:143–152
Farquhar MG, Palade GE (1964) Functional organization of amphibian skin. Proc Natl Acad Sci USA 51:569–577
Ferreira KTG (1978) The effect of copper ions on frog skin. The role of sulphydryl groups. Biochim Biophys Acta 510:298–304

Ferreira KTG (1979) The relationship of K^+ efflux at the inner surface of the isolated frog skin epithelium to the short circuit current. Biochim Biophys Acta 555:13–25
Ferreira KTG, Ferreira HG (1981) The regulation of volume and ion composition in frog skin. Biochim Biophys Acta 646:193–202
Finkelstein A, Mauro A (1963) Equivalent circuits as related to ionic systems. Biophys J 3:215–237
Foskett JK, Ussing HH (1986) Localization of chloride conductance to mitochondria-rich cells in frog skin epithelium. J Membr Biol 91:251–258
Frazier LW (1974) Interrelationship of H^+ and Na^+ reabsorption in the toad urinary bladder. J Membr Biol 19:267–276
Frazier LW, Vanatta JC (1981) Excretion of K^+ by frog with rate varying with K^+ load. Comp Biochem Physiol 69A:157–160
Frizzel RA, Schultz SG (1978) Effect of aldosterone on ion transport by rabbit colon in vitro. J Membr Biol 39:1–26
Fuchs W, Gebhardt U, Lindemann B (1972) Delayed voltage responses to fast changes of $(Na)_o$ at the outer surface of frog skin epithelium. In: Kreuzer F, Slegers JFG (eds) Biomembranes, vol 3. Passive permeability of cell membranes. Plenum Press, New York London, pp 483–498
Fuchs W, Larsen EH, Lindemann B (1977) Curerent-voltage curve of sodium channels and concentration dependence of sodium permeability in frog skin. J Physiol (London) 267:137–166
Fuhrman FA, Ussing HH (1951) A characteristic response of the isolated frog skin potential to neurohypophyseal principles and its relation to the transport of sodium and water. J Cell Comp Physiol 38:109–130
Garcia-Diaz JF, Baxendale LM, Klemperer G, Essig A (1985) Cell K in frog skin in the presence and absence of cell current. J Membr Biol 85:143–158
Garcia-Diaz JF, Klemperer G, Baxendale LM, Essig A (1986) Cell sodium and sodium pump function in frog skin. J Membr Biol 92:37–46
Garcia Romeu F, Salibian A, Pezzani-Hernandez S (1969) The nature of the in vivo sodium and chloride uptake mechanisms through the epithelium of Chilean frog *Calyptocephalella gayi* (Dum. et Bibr., 1841). *Exchanges of hydrogen against sodium and of bicarbonate against chloride.* J Gen Physiol 53:816–835
Garay RP, Garrahan PJ (1973) The interaction of sodium and potassium with the sodium pump in red cells. J Physiol (London) 231:297–325
Garty H (1986) Mechanisms of aldosterone action in tight epithelia. J Membr Biol 90:193–205
Garty H, Lindemann B (1984) Feedback inhibition of sodium uptake in K^+-depolarized toad urinary bladders. Biochim Biophys Acta 771:89–98
Geck P, Heinz E (1980) Coupling of ion flows in cell suspension systems. Ann NY Acad Sci 341:569–577
Giraldez F, Ferreira KTG (1984) Intracellular chloride activity and membrane potential in stripped frog skin (*Rana temporaria*). Biochim Biophys Acta 769:625–628
Gluck S, Al-Awqati Q (1984) An electrogenic proton translocating adenosine triphosphatase from bovine kidney medulla. J Clin Invest 73:1704–1710
Gluck S, Cannon C, Al-Awqati Q (1982) Exocytosis regulates urinary acidification in turtle bladder by rapid insertion of H^+ pumps into the luminal membrane. Proc Natl Acad Sci USA 79:4327–4331
Goldman DE (1943) Potential, impedance, and rectification in membranes. J Gen Physiol 27:37–60
Grinstein S, Erlij D (1978) Intracellular calcium and the regulation of sodium transport in frog skin. Proc R Soc London Ser B 202:353–360
Hanrahan JW, Alles WP, Lewis SA (1985) Single anion-selective channels in basolateral membrane of a mammalian tight epithelium. Proc Natl Acad Sci USA 82:7791–7795
Harck AF, Larsen EH (1986) Concentration dependence of halide fluxes and selectivity of the anion pathway in toad skin. Acta Physiol Scand 128:289–304
Harris EV, Burn GP (1949) The transfer of sodium and potassium ions between muscle and surrounding medium. Trans Faraday Soc 45:508–528
Harvey BJ, Ehrenfeld J (1986) Regulation of intracellular sodium and pH by the electrogenic H^+ pump in frog skin. Pfluegers Arch 406:362–366
Harvey BJ, Kernan RP (1984a) Intracellular ion activities in frog skin in relation to external sodium and effects of amiloride and/or ouabain. J Physiol (London) 349:501–517
Harvey BJ, Kernan RP (1984b) Sodium selective micro-electrode study of apical permeability in frog skin: effects of sodium, amiloride and ouabain. J Physiol (London) 356:359–374

Helman SI, Fisher RS (1977) Microelectrode studies of the active Na transport pathway of frog skin. J Gen Physiol 69:571 – 604
Helman SI, Miller DA (1974) Edge damage effect on measurements of urea and sodium in frog skin. Am J Physiol 226:1198 – 1203
Helman SI, Cox TC, Driessche W van (1983) Hormonal control of apical membrane Na transport in epithelia. *Studies with fluctuation analysis.* J Gen Physiol 82:201 – 220
Hodgkin AL (1951) The ionic basis of electrical activity in nerve and muscle. Biol Rev 26:339 – 409
Hodgkin AL, Huxley AF (1952) A quantitative description of membrane current and its application to conduction and excitation in nerve. J Physiol (London) 117:500 – 544
Hodgkin AL, Katz B (1949) The effect of sodium ions on the electrical activity of the giant axon of the squid. J Physiol (London) 108:37 – 77
Hoshiko T, Driessche W van (1986) Effect of sodium on amiloride- and triamterene-induced current fluctuations in isolated frog skin. J Gen Physiol 87:425 – 442
Huf EG (1972) The role of Cl^- and other anions in active Na^+ transport in isolated frog skin. Acta Physiol Scand 84:366 – 381
Ilic V, Brown D (1980) Modification of mitochondria-rich cells in different ionic conditions: changes in cell morphology and cell number in the skin of *Xenopus laevis.* Anat Rec 196: 153 – 161
Jørgensen CB (1949) Permeability of the amphibian skin II. Effect of moulting of the skin of anurans on the permeability to water and electrolytes. Acta Physiol Scand 18:171 – 180
Jørgensen CB, Levi H, Zerahn K (1954) On active uptake of sodium and chloride in anurans. Acta Physiol Scand 30:178 – 190
Katz U (1986) The role of amphibian epidermis in osmoregulation and its adaptive response to changing environment. In: Bereiter-Hahn J, Matoltsy GM, Richards A (eds) Biology of the integument, vol 2. Springer, Berlin Heidelberg New York Tokyo, pp 472 – 498
Katz U, Larsen EH (1984) Chloride transport in toad skin (*Bufo viridis*). The effect of salt adaptation. J Exp Biol 109:353 – 371
Katz U, Scheffey C (1986) The voltage-dependent chloride conductance of toad skin is localized to mitochondria-rich cells. Biochim Biophys Acta 861:480 – 482
Kirschner LB (1955) On the mechanism of active sodium transport across the frog skin. J Cell Comp Physiol 45:61 – 86
Kirschner LB (1970) The study of NaCl transport in aquatic animals. Am Zool 10:365 – 376
Kirschner LB (1973) Electrolyte transport across the body surface of fresh water fish and amphibians. In: Ussing HH, Thorn NA (eds) Transport mechanisms in epithelia. Proc Alfred Benzon Symp, vol 5. Munksgaard, Copenhagen, pp 447 – 463
Kirschner LB (1983) Sodium chloride absorption across the body surface: frog skin and other epithelia. Am J Physiol 244:R429 – R443
Kirschner LB, Greenwald L, Kerstetter TH (1973) Effect of amiloride on sodium transport across body surfaces of freshwater animals. Am J Physiol 224:832 – 837
Knauf PA, Fuhrman GF, Rothstein S, Rothstein A (1977) The relationship between anion exchange and net ion flow across the human red cell membrane. J Gen Physiol 69:363 – 386
Koefoed-Johnsen V (1957) The effect of g-strophantin (ouabain) on the active transport of sodium through the isolated frog skin. Acta Physiol Scand 42 (Suppl 145):87 – 88
Koefoed-Johnsen V (1979) Control mechanisms in amphibians. In: Gilles R (ed) Mechanisms of osmoregulation. Wiley, New York, pp 223 – 272 and appendix, pp 646 – 648
Koefoed-Johnsen V, Ussing HH (1958) The nature of the frog skin potential. Acta Physiol Scand 42:298 – 308
Koefoed-Johnsen V, Ussing HH (1974) Transport pathways in frog skin and their modification by copper ions. In: Thorn NA, Petersen OH (eds) Secretory mechanisms of exocrine glands. Proc Alfred Benzon Symp, vol 7. Munksgaard, Copenhagen, pp 411 – 419
Koefoed-Johnsen V, Levi H, Ussing HH (1952a) The mode of passage of chloride ions through the isolated frog skin. Acta Physiol Scand 25:150 – 163
Koefoed-Johnsen V, Ussing HH, Zerahn K (1952b) The origin of the short-circuit current in the adrenaline stimulated frog skin. Acta Physiol Scand 27:38 – 48
Kristensen P (1972) Chloride transport across isolated frog skin. Acta Physiol Scand 84:338 – 346
Kristensen P (1978) The effect of amiloride on chloride transport across amphibian epithelia. J Membr Biol 40S:167 – 185

Kristensen P (1981) Is chloride transfer in frog skin localized to a special cell type? Acta Physiol Scand 113:123 – 124
Kristensen P (1982) Chloride transport in frog skin. In: Zadunaisky JA (ed) Chloride transport in biological membranes. Academic Press, London New York, pp 310 – 332
Kristensen P (1983) Exchange diffusion, electrodiffusion and rectification in the chloride transport pathway of frog skin. J Membr Biol 72:141 – 151
Kristensen P, Larsen EH (1978) Relation between chloride exchange diffusion and a conductive chloride pathway across the isolated skin of the toad (*Bufo bufo*). Acta Physiol Scand 102:22 – 34
Kristensen P, Schousboe A (1968) The role of glycolysis in energy production in the isolated skin of the brown frog (*Rana temporaria* L.). Biochim Biophys Acta 153:132 – 137
Kristensen P, Schousboe A (1969) The influence of anaerobic conditions on sodium tranport and adenine nucleotide levels in isolated skin of the frog *Rana temporaria*. Biochim Biophys Acta 173:206 – 212
Kristensen P, Ussing HH (1985) Epithelial organization. In: Seldin W, Giebisch G (eds) The kidney: Physiology and pathophysiology, vol 2. Raven, New York, pp 173 – 188
Krogh A (1937) Osmotic regulation in the frog (*R. esculenta*) by active absorption of chloride ions. Skand Arch Physiol 76:60 – 74
Krogh A (1939) Osmotic regulation in aquatic animals. Cambridge Univ Press, Cambridge
Larsen EH (1971a) Effect of aldosterone and oxytocin on the active sodium transport across the isolated toad skin in relation to loosening of *stratum corneum*. Gen Comp Endocrinol 17:543 – 553
Larsen EH (1971b) The relative contribution of sodium and chloride ions to the conductance of toad skin in relation to shedding of the *stratum corneum*. Acta Physiol Scand 81:254 – 263
Larsen EH (1972) Characteristics of aldosterone stimulated transport in isolated skin of the toad, *Bufo bufo* (L.). J. Steroid Biochem 3:111 – 120
Larsen EH (1973) Effect of amiloride, cyanide and ouabain on the active transport pathway in toad skin. In: Ussing HH, Thorn NA (eds) Transport mechanisms in epithelia. Proc Alfred Benzon Symp, vol 5. Munksgaard, Copenhagen, pp 131 – 147
Larsen EH (1982) Chloride current rectification in toad skin epithelium. In: Zadunaisky JA (ed) Chloride transport in biological membranes. Academic Press, London New York, pp 333 – 364
Larsen EH, Kristensen P (1978) Properties of a conductive cellular chloride pathway in the skin of the toad (*Bufo bufo*). Acta Physiol Scand 102:1 – 21
Larsen EH, Rasmussen BE (1982) Chloride channels in toad skin. Philos Trans R Soc London Ser B 299:413 – 434
Larsen EH, Rasmussen BE (1983) Membrane potential plays a dual role for chloride transport across toad skin. Biochim Biophys Acta 728:455 – 459
Larsen EH, Rasmussen BE (1985) A mathematical model of amphibian skin epithelium with two types of transporting cellular units. Pfluegers Arch 405 (Suppl 1):S50 – S58
Larsen EH, Fuchs W, Lindemann B (1979) Dependence of Na-pump flux on intracellular Na-activity in frog skin epithelium (*R. esculenta*). Pfluegers Arch 382:R13
Larsen EH, Ussing HH, Spring KR (1986) Volume response of mitochondria-rich cells of toad skin to amiloride and Na-free outside solution. Fed Proc 45:746
Larsen EH, Ussing HH, Spring KR (1987) Ion transport by mitochondria-rich cells in toad skin. J Membr Biol 99:25 – 40
Larsen LO (1976) Physiology of molting. In: Lofts B (ed) Physiology of the amphibia, vol 3. Academic Press, London New York, pp 53 – 100
Leaf A (1956) On the mechanism of fluid exchange of tissues in vitro. Biochem J 62:241 – 248
Levi H, Ussing HH (1949) Resting potential and ion movements in frog skin. Nature (London) 164:928
Lew VL, Ferreira HG, Moura T (1979) The behavior of transporting epithelial cells. I Computer analysis of a basic model. Proc R Soc London Ser B 206:53 – 83
Li JH-Y, Lindemann B (1982) Movement of Na and Li across the apical membrane of frog skin. In: Emrich HM, Aldenhoff JB, Lux HD (eds) Basic mechanisms in the action of lithium. Excerpta Medica, Amsterdam Oxford Princeton, pp 28 – 35
Li JH-Y, Lindemann B (1983) Chemical stimulation of Na transport through amiloride-blockable channels in frog skin epithelium. J Membr Biol 75:179 – 192
Li JH-Y, Palmer LG, Edelman IS, Lindemann B (1982) The role of sodium-channel density in the natriferic response of the toad urinary bladder to antidiuretic hormone. J Membr Biol 64:77 – 89

Li JH-Y, Cragoe EJ, Lindemann B (1985) Structure-activity relationship of amiloride analogs as blockers of epithelial Na channels: I. Pyrazine-ring modifications. J Membr Biol 83:45–56
Li JH-Y, Cragoe EJ, Lindemann B (1987) Structure-activity relationship of amiloride analogs as blockers of epithelial Na channels: II. Side-chain modifications. J Membr Biol 95:171–185
Lindemann B (1984) Fluctuation analysis of sodium channels in epithelia. Annu Rev Physiol 46:497–515
Lindemann B, Driessche W van (1977) Sodium-specific membrane channels of frog skin are pores: current fluctuations reveal high turnover. Science 195:292–294
Lindemann B, Voûte CL (1976) Structure and function of the epidermis. In: Llinas R, Precht W (eds) Frog neurobiology. Springer, Berlin Heidelberg New York, pp 169–210
Lindemann B, Gebhardt U, Fuchs W (1972) A flow chamber for concentration-step experiments with epithelial membranes. T.-I.-T. J Life Sci 2:15–26
Linderholm H (1954) On the behaviour of the "sodium pump" in frog skin at various concentrations of Na ions in the solution on the epithelial side. Acta Physiol Scand 31:36–61
Ludens JH, Fanestil DD (1972) Acidification of urine by the isolated urinary bladder of the toad. Am J Physiol 233:1338–1344
Macey RI, Meyers S (1963) Dependence of chloride permeability on sodium in the isolated frog skin. Am J Physiol 204:1095–1099
MacRobbie EAC, Ussing HH (1961) Osmotic behaviour of the epithelial cells of frog skin. Acta Physiol Scand 53:348–365
Madsen K, Tisher CC (1986) Structural-functional relatonships along the distal nephron. Am J Physiol 250:F1–F15
Mandel LJ (1975) Actions of external hypotonic urea, ADH, and theophylline on transcellular and extracellular solute permeabilities in frog skin. J Gen Physiol 55:599–615
Mandel LJ, Curran PF (1972) Chloride flux via a shunt pathway in frog skin: apparent exchange diffusion. Biochim Biophys Acta 282:258–264
Mills JW (1985) Ion transport across the exocrine glands of the frog skin. Pfluegers Arch 405 (Suppl 1):S44–S49
Mills JW, Ernst SA, DiBona DR (1977) Localization of Na^+-pump sites in frog skin. J Cell Biol 73:88–110
Morel F, Leblanc G (1975) Transient current changes and Na compartmentalization in frog skin epithelium. Pfluegers Arch 358:135–157
Nagel W (1976) The intracellular electrical potential profile of the frog skin epithelium. Pfluegers Arch 365:135–143
Nagel W (1977) The dependence of the electrical potentials across the membranes of the frog skin upon the concentration of sodium in the mucosal solution. J Physiol (London) 269: 777–796
Nagel W (1978) Effects of antidiuretic hormone upon electrical potential and resistance of apical and basolateral membranes of frog skin. J Membr Biol 42:99–122
Nagel W (1979) Inhibition of potassium conductance by barium in frog skin epithelium. Biochim Biophys Acta 552:346–357
Nagel W (1980) Rheogenic sodium transport in a tight epithelium, the amphibian skin. J Physiol (London) 302:281–295
Nagel W, Crabbé J (1980) Mechanism of action of aldosterone on active sodium transport across toad skin. Pfluegers Arch 385:181–187
Nagel W, Essig A (1982) Relationship of transepithelial electrical potential to membrane potentials and conductance ratios in frog skin. J Membr Biol 69:125–136
Nagel W, Hirschmann W (1980) K^+-permeability of the outer border of the frog skin (*R. temporaria*). J Membr Biol 62:107–113
Nagel W, Garcia-Diaz JF, Armstrong W McD (1981) Intracellular ion activities in frog skin. J Membr Biol 61:127–134
Nagel W, Garcia-Diaz JF, Essig A (1983) Contribution of junctional conductance to the cellular voltage-divider ratio in frog skins. Pfluegers Arch 399:336–341
Nagel W, Beauwens R, Crabbé J (1985) Opposite effects of indacrinone (MK-196) on sodium and chloride conductance of amphibian skin. Pfluegers Arch 403:337–343
Nelson DJ, Tang JM, Palmer LG (1984) Single-channel recordings of apical membrane chloride conductance in A6 epithelial cells. J Membr Biol 80:81–89

Nielsen R (1969) The effect of aldosterone in vitro on the active sodium transport and moulting in frog skin. Acta Physiol Scand 77:85–94

Nielsen R (1979a) Coupled transepithelial sodium and potassium transport across isolated frog skin: effect of ouabain, amiloride and the polyene antibiotic filipin. J Membr Biol 51:161–184

Nielsen R (1979b) A 3 to 2 coupling of the Na-K pump responsible for transepithelial Na transport in frog skin disclosed by the effect of Ba. Acta Physiol Scand 107:189–191

Nielsen R (1982) Effect of ouabain, amiloride, and antidiuretic hormone on the sodium transport pool in isolated epithelia from frog skin (*Rana temporaria*). J Membr Biol 65:221–226

Nielsen R (1984) Active transepithelial potassium transport in frog skin via specific potassium channels in the apical membrane. Acta Physiol Scand 120:287–296

Nielsen R (1985) Ba^{++}-induced changes in the Na^+- and K^+-permeability of the isolated frog skin. Acta Physiol Scand 124:61–70

Palmer LG, Li JH-Y, Lindemann B, Edelman IS (1982) Aldosterone control of the density of sodium channels in the toad urinary bladder. J Membr Biol 64:91–102

Philpott CW (1965) Halide localization in the teleost chloride cell and its identification by selected area electron diffraction. Direct evidence supporting an osmoregulatory function for the seawater adapted chloride cell of *Fundulus*. Protoplasma 60:317–330

Ques-von Petery MV, Rotunno, CA, Cereijido M (1978) Studies on chloride permeability of the skin of *Leptodactylus ocellatus* I. Na^+ and Cl^- effect on passive movement of Cl^-. J Membr Biol 42:317–330

Rhodin J (1958) Anatomy of kidney tubules. Int Rev Cytol 7:485–534

Rick R, Dörge A, Arnim E von, Thurau K (1978) Electron microprobe analysis of frog skin epithelium: evidence for a syncytial sodium transport compartment. J Membr Biol 39:313–331

Rick R, Dörge A, Katz U, Bauer R, Thurau K (1980) The osmotic behaviour of toad skin epithelium (*Bufo viridis*). An electron microprobe analysis. Pfluegers Arch 385:1–10

Rick R, Roloff C, Dörge A, Beck FX, Thurau K (1984) Intracellular electrolyte concentrations in the frog skin epithelium: effect of vasopressin and dependence on the Na concentration in the bathing media. J Membr Biol 78:129–145

Rosen S, Friedley NJ (1973) Carbonic anhydrase activity in *Rana pipiens* skin: biochemical and histochemical analysis. Histochemie 36:1–4

Rotunno CA, Pouchan MT, Cereijido M (1966) Location of the mechanism of active transport of sodium across frog skin. Nature (London) 210:597

Rotunno CA, Vilallonga FA, Fernandez M, Cerreijido M (1970) The penetration of sodium into the epithelium of the frog skin. J Gen Physiol 55:716–735

Rudneff M (1865) Über die epidermoidale Schicht der Froschhaut. Arch Mikrosk Anat 1:295–298

Sariban-Sohraby S, Benos DJ (1986) The amiloride-sensitive sodium channel. Am J Physiol 250:C175–C190

Schoen HF, Erlij D (1985a) Basolateral response to transport modifiers in frog skin epithelium. Pfluegers Arch 405 (Suppl 1):S33–S38

Schoen HF, Erlij D (1985b) Current-voltage relations of the apical and basolateral membranes of the frog skin. J Gen Physiol 86:257–287

Schultz SG (1972) Electrical potential difference and electromotive forces in epithelial tissues. J Gen Physiol 59:794–825

Schultz SG (1981) Homocellular regulatory mechanisms in sodium-transporting epithelia: avoidance of extinction by "flush-through". Am J Physiol 241:F579–F590

Schwartz GJ, Al-Aqwati Q (1985) Carbon dioxide causes exocytosis of vesicles containing H^+ pumps in isolated perfused proximal and collecting tubule. J Clin Invest 75:1638–1644

Sharp GWG, Leaf A (1964) Biological action of aldosterone in vitro. Nature (London) 202: 1185–1188

Smith PG (1971) The low frequency impedance of the isolated frog skin. Acta Physiol Scand 81:355–366

Spring KR, Hope A (1978) Size and shape of the lateral intercellular spaces in a living epithelium. Science 200:54–57

Spring KR, Ussing HH (1986) The volume of mitochondria-rich cells of frog skin epithelium. J Membr Biol 92:21–26

Steinmetz, PR, Andersen OS (1982) Electrogenic proton transport in epithelial membranes. J Membr Biol 65:155–174

Sten-Knudsen O (1978) Passive transport processes. In: Giebisch G, Tosteson DC, Ussing HH (eds) Membrane transport in biology, vol 1, Concepts and models. Springer, Berlin Heidelberg New York, pp 5–113

Sten-Knudsen O, Ussing HH (1981) The flux ratio equation under nonstationary conditions. J Membr Biol 63:233–242

Stoddard JS, Helman SI (1985) Dependence of intracellular Na^+ concentration on apical and basolateral membrane Na^+ influx in frog skin. Am J Physiol 249:F662–F671

Stoddard JS, Jacobsson E, Helman SI (1985) Basolateral membrane chloride transport in isolated epithelia of frog skin. Am J Physiol 249:C318–C329

Taylor A, Windhager EE (1979) Possible role of cytosolic calcium and Na-Ca exchange in regulation of transepithelial sodium transport. Am J Physiol 236:F505–F512

Teorell T (1949) Membrane electrophoresis in relation to bioelectrical polarization effects. Arch Sci Physiol 3:205

Thomas RC (1972) Electrogenic sodium pump in nerve and muscle cells. Physiol Rev 52:563–593

Tosteson DC (1959) Halide transport in red blood cells. Acta Physiol Scand 46:19–41

Tostesen DC, Hoffman JF (1960) Regulation of cell volume by active cation transport in high and low potassium sheep red cells. J Gen Physiol 44:169–194

Ussing HH (1949a) The active ion transport through the isolated frog skin in the light of tracer studies. Acta Physiol Scand 17:1–37

Ussing HH (1949b) The distinction by means of tracers between active transport and diffusion. Acta Physiol Scand 17:1–37

Ussing HH (1978) Interpretation of tracer fluxes. In: Giebisch G, Tosteson DC, Ussing HH (eds) Membrane transport in biology, vol 1. Concepts and models. Springer, Berlin Heidelberg New York, pp 115–140

Ussing HH (1982a) Volume regulation in frog skin epithelium. Acta Physiol Scand 114:363–369

Ussing HH (1982b) Pathways for transport in epithelia. In: Corradino RA (ed) Functional regulation at the cellular and molecular level. Elsevier, Amsterdam, pp 285–297

Ussing HH (1985) Volume regulation and basolateral co-transport of sodium, potassium, and chloride ions in frog skin epithelium. Pfluegers Arch 405 (Suppl 1):S1–S7

Ussing HH (1986) Epithelial cell volume regulation illustrated by experiments in frog skin. Renal Physiol 9:38–46

Ussing HH, Andersen B (1956) The relation between solvent drag and active transport of ions. Proc 3rd Int Congr Biochem, Bruxelles, pp 434–440

Ussing HH, Windhager EE (1964) Nature of shunt path and active sodium transport path through frog skin epithelium. Acta Physiol Scand 61:484–504

Ussing HH, Zerahn K (1951) Active transport of sodium as the source of electric current in the short-circuited isolated frog skin. Acta Physiol Scand 23:110–127

Veech RL, Lawson JWR, Cornell NW, Krebs HA (1979) Cytosolic phosphorylation potential. J Biol Chem 251:6538–6547

Vestergaard-Bogind B, Lassen UV (1974) Membrane potential of *Amphiuma* red cells: hyperpolarizing effect of phloretin. In: Bolis L, Bloch K, Luria SE, Byneu F (eds) Comparative biochemistry and physiology of transport. Elsevier North-Holland Biomed Press, Amsterdam, pp 346–353

Voûte CL, Meier W (1978) The mitochondria rich cell of frog skin as hormone sensitive 'shunt path'. J Membr Biol. 40:141–165

Voûte CL, Ussing HH (1968) Some morphological aspects of active sodium transport. The epithelium of the frog skin. J Cell Biol 36:625–638

Voûte CL, Thummel J, Brenner M (1975) Aldosterone effect in the epithelium of the frog skin – a new story about an old anzyme. J Steroid Biochem 6:1175–1179

Warncke J, Lindemann B (1985) Voltage dependence of the blocking rate constants of amiloride at apical Na channels. Pfluegers Arch 405 (Suppl 1):S89–S94

Whitear M (1975) Flask cells and epidermal dynamics in frog skin. J Zool 175:107–149

Wieth JO (1979) Bicarbonate exchange through the red cell membrane determined with $^3(H)HCO_3$. J Physiol (London) 294:521–539

Wieth JO, Bjerrum PJ (1982) Titration of transport and modifier sites in red cell anion transport system. J Gen Physiol 79:253–282

Wieth JO, Brahm J (1985) Cellular anion transport. In: Seldin DW, Giebiesch G (eds) The kidney: physiology and pathophysiology. Raven, New York, pp 49–89

Willumsen NJ, Larsen EH (1985) Passive Cl^- currents in toad skin: potential dependence and relation to mitochondria-rich cell density. In: Gilles R, Gilles-Baillien M (eds) Transport processes, iono- and osmoregulation. Springer, Berlin Heidelberg New York Tokyo, pp 20–30
Willumsen NJ, Larsen EH (1986) Membrane potentials and intracellular Cl^- activity of toad skin epithelium in relation to activation and deactivation of the transepithelial Cl^- conductance. J Membr Biol 94:173–190
Zaccone G, Fasulo S, Lo Cascio P, Licata A (1986) Enzyme cytochemical and immunocytochemical studies of flask cells in the amphibian epidermis. Histochemistry 84:5–9
Zadunaisky JA, Candia OA, Chiarandini DJ (1963) The origin of the short-circuit current in the isolated skin of the South American frog *Leptodactyllus ocellatus*. J Gen Physiol 47:393–402
Zeiske W, Driessche W van (1979) Saturable K^+ pathway across the outer border of frog skin (*Rana temporaria*): kinetics and inhibition by Cs^+ and other cations. J Membr Biol 47:77–96
Zeiske W, Driessche W van (1983) The interaction of "K^+-like" cations with the apical K^+ channel in frog skin. J Membr Biol 76:57–72
Zeiske W, Lindemann B (1974) Chemical stimulation of Na^+ current through the outer surface of frog skin epithelium. Biochim Biophys Acta 352:323–326
Zerahn K (1955) Studies of the active transport of lithium in the isolated frog skin. Acta Physiol Scand 33:347–358
Zerahn K (1958) Oxygen consumption and active sodium transport in isolated amphibian skin under varying experimental conditions. Diss, Univ Copenhagen, Universitetsforlaget, Aarhus, pp 1–40

Chapter 6

NaCl Transport in Tracheal Epithelium

M. J. Welsh

Contents

1 Introduction

In this review I will focus on the mechanism and regulation of Na^+ and Cl^- transport by tracheal epithelium. Some features of salt transport that I will discuss are unique to trachea, however, the mechanisms of ion transport are in many ways similar to those in other epithelia. For example at the cellular level, a

Laboratory of Epithelial Transport and Pulmonary Disease Division, Department of Internal Medicine, University of Iowa College of Medicine, Iowa City, IA 52242, USA

Advances in Comparative and Environmental Physiology
Vol. 1. Edited by R. Greger

diverse group of epithelia (including shark rectal gland, cornea, and large and small intestinal epithelia) have a similar mechanism of Cl^- secretion. Likewise at the molecular level, many epithelia contain similar ion transport processes, although the individual processes may be located on different membranes than in the trachea. In this review I will note similarities to and differences from other epithelia in the mechanisms and regulation of electrolyte transport. Such comparison and contrast may be helpful in understanding common themes of epithelial function and the evolution of ion transport processes. I will also point out uncertainties and gaps in our knowledge. This approach may help focus important but unanswered questions. I will only discuss the histology in passing and will not cover the mechanism or regulation of mucus secretion.

2 Relation of Sodium and Chloride Transport to Mucociliary Clearance

Electrolyte transport by the airway epithelium controls, in part, the quantity and composition of respiratory tract fluid. Thus, it is important for mucociliary clearance, the pulmonary defense mechanism that removes inhaled particulate material from the airways.

Several important components combine to effect normal mucociliary clearance (Wanner et al. 1977; Nadel et al.1985). First, the mucus must trap and carry inhaled particulates. The efficiency with which mucus performs this task depends on its quantity and its viscoelastic properties. Second, the cilia provide the engine to propel the mucus up out of the lungs. The characteristic structure, function, and coordinated beating of airway cilia share several features with cilia from other locations. Third, a thin layer of periciliary fluid is required to provide a less viscous watery environment in which the cilia can beat freely, with only their tips projecting up to contact the mucus layer. The periciliary fluid allows for efficient transmission of the mechanical shear force generated by the cilia to the overlying mucus (Fulford and Blake 1986). The depth of periciliary fluid seems critical to the overall success of clearance. If the periciliary fluid were too deep, the cilia and mucus might not interact effectively; if the fluid were too shallow, the mucus might prevent the optimal pattern of ciliary beating with a power and recovery stroke. The periciliary fluid may also be important in hydrating the overlying mucus. Although our notions of the function of the periciliary fluid are reasonable, these notions have not been rigorously proven.

What is the role of electrolyte transport in production and regulation of the periciliary fluid and in mucociliary clearance? Again, we have little direct experimental evidence. However, the most widely held opinion is that the epithelium secretes the periciliary fluid and hydrates the mucus (Nadel et al. 1979, 1985). Certainly the epithelium can actively secrete Cl^-, which can drive fluid secretion (Olver et al. 1975; Welsh et al. 1980). Moreover, hormones known to control Cl^- secretion also regulate mucociliary clearance (Foster et al. 1976; Gerrity et al. 1983; Wanner et al. 1977). However, the airway epithelium can also actively absorb Na^+ and presumably absorb fluid. Thus, the airway epithelium may precisely control the amount of airway fluid by balancing fluid loss by evaporation with fluid production and fluid absorption.

Although the physiological details are not yet certain, there is no question about the importance of electrolyte transport for normal airway function. This is dramatically demonstrated by the pathophysiology caused by abnormal electrolyte transport in the disease, cystic fibrosis (Knowles et al. 1981; Widdicombe et al. 1985). In cystic fibrosis abnormally decreased epithelial Cl^- secretion contributes to the dehydrated secretions, impaired mucociliary clearance, and susceptibility to recurrent pulmonary infections that make cystic fibrosis a fatal disease.

3 Transport Properties of Dog Tracheal Epithelium

3.1 Transepithelial Properties

I will focus primarily on dog tracheal epithelium; it was the first and continues to be the most extensively studied airway epithelium, at least in part because of its large size and accessibility. The posterior membranous portion provides a flat sheet of epithelium that can be dissected free from the underlying connective tissue and smooth muscle. When mounted in an Ussing chamber the epithelium generates a transepithelial electrical potential difference (Ψ_t) of +10 to +40 mV (referenced to the mucosal surface) (Al-Bazzaz and Cheng 1979; Boucher et al. 1980; Olver et al. 1975; Widdicombe et al. 1979b). The values vary substantially between individual animals and depend on factors in the neurohumoral environment. The transepithelial resistance (R_t) ranges from 200 to 700 $\Omega \cdot cm^2$; again there is substantial variability.

When Ψ_t is clamped to zero, the short-circuit current (Isc) is 30 to 140 $\mu A\ cm^{-2}$. Transepithelial ion fluxes measured under short-circuit conditions have shown that the I_{sc} is produced by a net secretory flux of Cl^- from submucosa to mucosa and a net absorptive flux of Na^+ in the opposite direction. In canine trachea, the fluxes vary in absolute size; however, under baseline conditions, the rate of Cl^- secretion is usually two to four times that of Na^+ absorption.

In addition to Cl^- and Na^+ transport, small secretory fluxes of Ca^{2+} (1 to 2 nEq $cm^{-2} h^{-1}$) (Al-Bazzaz et al. 1985) and K^+ (Smith and Frizzell 1984; Welsh 1983a) have been measured under some conditions. However, these fluxes make no appreciable contribution to the I_{sc}. The physiological significance and function of the small Ca^{2+} and K^+ fluxes are unknown.

3.2 Fluid Transport

Active electrolyte transport can drive fluid movement in canine trachea. Under baseline conditions no net fluid movement was observed, however, stimulation of Cl^- secretion induced a small secretion volume flow (Welsh et al. 1980). Fluid absorption has only been observed when the trachea was treated with mucosal amphotericin to increase the rate of Na^+ absorption (Nathanson et al. 1983b).

A significant part of the fluid flowing across the epithelium appears to move through the lateral intercellular space. This is evidenced by changes in structure of the lateral intercellular spaces that accompany fluid transport. On the one hand, when epinephrine stimulated Cl^- and fluid secretion, these spaces narrowed and collapsed (Al-Bazzaz and Veech 1984). On the other hand, when amphotericin stimulated Na^+ and fluid absorption, the lateral intercellular spaces dilated (Widdicombe et al. 1986), an effect that has also been observed when an osmotic gradient induced mucosal to submucosal fluid flow (Man et al. 1984).

3.3 Identity of the Transporting Cell

The airway epithelium is a pseudostratified columnar epithelium containing three main cell types: ciliated cells, goblet cells, and basal cells. Ciliated cells are roughly columnar, approximately 20 to 60 μm tall, 2 – 7 μm wide, and extend from the apical surface to the basement membrane (Breeze and Wheeldon 1977; Frasca et al. 1968). Approximately 250 cilia and numerous microvilli emerge from the apical surface. Mucus-containing goblet cells are interspersed with the ciliated cells at a ratio of approximately one for each five or more ciliated cells. Basal cells and intermediate cells contact the basement membrane but do not stretch up to the luminal surface, thus it is unikely that they are directly involved in transepithelial transport. Basal cells appear to be poorly differentiated germinal cells that develop into ciliated and goblet cells via intermediate cells. The epithelium also contains several other cell types, however, they are not common and their significance in transepithelial electrolyte transport is unknown.

Which of these cells secrete Cl^- and which absorb Na^+? In many epithelia heterogeneity in ion transport properties is paralleled by cellular heterogeneity. For example, in intestinal epithelium, crypt cells secrete Cl^- and surface cells absorb Na^+ (Welsh et al. 1982b). In cortical collecting tubules, principle cells absorb Na^+ and intercalated cells secrete H^+ or HCO_3^- (Schwartz 1985; Wade et al. 1979). However, in airway epithelia the ciliated cell probably both secretes Cl^- and absorbs Na^+. The most direct evidence to support this contention comes from intracellular microelectrode studies. During impalement of a single cell, one observes the characteristic electrophysiological responses to amiloride (inhibition of Na^+ absorption) and to epinephrine (stimulation of Cl^- secretion) (Welsh et al. 1983). However, this observation could result from electrical coupling between absorptive and secretory cells.

Other, less direct evidence also suggests that the same cell secretes Cl^- and absorbs Na^+. First, Cl^- secretion has been localized to ciliated cells with immunocytochemical techniques and anti-cAMP antibodies: secretagogues increase cAMP in ciliated cells (Lazarus et al. 1984b). Moreover, in the absence of Cl^- the same secretagogues regulate Na^+ absorption (Cullen and Welsh 1987). Second, even though cultured airway cells appear much more homogeneous (lacking goblet cells) than cells in the native epithelium, cultured cells both secrete Cl^- and absorb Na^+.

3.4 Cultured Tracheal Epithelium

The recent ability to culture tracheal epithelial cells that retain their ion transport properties provides an advance for studies of the mechanism and regulation of electrolyte transport, as well as for studies of human airways in which the availability of tissue is limited. Epithelial cells are enzymatically and mechanically isolated and then grown on permeable supports (collagen-coated filters). The cultured monolayers secrete Cl^-, absorb Na^+, respond to hormones, and transport ions by the same cellular mechanisms as native epithelia, indicating that they provide a suitable model for study (Coleman et al. 1984; Welsh 1985). The morphology of the cultured cells resembles that of the native epithelium, although cultured cells have a flatter, more cuboidal shape than cells in the native epithelium. This difference may account for the lower rate of transepithelial transport per cm^2 area and the tendency toward higher values of R_t in cultured monolayers. At present, we are limited to the study of primary cultures; the development of immortal cell lines would provide a significant advantage for future studies.

4 Cellular Mechanism of Chloride Secretion

Figure 1 shows a model that describes out current understanding of the mechanism of Cl^- secretion and Na^+ absorption by canine tracheal epithelium. I will consider each of the transport process in turn.

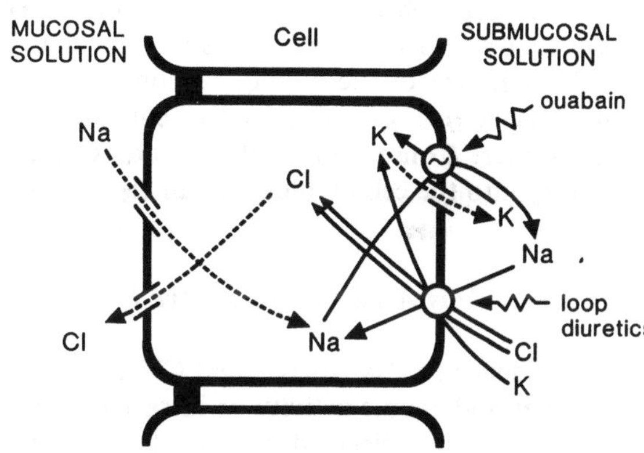

Fig. 1. Cellular model of the mechanism of electrolyte transport in canine tracheal epithelium

4.1 Chloride Entry at the Basolateral Membrane

4.1.1 Chloride Entry is Sodium-Dependent

In some of the first Ussing chamber studies directed at understanding the mechanism of Cl^- secretion, investigators deleted Na^+ from the bathing media (Al-Bazzaz and Al-Awqati 1979; Marin and Zaremba 1979; Widdicombe et al.

1979b). When choline replaced Na^+ in either the submucosal or both bathing solutions, Cl^- secretion was abolished. When Na^+ was replaced in the mucosal solution, Cl^- secretion did not change. Those studies showed that Cl^- secretion required submucosal Na^+ and suggested that Cl^- entry is coupled to Na^+.

4.1.2 Chloride Entry is Electrically Neutral

Studies with intracellular microelectrodes indicated that Cl^- entry was electrically neutral. When SO_4^{2-} or gluconate replaced submucosal Cl^-, the electrical potential difference across the basolateral membrane (Ψ_b) did not change (Welsh et al. 1982a). This result suggests that the chemical potential difference for Cl^- plays little or no part in determining basolateral voltage. The same ion substitution did not decrease the ratio of membrane resistances, R_a/R_b (where R_a and R_b refer to apical and basolateral resistances). If the basolateral membrane were Cl^--conductive, R_a/R_b should decrease when an impermeant anion replaces submucosal Cl^-.

4.1.3 Loop Diuretics Inhibit Chloride Entry

Submucosal addition of the loop diuretics, furosemide, piretanide, and bumetanide (in order of increasing potency), inhibited net Cl^- secretion by decreasing the unidirectional flux of Cl^- from submucosal to mucosal solution (Welsh 1983b; Widdicombe et al. 1983). By analogy with the effect of loop diuretics in other cells, this observation suggested basolateral Na^+-coupled Cl^- uptake. Subsequent microelectrode studies showed that furosemide did not alter basolateral resistance or electromotive force, indicating that furosemide did not inhibit an electrically conductive Cl^- entry step, nor did it alter other basolateral conductances. Consequently, these findings indicate that furosemide inhibited Cl^- entry into the cell by blocking an electrically neutral Cl^- entry step.

4.1.4 Chloride is Accumulated Above Electrochemical Equilibrium

Direct evidence for Na^+-dependent Cl^- accumulation came from measurements of the thermodynamic activity of intracellular Cl^-, a_c^{Cl}, made with intracellular Cl^--selective microelectrodes (Shorofsky et al. 1984; Welsh 1983c). Table 1 shows that under both secreting and nonsecreting conditions, a_c^{Cl} is two to four times the value predicted for electrochemical equilibrium. The absolute value of a_c^{Cl} varies, depending upon whether the epithelium was studied under short- or open-circuit conditions. These measurements of a_c^{Cl} are in excellent agreement with estimates of a_c^{Cl} based on equivalent electrical circuit analysis (Welsh et al. 1983) and measurements of Cl^- concentration (50 mM) in isolated tracheal epithelial cells (Widdicombe et al. 1981).

Two studies established that Cl^- accumulation depended on Na^+. First, when choline replaced Na^+, a_c^{Cl} decreased to 17 mM, the value predicted for a

Table 1. Intracellular ion activities in canine tracheal epithelium[a]

Condition	Ψ_a (mV)	a_c^{Cl} (mM)	$\Delta\tilde{\mu}_a^{Cl}$ (mV)	$\Delta\tilde{\mu}_b^{Cl}$ (mV)	Ψ_b (mV)	a_c^K (mM)	$\Delta\tilde{\mu}_b^K$ (mV)	Ψ_b (mV)	a_c^{Na} (mM)	$\Delta\tilde{\mu}_b^{Na}$ (mV)
Baseline										
Short circuit	−60	37	−35	−35	−58	69	−19			
Open circuit	−49	47	−27	−41	−61	83	−19	−64	11	+130
Epinephrine										
Short circuit	−46*	39	−22*	−22	−54*	70	−24*			
Open circuit	−44*	32*	−11*	−42	−58	69	−18	−65	21*	+110*

[a] Indomethacin (10^{-6}) was present during baseline conditions to minimize the rate of Cl^- secretion and epinephrine (10^{-6} M) was added during the second period.
* indicates significant change from baseline. (After data in Shorofsky et al. 1984, 1986; Smith and Frizzell 1984; Welsh 1983c, d).

passive distribution (Welsh 1983c). Second, following submucosal addition of bumetanide, a_c^{Cl} decreased from 26 to 16 mM.

4.1.5 *Chloride Probably Enters on an Na−K−2Cl Cotransporter*

There are several known molecular mechanisms in other epithelia that could effect NaCl entry: parallel Na−H and $Cl-HCO_3$ exchangers (Liedtke and Hopfer 1982a, b), coupled Na−Cl entry (Stokes 1984), and Na−K−2Cl cotransport (Geck and Heinz 1986). Although there is no definitive proof in trachea, several observations favor the last alternative. Maneuvers expected to inhibit parallel $Cl-HCO_3$ and Na−H cotransport did not inhibit Cl^- secretion: neither the removal of HCO_3^- and CO_2 from the bathing solutions, the application of substituted stilbenes (4-acetamido-4′-isothiocyano-2,2′-stilbene disulfonate, SITS, or 4,4′-diisothiocyanostilbene-2,2′-disulfonate, DIDS), nor the addition of acetazolamide inhibited Cl^- secretion. Hydrochlorothiazide, which blocks NaCl cotransport in flounder bladder (Stokes 1984) did not block Cl^- secretion. In contrast, loop diuretics inhibited Cl^- transport with a half-maximal concentration, similar to that required to inhibit Na−K−2Cl cotransport in other tissues.

Further support for Na−K−2Cl cotransport comes from the observation that the oxygen cost of Cl^- transport is less than the oxygen cost of Na^+ transport (Welsh 1984a), suggesting that more than one Cl^- ion enters coupled to each Na^+ ion. This finding is also consistent with energetic measurements in other epithelia that conform to the same mechanism of Cl secretion as canine trachea (Silva et al. 1980; Smaje et al. 1986). Table 1 shows that the Na^+ gradient contains more than enough energy to drive Cl^- uptake via an Na−K−2Cl cotransporter. Thus, in trachea, and in several other epithelia, Cl^- transport may be coupled to Na^+ transport in a way that maximizes use of the metabolic energy that generates the Na^+ gradient.

4.1.6 Chloride Entry is Regulated

When Cl^- secretion is stimulated, the rate of Cl^- entry increases. Table 1 shows that such an increase cannot be attributed to an increase in the electrochemical driving force, no matter what the molecular mechanism. In fact, under open-circuit conditions the net electrochemical gradient for Cl^- entry may actually decrease. Thus, the entry step must be regulated. Although the mediator is unknown, cAMP is a reasonable candidate since many agents that stimulate secretion increase cAMP and there is a precedent for cAMP mediation of Na–K–2Cl cotransport (Geck and Heinz 1986). Thus, cAMP might control both the rate of Cl^- entry and the rate of Cl^- exit.

4.2 Chloride Exit at the Apical Membrane

4.2.1 Chloride Exit is Electrically Conductive

Stimulation of Cl^- secretion increases I_{sc} and decreases R_t. This was the first observation suggesting that Cl movement was electrically conductive. Subsequent studies with intracellular microelectrodes clearly established the Cl^- conductive nature of the apical membrane (Shorofsky et al. 1983, 1984; Welsh 1983a, c, 1985; Welsh et al. 1982a, 1983). First, when nontransported anions replace mucosal Cl^-, Ψ_a depolarizes. Second, when mucosal Cl^- concentration decreases, R_a/R_b increases. Third, under secreting conditions, the chemical gradient for Cl^- primarily determines the apical membrane EMF. In contrast, changes in the mucosal Na^+ or K^+ concentration have little influence on Ψ_a or R_a/R_b.

4.2.2 Carboxylic Acid Analogs Block the Chloride Conductance

Two carboxylic acid analogs, anthracene-9-carboxylic acid (9-AC) and diphenylamine-2-carboxylate (DPC) inhibit apical Cl^- conductance (Welsh 1984b, c). When added to the mucosal solution, 9-AC rapidly and reversibly decreased G_t and inhibited Cl^- secretion. This effect was localized to the apical membrane: 9-AC hyperpolarized Ψ_a, increased R_a/R_b, and prevented the characteristic electrical changes produced by lowering mucosal Cl^- concentration. The synthesis and testing of new agents should lead to more specific and potent inhibitors of the apical Cl^- conductance. It is encouraging that such agents have recently been described for the thick ascending limb of the loop of Henle (Wangemann et al. 1986).

4.2.3 The Apical Chloride Permeability Regulates the Rate of Secretion

The first effect of a Cl^- secretagogue, such as epinephrine or isoproterenol, is to decrease R_a, often to a level one-tenth the initial value (Welsh et al. 1983;

Shorofsky et al. 1983). The decrease in R_a results from an increase in apical Cl permeability. As R_a decreases, Ψ_a depolarizes and approaches the electromotive force for Cl^- at the apical membrane. The quantitative and temporal relation between an increase in Cl^- secretory rate and an increase in apical Cl^- permeability suggest that the Cl^- exit step is a major regulatory site controlling secretion.

4.2.4 Chloride Exits via Apical Chloride Channels

Recent studies using the patch-clamp technique (Hamill et al. 1981) have demonstrated apical membrane Cl^- channels in dog and human airways (Frizzell et al. 1986; Shoemaker et al. 1986; Welsh 1986b, c). The properties of this channel are different from those described in other epithelia or excitable cells. Figure 2 shows the current-voltage relation of a Cl^- channel in an excised, inside-out patch with 145 mM NaCl solutions bathing both surfaces of the patch. Despite the symmetrical solutions, there is outward rectification: conductance increases at depolarizing voltages. Figure 2 also shows the effect of anion gradients on the current-voltage relation: this channel was seven to ten times more selective for Cl^- than for Na^+. Current was not carried by SO_4^{2-} or gluconate and substitution of Cs^+ or K^+ for Na^+ had little effect on the conductive properties. The single-channel conductance varies somewhat from channel to channel: there may be more than one class of Cl^- channel (Shoemaker et al. 1986) or the size of the single class of channel may vary (Welsh 1986c). In either case, at this time the properties and regulation of all the Cl^- channels appear similar. However, further work is needed.

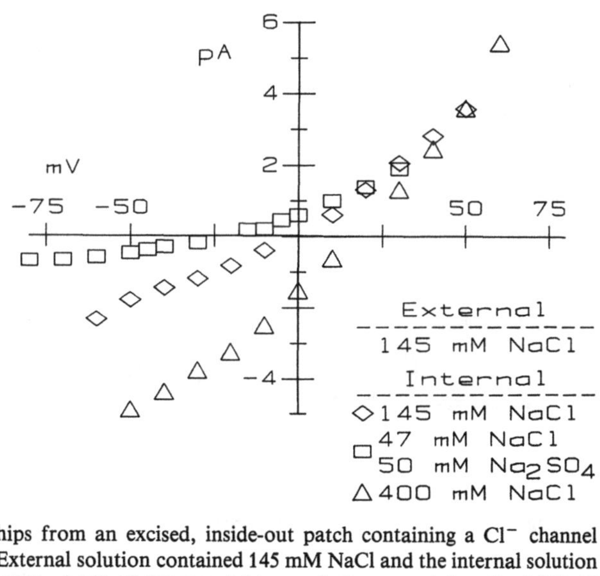

Fig. 2. Current-voltage relationships from an excised, inside-out patch containing a Cl^- channel from canine tracheal epithelium. External solution contained 145 mM NaCl and the internal solution was changed from one containing 145 mM NaCl (*diamonds*) to a solution containing 47 mM NaCl and 50 mM Na_2 SO_4 (*squares*) and then to a solution containing 400 mM NaCl (*triangles*). (After Welsh 1986c)

This channel could be found in patches of membrane obtained from single cells, cells growing at the edge of a group of cells, and cells in confluent sheets, where only the apical membrane is accessible to the recording pipette. In contrast, the basolateral K^+ channel was not observed in the apical membrane of confluent sheets of cells (Welsh and McCann 1985). This observation suggests that the channel is located in the apical membrane and produces the apical Cl^- permeability. Additional evidence is that both DPC and 9-AC decreased single-channel current with a rank-order potency and at inhibitor concentrations similar to those required to block Cl^- secretion in native epithelia (Welsh 1986c).

4.2.5 Intracellular Second Messengers Regulate the Chloride Channel

Figure 3 shows the effect of isoproterenol on a Cl^- channel recorded from a cell-attached patch. Before addition of secretagogue, the patch was quiet with no channel openings; afterwards a Cl^- channel frequently opened. These results correspond with the isoproterenol-induced increase in apical Cl^- conductance in native and cultured epithelia.

Although the identity of second messengers that directly regulate the Cl^- channel is uncertain, cAMP must be considered a strong candidate. Studies of ion channels in excitable cells have shown that channel phosphorylation by cAMP-dependent protein kinase modulates function, increasing or decreasing the probability that ion channels are open (Levitan 1985). By analogy, cAMP might trigger Cl^- channel phosphorylation in airway epithelium, leading to a change in conformation and increased opening. However, other scenarios should not be ruled out. For example, the agent that directly regulates the channel might lie several steps distal to cAMP or a cAMP-dependent kinase. Moreover, since secretagogues do not open all channels in the cell membrane (some Cl^- channels do not open until they are excised from the cell), some intracellular substance

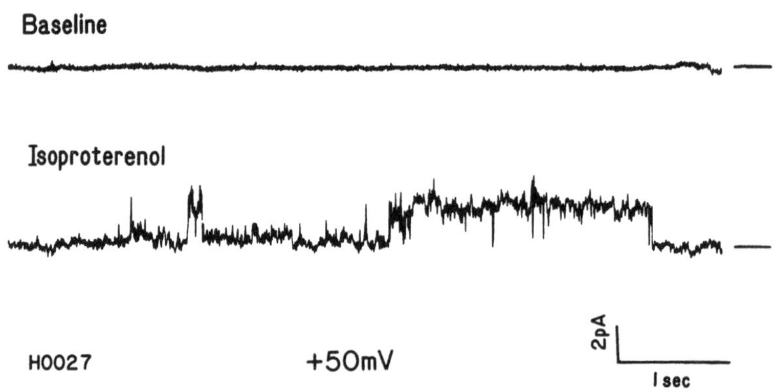

Fig. 3. Effect of isoproterenol on a cell-attached Cl^- channel. The top sweep was taken before and the bottom sweep 1 min after addition of isoproterenol (5 μM). Solid line on right indicates zero-current level

may tonically inhibit the channel (Welsh 1986c); when the channel is removed from the cell, the channel is no longer inhibited.

The intracellular Ca^{2+} concentration mediates stimulus-secretion coupling in a variety of cells (Berridge 1980) and has been proposed as a regulator of the Cl^- channel in airway cells. However, its precise regulatory role is unresolved. In some studies, changes in the Ca^{2+} concentration bathing the cytosolic surface of excised, inside-out patches did not acutely regulate the Cl^- channel (Welsh 1986b, c). However, in another study (Frizzell et al. 1986), an increase of internal Ca^{2+} did increase channel opening. In that study the response to Ca^{2+} was delayed: channel activity did not change for several minutes after bath Ca^{2+} concentration changed or after quiescent patches were excised into a bath containing high concentrations of Ca^{2+}. These observations suggest that Ca^{2+} may not be the primary or sole regulator of the Cl^- channel, although it may be involved at some point in regulation.

An understanding of the direct and indirect metabolic steps that lead to regulation of the Cl^- channel is very important goal. Both cAMP and Ca^{2+} may play important roles, but their precise functions are not yet defined.

4.3 The Na^+-K^+ ATPase

4.3.1 The Na^+-K^+ ATPase is Located on the Basolateral Membrane

Na^+ that enters the cell coupled to Cl^- is pumped out at the basolateral membrane by the Na^+-K^+ ATPase. Many observations point to the pump's basolateral location. First, ouabain inhibited transport only when added to the submucosal solution, not when added to the mucosal solution (Widdicombe et al. 1979b). Second, when K^+ was removed from the submucosal solution, transport was inhibited; removing K^+ from the mucosal solution had no effect (Widdicombe et al. 1979b; Westenfelder et al. 1980). Third, binding of ^{3}H-ouabain was much greater from the submucosal than from the mucosal solution (Widdicombe et al. 1979a). Fourth, autoradiographic ^{3}H-ouabain binding was localized to the basolateral cell membrane (Widdicombe et al. 1979a).

4.3.2 The Na^+-K^+ ATPase Provides the Energy for Secretion

The Na^+-K^+ ATPase in airway epithelia has biochemical and functional properties like those of Na^+-K^+ ATPase in a variety of other tissues (Westenfelder et al. 1980). Table 1 shows that the Na pump accumulates K^+ in the cell and maintains a low intracellular Na^+ activity. Ouabain addition to the submucosal solution inhibits the Na^+-K^+ ATPase and abolishes Cl^- secretion and Na^+ absorption. Na-pump activity at the basolateral membrane does not appear to determine directly the electrical properties of the basolateral membrane.

4.4 Potassium Exit at the Basolateral Membrane

Potassium that enters the cell at the basolateral membrane on the Na^+-K^+ ATPase (and possibly on an Na-K-2Cl cotransporter) must also exit across the basolateral membrane since the epithelium secretes very little K^+.

4.4.1 Potassium Exit is Electrically Conductive

Three observations support this conclusion. First, when submucosal K^+ increases, Ψ_b and Ψ_t depolarize (Smith and Frizzell 1984; Welsh 1983a). In contrast, neither Na^+ nor Cl^- substitutions change Ψ_b. Second, submucosal addition of Ba^{2+}, which blocks many types of K^+ channels, inhibited Cl^- secretion, depolarized Ψ_b, doubled R_t, and attenuated the K^+-induced depolarization of Ψ_b (Smith and Frizzell 1984; Welsh 1983e). Third, as shown in Table 1, the K^+ electrochemical gradient is more than sufficient to account for the estimated rate of K^+ exit via the basolateral K^+ conductance.

4.4.2 Potassium Channels Make the Basolateral Membrane K^+ Permeable

The patch-clamp technique has demonstrated K^+ channels in the epithelium; Fig. 4 shows the current-voltage relationship of a single K^+ channel in an excised, inside-out path (Welsh and McCann 1985). The channel is highly selective for K^+ over anions and over Na^+, the I-V relation shows inward rectification (conductance increases at hyperpolarizing voltages) even in the presence of symmetrical

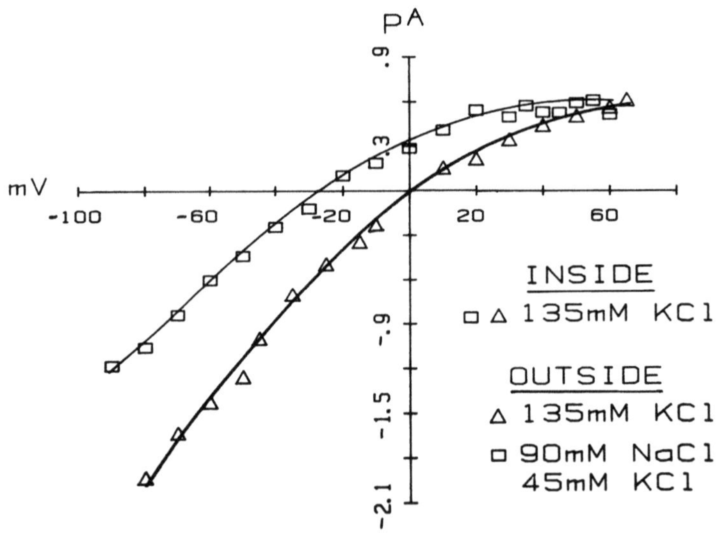

Fig. 4. Current-voltage relations of K^+ channels in excised, inside-out patches from canine tracheal epithelium. Internal solution contained 135 mM KCl and external solution contained either 135 mM KCl (*triangles*) or 90 mM NaCl and 45 mM KCl (*squares*). (After Welsh and McCann 1985)

K^+ solutions, and the channel was not voltage-gated. These properties are different from those of previously described K^+ channels, although they somewhat resemble those of the inward rectifier in cardiac ventricular cells (Sakmann and Trube 1984).

The K^+ channel was found in patches from isolated cells, from freshly dissociated cells, and from cells at the edge of a cluster of confluent cells but not in the apical membrane of confluent cells. These observations suggest that the K^+ channel is normally located on the basolateral membrane of confluent cells and is therefore inaccessible to the pipette tip.

4.4.3 Intracellular Calcium Regulates the Potassium Channel

When secretion is stimulated, basolateral K^+ conductance increases, so that during steady-state stimulation the apical membrane becomes more permeable to Cl^- and the basolateral membrane becomes more permeable to K^+ (Shorofsky et al. 1983; Welsh et al. 1982a, 1983). Figure 5 shows that when the K^+ channel is studied in cell-attached patches, addition of epinephrine or isoproterenol to the bathing solution increased the frequency of channel opening, a result which indicates that an intracellular second messenger regulates the channel.

Two studies indicate that the second messenger is cytosolic Ca^{2+} (Welsh and McCann 1985). First, in cell-attached patches, the calcium ionophore, A-23187, added to the bath solution increased the probability of channel opening in a Ca^{2+}-dependent manner. Second, in cell-free, excised patches an increase in the Ca^{2+} concentration on the cytosolic surface increased the probability of finding the channel open, most markedly at concentrations of Ca^{2+} between 100 nM and

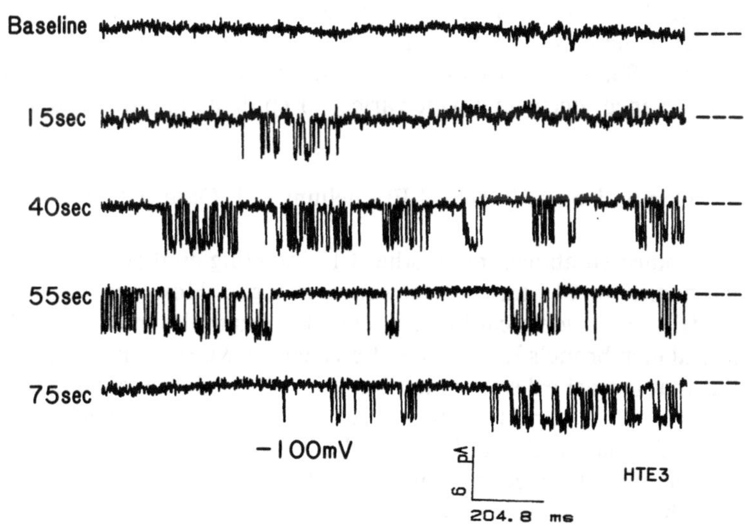

Fig. 5. Effect of isoproterenol on K^+ channel activity in a cell-attached patch. Tracings were obtained before and then at the times indicated after addition of isoproterenol to the bathing solution. *Dashed lines* indicate zero-current level

1 μM. The duration of the channel closed state decreased; the duration of the open state changed little.

Other experiments also suggest that Ca^{2+} regulates the basolateral K^+ channel. First, when A-23187 was added to cell monolayers, the first change was an increase in basolateral K^+ conductance (Welsh 1986a). Second, α_1-adrenergic agonists, whose effect is usually mediated by an increase in cell Ca^{2+} (Exton 1981) increased basolateral K^+ conductance (Welsh 1986a). Third, in a suspension of cells loaded with a fluorescent Ca^{2+} indicator (quin2), epinephrine increased cell Ca^{2+} concentration (Welsh and McCann 1985). Although Ca^{2+} regulates the K^+ channel, it might also be regulated by other intracellular second messengers because the epinephrine-induced increase in cell Ca^{2+} is transient. We need further studies to completely understand the regulation.

4.4.4 Regulation of Potassium Permeability is Critical for Secretion

Parallel increases in apical Cl^- and basolateral K^+ permeabilities have two important physiological consequences. First, the parallel increases prevent large transport-related alterations in intracellular K^+ activity. When the rate of Cl^- secretion increases, the rate of K^+ entry into the cell must also increase. To drive the K^+ back across the basolateral membrane, either the driving force must increase or the basolateral membrane K^+ permeability must increase. If the driving force increased, the K^+ concentration in the cell would increase, possibly causing large alterations in cell volume. Instead, the K^+ permeability increases and intracellular K^+ activity remains nearly constant. Second, an increase in basolateral K^+ permeability prevents a sustained depolarization of intracellular voltage. If K^+ permeability did not increase with the Cl^- permeability, the cell would remain depolarized when secretagogues were added. When basolateral K^+ conductance increases, intracellular voltage shifts back toward the chemical gradient for K^+. Similar homeostatic mechanisms for coupling apical and basolateral conductances occur in a wide variety of epithelia (Schultz 1981).

4.5 Comparison of Tracheal Epithelium with Other Cl^--Secreting Epithelia

As I indicated above, many other Cl^- secreting epithelia share the same cellular mechanism of Cl^- transport as shown in Fig. 1. However, the components of the model are not identical in all tissues. For example, in salivary glands the basolateral membrane's K^+ channel (Petersen and Maruyama 1984) is Ca^{2+}-activated, like that in tracheal epithelium, but it is voltage-dependent and the single-channel conductance is almost ten times greater than that in tracheal epithelium. The K^+ channel in lacrimal glands is like the large conductance Ca^{2+}-activated "maxi" K^+ channel observed in many cells (Petersen and Maruyama 1984). The apical Cl^- channel may also vary in different cells. Lacrimal glands contain a low conductance (1 – 2 pS) Ca^{2+}-activated Cl^- channel, as indicated by noise analysis (Marty et al. 1984). Two cultured cell lines that secrete Cl^-, MDCK (Kolb et al. 1985) and A6 (Nelson et al. 1984) epithelial cells, contain a large conductance

(400 pS) voltage-dependent Cl^- channel that resembles the voltage-dependent anion channel in the outer mitochondrial membrane (Colombini 1979).

Thus, the use of the Na^+ gradient for "secondary-active" Cl^- transport and the use of the Na^+ pump to supply the energy to Cl^- transport is a common feature of several cells. However, the individual transport process may vary depending on the requirements and evolution of the cell.

5 Cellular Mechanism of Sodium Absorption

Airwar epithelia absorb Na^+ by an electrogenic process that conforms to the classic Koefoess-Johnsen and Ussing model (Koefoed-Johnsen and Ussing 1958). Figure 1 shows the details.

5.1 Sodium Entry at the Apical Membrane

Sodium enters the cell via an amiloride-sensitive Na^+ channel. Amiloride added to the mucosal solution inhibits both Na^+ absorption and I_{sc} (Boucher and Gatzy 1981; Widdicombe and Welsh 1980), approximately 100 nM amiloride is required for half-maximal inhibition (Cullen and Welsh 1987). Amiloride also increases R_a and hyperpolarizes Ψ_a (Welsh et al. 1983), changes idential to those observed in a variety of Na^+-absorbing epithelia. However, electrogenic Na^+ absorption is not completely amiloride-sensitive: following addition of maximal concentrations of amiloride some residual electrogenic Na^+ absorption often remains (Cullen and Welsh 1987; Langridge-Smith 1986; Widdicombe and Welsh 1980). Sodium entry is a passive process, with Na^+ moving down a favorable electrochemical gradient as shown in Table 1.

5.2 Sodium Exit at the Basolateral Membrane

Addition of submucosal ouabain abolishes Na^+ absorption (Widdicombe et al. 1979b), indicating that the basolateral membrane Na^+-K^+ ATPase pumps Na^+ out the cell. Thus, the Na^+ pump provides the energy for both Cl^- secretion and Na^+ absorption.

6 Regulation of Chloride Secretion

In discussing the mechanisms of Cl^- secretion, I have indicated our knowledge of how the individual ion transport properties are regulated. In this section I will briefly review our knowledge of how neurotransmitters, hormones, and autocoids control Cl^- secretion. Table 2 shows a list of agents reported to stimulate secretion in canine tracheal epithelium. Rather than discuss each agent individually, I will make a few general points.

Table 2. Agents that stimulate chloride secretion

Agent	Surface[a]	cAMP[b]	Reference
β-Adrenergic agonists	S	↑	Al-Bazzaz and Cheng (1979), Boucher and Gatzy (1982), Davis et al. (1979), Lazarus et al. (1984b), Smith et al. (1982), Welsh (1986a)
Prostaglandin E_1 and E_2	M or S	↑	Al-Bazzaz et al. (1981) Lazarus et al. (1984a), Smith et al. (1982)
Prostaglandin $F_{2\alpha}$	M or S	–	Al-Bazzaz et al. (1981), Smith et al. (1982)
Leukotrienes C_4 and D_4	S		Leikauf et al. (1986)
Vasoactive intestinal peptide	S	↑	Lazarus et al. (1986a), Nathanson et al. (1983a)
Adenosine	M	↑	Pratt et al. (1986)
Bradykinin	M > S		Leikauf et al. (1985)
Substance P	M		Al-Bazzaz and Jayaram (1985)
Mast cell mediators	M and S		Lazarus et al. (1986b)
Eosinophil major basic protein	M		Jacoby et al. (1986)

[a] M and S refer to mucosal and submucosal surface.
[b] ↑ increase; – no change. A blank space indicates that data is unavailable.

6.1 Stimulation from the Mucosal and Submucosal Surface

In many epithelia, signals that regulate ion transport come from the submucosal or basolateral surface of the cell. Table 2 shows a similar situation in airway epithelia: most mediators have their receptors on the submucosal surface. However, the location of the tracheal epithelium at the interface between the external and internal environment and the variable demands on the mucociliary clearance apparatus might lead one the expect some regulation of transport from the mucosal surface. In fact, receptors to the local transmitter adenosine are located on the apical membrane (Pratt et al. 1986) and bradykinin is much more potent when added to the mucosal surface than to the submucosal surface (Leikauf et al. 1985). Hormone receptors located on the mucosal surface would allow regional regulation of transport; inhaled material might trigger resident cells in the airway lumen to release transmitters, thereby stimulating secretion and mucociliary clearance.

6.2 Regulation by Mediators of Inflammation

Because airways sit at the interface between the external environment and the organism, they are frequently involved with the inflammatory response. Thus, it is not surprising that a variety of inflammatory mediators regulate secretion. For example: bradykinin (Leikauf et al. 1985), leukotrienes C_4 and D_4 (Leikauf et al. 1986), and eosinophilic major basic protein (Jacoby et al. 1986) induce Cl^- secretion. The physiological function of these mediators may be to increase mucociliary clearance.

6.3 Regulation by Endogenous Prostaglandins

Several observations indicate that the epithelium produces prostaglandins to regulate secretion. First, exogenous prostaglandin E_1 and E_2 stimulate the epithelium to secrete Cl^- and increase cellular levels of cAMP (Al-Bazzaz et al. 1981; Lazarus et al. 1984a; Smith et al. 1982). Second, indomethacin and other cyclooxygenase inhibitors decrease the baseline rate of secretion without attenuating the response to exogenous prostaglandins (Al-Bazzaz et al. 1981). Third, indomethacin decreases baseline levels of cellular cAMP (Lazarus et al. 1984a; Smith et al. 1982). Fourth, the epithelium produces prostaglandins and indomethacin inhibits their production (Eling et al. 1986; Smith et al. 1982; Xu et al. 1986).

Moreover, agents may induce secretion by stimulating the epithelium to produce prostaglandins. The secretory effects of bradykinin (Leikauf et al. 1985), leukotriene C_4 and D_4 (Leikauf et al. 1986), mast cell mediators (Lazarus et al. 1986b), and eosinophilic major basic protein (Jacoby et al. 1986) are all abolished or attenuated by prostaglandin synthesis inhibitors. In addition, bradykinin and leukotriene C_4 and D_4 have been shown to increase the production of prostaglandins. Thus, prostaglandins play a central role in controlling the baseline rate of Cl^- secretion and the secretory response to neurohumoral agents.

6.4 Regulation by Cyclic-AMP

There is considerable evidence that cAMP is an important second messenger regulating Cl^- secretion. Table 2 shows that many agents which stimulate secretion also increase intracellular levels of cAMP. The phosphodiesterase inhibitor, theophylline, and the adenylate cyclase activator, forskolin, also increase cellular cAMP levels and stimulate secretion (Al-Bazzaz and Al-Awqati 1979; Smith et al. 1982). Dibutyryl cAMP or 8-bromo-cAMP mimics the secretory effect of hormones (Al-Bazzaz 1981). However, we do not know the metabolic steps between an increase in cAMP and an increase in Cl^- secretion: the pathways may be direct or indirect and cAMP might not be the only important physiological regulator. Studies of the cellular regulatory pathways are a particularly important focus for future work.

6.5 Regulation by Calcium

In exocrine glands, an increase in the concentration of cytosolic Ca^{2+} stimulates fluid secretion, presumably mediated by Cl^- transport (Petersen 1986). In canine trachea, there is evidence for a secretory effect of raising cell Ca^{2+}: the calcium ionophore, A-23187, stimulated secretion in a Ca^{2+}-dependent manner (Al-Bazzaz and Jayaram 1981). However, recent data indicates that the stimulatory effect of A-23187 may not be entirely a direct result of an increase in cell Ca^{2+} in dog tracheal epithelium: indomethacin attenuated the epithelium's secretory response to A-23187, and A-23187 stimulated prostaglandin production (Eling et al. 1986; Welsh 1987). However, there may be species differences in this effect. New techniques may help us understand how cytoplasmic Ca^{2+} concentrations are regulated and how Ca^{2+} effects transepithelial Cl^- secretion and the individual ion transport processes.

6.6 Comparison with Other Secretory Epithelia

The regulation of secretion varies widely. For example, in many glandular epithelia muscarinic acetylcholine receptors regulate secretion. In trachea, these receptors regulate submucosal gland secretion but have little if any effect on transport by the surface epithelium. Because of their differing receptors, secretory cells adapt to the specific physiological requirements of different organs and tissues yet maintain a common mechanism of secretion. Unfortunately, we have little information on the differences and similarities in the second messenger systems that regulate secretion in this diverse group of epithelia.

7 Regulation of Sodium Absorption

Sodium absorption may be just as important as Cl^- secretion for controlling the quantity and composition of the respiratory tract fluid, but we know much less about it.

7.1 Acute Regulation of Sodium Absorption

Agents that stimulate Cl^- secretion frequently decrease Na^+ absorption; prostaglandin E_1 (Al-Bazzaz et al. 1981) and dibutyryl cAMP (Al-Bazzaz 1981) are two examples. Many of the other agents listed in Table 2 also tended to inhibit Na^+ absorption, but the decrease was often not statistically significant. In contrast, when canine tracheal epithelium was bathed in a Cl^--free gluconate solution, prostaglandin E_2, 2-chloroadenosine, isoproterenol, and 8-bromo-cAMP stimulated Na^+ absorption (Cullen and Welsh 1987). These observations suggest that cAMP regulates Na^+ absorption as well as Cl^- secretion. However, the contrast between the effects of cAMP on Na^+ absorption in the presence and absence of Cl^- raises interesting questions about the regulation and interaction between Na^+ and Cl^-. It is interesting the cAMP also stimulates Na^+ absorption in another lung epithelium: cultured monolayers of alveolar type-2 cells (Cott et al. 1983; Crandall et al. 1983).

Both in the presence and absence of Cl^-, A-23187 inhibited Na^+ absorption (Al-Bazzaz and Jayaram 1981a; Cullen and Welsh 1987), suggesting that an increase in cell Ca^{2+} inhibits absorption.

7.2 Chronic Regulation of Sodium Absorption

Aldosterone chronically regulates Na^+ absorption in many epithelia and may also do so in trachea. In monolayers of epithelia grown on permeable supports in serum-free media, 2-day exposure to aldosterone stimulated the amiloride-sensitive current by 50% without altering Cl^- secretion (Cullen and Welsh 1987). In contrast, in vivo studies in humans showed that the aldosterone antagonist, spironolactane, did not alter transepithelial nasal voltage, although voltage across rectal mucosa decreased (Knowles et al. 1985).

7.3 Interaction Between Chloride Secretion and Sodium Absorption

As indicated above, the same airway cell probably both secretes Cl^- and absorbs Na^+. An important but unanswered question is whether the pathways which regulate these processes interact. There is no direct evidence on this in canine tracheal epithelium, although several agents that stimulate secretion tend to decrease the rate of Na^+ absorption. In bovine tracheal epithelium, interactions between the two transport processes have been observed under some conditions (Langridge-Smith et al. 1984). Given the importance of both processes in controlling the respiratory tract fluid, further study is clearly warranted.

Acknowledgments. I thank Sandy Wagner for secretarial assistance. Work from the author's lab was supported by grants from the NIH (HL-29851 and HL-14388), American Heart Association, and Council for Tobacco Research. The author is an Established Investigator of the American Heart Association.

References

Al-Bazzaz FJ (1981) Role of cyclic AMP in regulation of chloride secretion by canine tracheal mucosa. Am Rev Respir Dis 123:295–298

Al-Bazzaz FJ, Al-Awqati Q (1979) Interaction between sodium and chloride transport in canine tracheal mucosa. J Appl Physiol 46(1):111–119

Al-Bazzaz FJ, Cheng E (1979) Effect of catecholamines on ion transport in dog tracheal epithelium. J Appl Physiol Respirt Environ Physiol 47(2):397–403

Al-Bazzaz FJ, Jayaram T (1981) Ion transport by canine tracheal mucosa: effect of elevation of cellular calcium. Exp Lung Res 2:121–130

Al-Bazzaz FJ, Jayaram T (1985) Calcium secretion in canine tracheal mucosa. J Appl Physiol 59(4):1191–1195

Al-Bazzaz F, Veech J (1984) Narrowing of intercellular spaces by epinephrine in canine tracheal mucosa. Am Rev Respir Dis 129:A315

Al-Bazzaz FJ, Yadava VP, Westenfelder C (1981) Modification of Na and Cl transport in canine tracheal mucosa by prostaglandins. Am J Physiol 240 (Renal Fluid Electrolyte Physiol 9):F101 – F105
Al-Bazzaz FJ, Kelsey JG, Kaage WD (1985) Substance P stimulation of chloride secretion by canine tracheal mucosa. Am Rev Respir Dis 131:86 – 89
Berridge MJ (1980) The role of cyclic nucleotides and calcium in the regulation of chloride transport. Ann N Y Acad Sci 341:156 – 171
Boucher RC, Gatzy JT (1981) Effect of amiloride (Am) and mucosal sodium removal on canine bronchial ion transport. Fed Proc 40:447
Boucher RC, Gatzy JT (1982) Regional effects of autonomic agents on ion transport across excised canine airways. J Appl Physiol 52(4):893 – 901
Boucher RC, Bromberg PA, Gatzy JT (1980) Airway transepithelial electrical potential in vivo: species and regional differences. J Appl Physiol 48(1):169 – 176
Breeze RG, Wheeldon EB (1977) The cells of the pulmonary airways. Am Rev Respir Dis 116:705 – 777
Coleman DL, Tuet IK, Widdicombe JH (1984) Electrical properties of dog tracheal epithelial cells grown in monolayer culture. Am J Physiol 246 (Cell Physiol 15):C355 – C359
Colombini M (1979) A candidate for the permeability pathway of the outer mitochondrial membrane. Nature (London) 279:643 – 645
Cott GR, Sugahara K, Mason RJ (1983) Cyclic adenosine monophosphate is a modulator of the bioelectric properties of alveolar type II cell monolayers. Physiologist 26:A125
Crandall ED, Brown SES, Goodman BE (1983) Regulation of transport across alveolar epithelial cell monolayers by cAMP. Am Rev Respir Dis 127:306
Cullen JJ, Welsh MJ (1987) Regulation of sodium absorption by canine tracheal epithelium. J Clin Invest 79:73 – 79
Davis B, Marin MG, Yee JW, Nadel JA (1979) Effect of terbutaline on movement of Cl^- and Na^+ across the trachea of the dog in vitro. Am Rev Respir Dis 120:547 – 552
Eling TE, Danilowica RM, Henke DC, Sivarajah K, Yankaskas JR, Boucher RC (1986) Arachidonic acid metabolism by canine tracheal epithelial cells. J Biol Chem 261:12841 – 12849
Exton JH (1981) Molecular mechanisms involved in α-adrenergic responses. Mol Cell Endocrinol 23:233 – 264
Foster WM, Bergofsky EH, Bohning DE, Lippmann M, Albert RE (1976) Effect of adrenergic agents and their mode of action on mucociliary clearance in man. J Appl 41:146 – 152
Frasca JM, Auerbach O, Parks VR, Jamieson JD (1968) Electron microscopic observations of the bronchial epithelium of dogs. Exp Mol Pathol 9:363 – 379
Frizzell RA, Rechkemmer G, Shoemaker RL (1986) Altered regulation of airway epithelial cell chloride channels in cystic fibrosis. Science 233:558 – 560
Fulford GR, Blake JR (1986) Muco-ciliary transport in the lung. J Theor Biol 121:381 – 402
Geck P, Heinz E (1986) The Na-K-2Cl co-transport system. J Membr Biol 91:97 – 105
Gerrity TR, Cotromanes E, Garrard CS, Yeates DB, Lourensco RV (1983) The effect of aspirin on lung mucociliary clearance. J Engl J Med 308:139 – 141
Hamill OP, Marty A, Neher E, Sakmann B, Sigworth FJ (1981) Improved patch-clamp techniques for high-resolution current recording from cells and cell-free membrane patches. Pfluegers Arch 391:85 – 100
Jacoby DB, Ueki IF, Gleich GJ, Nadel JA (1986) Effects of human eosinophil major basic protein on ion transport in canine tracheal epithelium. Clin Res 34:108A
Knowles M, Gatzy J, Boucher R (1981) Increased bioelectric potential difference across respiratory epithelia in cystic fibrosis. N Engl J Med 305(25):1489 – 1494
Knowles MR, Gatzy JT, Boucher RC (1985) Aldosterone metabolism and transepithelial potential difference in normal and cystic fibrosis subjects. Pediatr Res 19(7):676 – 679
Koefoed-Johnsen V, Ussing HH (1958) The nature of the frog skin potential. Acta Physiol Scand 42:298 – 308
Kolb HA, Brown CDA, Murer H (1985) Identification of a voltage-dependent anion channel in the apical membrane of a Cl^--secretory epithelium (MDCK). Pfluegers Arch 403:262 – 265
Langridge-Smith JE (1986) Interaction between sodium and chloride transport in bovine tracheal epithelium. J Physiol (London) 376:299 – 319
Langridge-Smith JE, Rao MC, Field M (1984) Chloride and sodium transport across bovine tracheal epithelium. Pfluegers Arch 402:42 – 47

Lazarus SC, Basbaum CB, Gold WM (1984a) Prostaglandins and intracellular cyclic AMP in respiratory secretory cells. Am Rev Respir Dis 130:262–266
Lazarus SC, Basbaum CB, Gold WM (1984b) Localization of cAMP in dog and cat trachea: effects of β-adrenergic agonists. Am J Physiol 247 (Cell Physiol 16):C327–C334
Lazarus SC, Basbaum CB, Barnes PJ, Gold WM (1986a) cAMP immunocytochemistry provides evidence for functional VIP receptors in trachea. J Physiol (London) 251 (Cell Physiol 20): C115–C119
Lazarus SC, McCabe LJ, Nadel JA, Gold WM, Leikauf GD (1986b) Effects of mast cell-derived mediators on epithelial cells in canine trachea. Am J Physiol 251 (Cell Physiol 20): C387–C394
Leikauf GD, Ueki IF, Nadel JA, Widdicombe JH (1985) Bradykinin stimulates Cl secretion and prostaglandin E2 release by canine tracheal epithelium. Am J Physiol 248:F48–F55
Leikauf GD, Ueki IF, Widdicombe JH, Nadel JA (1986) Alteration of chloride secretion across canine tracheal epithelium by lipoxygenase products of arachidonic acid. Am J Physiol 250(19):F47–F53
Levitan IB (1985) Phosphorylation of ion channels. J Membr Biol 87:177–190
Liedtke CM, Hopfer U (1982a) Mechanism of Cl^- translocation across small intestinal brush border membrane. II. Demonstration of Cl^- OH^- exchange and Cl^- conductance. Am J Physiol 242:G272–G280
Liedtke CM, Hopfer U (1982b) Mechanism of Cl^- translocation across small intestinal brush border membrane. I. Absence of Na^+-Cl^- cotransport. Am J Physiol 242:G263–G271
Man SFP, Hulbert WC, Park DSK, Thomson ABR, Hogg JC (1984) Asymmetry of canine tracheal epithelium: osmotically induced changes. J Appl Physiol Respir Environ Physiol 57:1338–1346
Marin MG, Zaremba MM (1979) Interdependence of Na^+ and Cl^- transport in dog tracheal epithelium. J Appl Physiol 44(6):900–904
Marty A, Tan YP, Trautmann A (1984) Three types of calcium-dependent channels in rat lacrimal glands. J Physiol (London) 357:293–325
Nadel JA, Davis B, Phipps RJ (1979) Control of mucus secretion and ion transport in airways. Annu Rev Physiol 41:369–381
Nadel JA, Widdicombe JH, Peatfield AC (1985) Regulation of airway secretion, ion transport and water movement. In: Fishman AP, Fisher AB (eds) Handbook of physiology, Sect 3. The respiratory system, vol 1. Circulation and nonrespiratory functions. Am Physiol Soc, Bethesda Maryland
Nathanson I, Widdicombe JH, Barnes PJ (1983a) Effect of vasoactive intestinal peptide on ion transport across dog tracheal epithelium. J Appl Physiol 55(6):1844–1848
Nathanson I, Widdicombe JH, Nadel JA (1983b) Effects of amphotericin B on ion and fluid movement across dog tracheal epithelium. J Appl Physiol 55(4):1257–1261
Nelson DJ, Tang JM, Palmer LG (1984) Single-channel recordings of apical membrane chloride conductance in A6 epithelial cells. J Membr Biol 80:81–89
Olver RE, Davis B, Marin MG, Nadel JA (1975) Active transport of Na^+ and Cl^- across the canine tracheal epithelium in vitro 1–3. Am Rev Respir Dis 112:811–815
Petersen OH (1986) Calcium-activated potassium channels and fluid secretion by exocrine glands. Am J Physiol 251:G1–G13
Petersen OH, Maruyama Y (1984) Calcium-activated potassium channels and their role in secretion. Nature (London) 307:693–696
Pratt AD, Clancy G, Welsh MJ (1986) Mucosal adenosine stimulates chloride secretion in canine tracheal epithelium. Am J Physiol 251 (Cell Physiol 20):C167–C174
Sakmann B, Trube G (1984) Conductance properties of single inwardly rectifying potassium channels in ventricular cells from guinea-pig heart. J Physiol (London) 347:641–657
Schultz SG (1981) Homocellular regulatory mechanisms in sodium-transporting epithelia: avoidance of extinction by "flush-through". Am J Physiol 241:F579–F590
Schwartz GH, Barasch J, Al-Awqati Q (1985) Plasticity of functional epithelial polarity. Nature (London) 318:368–371
Shoemaker RL, Frizzell RA, Dwyer TM, Farley JM (1986) Single chloride channel currents from canine tracheal epithelial cells. Biochim Biophys Acta 858:235–242
Shorofsky SR, Field M, Fozzard HA (1983) Electrophysiology of Cl secretion in canine trachea. J Membr Biol 72:105–115

Shorofsky SR, Field M, Fozzard HA (1984) Mechanism of Cl secretion in canine trachea: changes in intracellular chloride activity with secretion. J Membr Biol 81:1 – 8

Shorofsky SR, Field M, Fozzard HA (1986) Changes in intracellular sodium with chloride secretion in dog tracheal epithelium. Am J Physiol 250 (Cell Physiol 19):C646 – C650

Silva P, Stoff J, Solomon RJ, Rosa R, Stevens A, Epstein J (1980) Oxygen cost of chloride transport in perfused rectal gland of *Squalus acanthias*. J Membr Biol 53:215 – 221

Smaje LH, Hedemark Poulsen J, Ussing HH (1986) Evidence from O_2 uptake measurements for $Na^+-K^+-2Cl^-$ co-transport in the rabbit submandibular gland. Pfluegers Arch 406: 492 – 496

Smith PL, Frizzel RA (1984) Chloride secretion by canine tracheal epithelium: IV. Basolateral membrane K permeability parallels secretion rate. J Membr Biol 77:187 – 199

Smith PL, Welsh MJ, Stoff JW, Frizzell RA (1982) Chloride secretion by canine tracheal epithelium: I. Role of intracellular cAMP levels. J Membr Biol 70:217 – 226

Stokes JB (1984) Sodium chloride absorption by the urinary bladder of the winter flounder. A thiazide-sensitive, electrically neutral transport system. J Clin Invest 74:7 – 16

Wade JB, O'Neil RG, Pryor JL, Boulpaep EL (1979) Modulation of cell membrane area in renal collecting tubules by corticosteroid hormones. J Cell Biol 81:434 – 445

Wangemann P, Wittner M, Di Stefano A, Englert HC, Lang HJ, Schlatter E, Greger R (1986) Cl-channel blockers in the thick ascending limb of the loop of Henle. Structure activity relationship. Pfluegers Arch 407:S128 – S141

Wanner A (1977) Clinical aspects of mucociliary transport. Am Rev Respir Dis 116:73 – 125

Welsh MJ (1983a) Evidence for basolateral membrane potassium conductance in canine tracheal epithelium. Am J Physiol 244 (Cell Physiol 13):C377 – C384

Welsh MJ (1983b) Inhibition of chloride secretion by furosemide in canine tracheal epithelium. J Membr Biol 71:219 – 226

Welsh MJ (1983c) Intracellular chloride activities in canine tracheal epithelium: direct evidence for sodium-coupled intracellular chloride accumulation in a chloride-secreting epithelium. J Clin Invest 71:1392 – 1401

Welsh MJ (1983d) Intracellular potassium activities in canine tracheal epithelium. Am J Physiol 245 (Cell Physiol 14):C248 – C254

Welsh MJ (1983e) Barium inhibition of basolateral membrane potassium conductance in tracheal epithelium. Am J Physiol 244 (Renal Fluid Electrolyte Physiol 13):F639 – F645

Welsh MJ (1984a) Energetics of chloride secretion in canine tracheal epithelium: comparison of the metabolic cost of chloride transport with the metabolic cost of sodium transport. J Clin Invest 74:262 – 268

Welsh MJ (1984b) Anthracene-9-carboxylic acid inhibits an apical membrane chloride conductance in canine tracheal epithelium. J Membr Biol 78:61 – 71

Welsh MJ (1985) Ion transport by primary cultures of canine tracheal epithelium: methodology, morphology, and electrophysiology. J Membr Biol 88:149 – 163

Welsh MJ (1986a) Adrenergic regulation of ion transport by primary cultures of canine tracheal epithelium: cellular electrophysiology. J Membr Biol 91:121 – 128

Welsh MJ (1986b) An apical-membrane chloride channel in human tracheal epithelium. Science 232:1648 – 1650

Welsh MJ (1986c) Single apical membrane anion channels in primary cultures of canine tracheal epithelium. Pfluegers Arch 407:S116 – S122

Welsh MJ (1987) Effect of phorbol ester and calcium ionophore on chloride secretion in canine tracheal epithelium. Am J Physiol 253 (Cell Physiol 22):C828 – C834

Welsh MJ, McCann JD (1985) Intracellular calcium regulates basolateral potassium channels in a chloride-secreting epithelium. Proc Natl Acad Sci USA 82:8823 – 8826

Welsh MJ, Widdicombe JH, Nadel JA (1980) Fluid transport across the canine tracheal epithelium. J Appl Physiol 49(5):905 – 909

Welsh MJ, Smith PL, Frizzell RA (1982a) Chloride secretion by canine tracheal epithelium: II. The cellular electrical potential profile. J Membr Biol 70:227 – 238

Welsh MJ, Smith PL, Fromm M, Frizzell RA (1982b) Crypts are the site of intestinal fluid and electrolyte secretion. Science 218:1219 – 1221

Welsh MJ, Smith PL, Frizzell RA (1983) Chloride secretion by canine tracheal epithelium: III. Membrane resistances and electromotive forces. J Membr Biol 71:208 – 218

Westenfelder C, Earnest WR, Al-Bazzaz FJ (1980) Characterization of Na-K-ATPase in dog tracheal epithelium: Enzymatic and ion transport measurements. J Appl Physiol 48(6):1008–1019

Widdicombe JH, Basbaum CB, Highland E (1981) Ion contents and other properties of isolated cells from dog tracheal epithelium. Am J Physiol 241 (Cell Physiol 10):C184–C192

Widdicombe JH, Basbaum CB, Yee JY (1979a) Localization of Na pumps in the tracheal epithelium of the dog. J Cell Biol 82:380–390

Widdicombe JH, Gashi AA, Basbaum CB, Nathanson IT (1986) Structural changes associated with fluid absorption by dog tracheal epithelium. Exp Lung Res 10:57–69

Widdicombe JH, Nathanson IT, Highland E (1983) Effects of "loop" diuretics on ion transport by dog tracheal epithelium. Am J Physiol 245 (Cell Physiol 14):C388–C396

Widdicombe JH, Ueki IF, Bruderman I, Nadel JA (1979b) The effects of sodium substitution and ouabain on ion transport by dog tracheal epithelium. Am Rev Respir Dis 120:385–392

Widdicombe JH, Welsh MJ (1980) Ion transport by dog tracheal epithelium. Fed Proc 39: 3062–3066

Widdicombe JH, Welsh MJ, Finkbeiner WE (1985) Cystic fibrosis decreases the apical membrane chloride permeability of monolayers cultured from cells of tracheal epithelium. Proc Natl Acad Sci USA 82:6167–6171

Xu GL, Sivarajah K, Wu R, Nettesheim P, Eling T (1986) Biosynthesis of prostaglandins by isolated and cultured airway epithelial cells. Exp Lung Res 10:101–114

Chapter 7

NaCl Transport in Salt Glands

E. Schlatter and *R. Greger*

Contents

1 Introduction

Thirty years ago Schmidt-Nielsen and his co-workers described a salt-secreting supraorbital nasal gland in marine birds (K. Schmidt-Nielsen et al. 1957, 1958). The avian salt gland was the second extrarenal organ after the gills of fish that was found to participate in the osmoregulation in marine animals. Later, other homologous and nonhomologous exocrine salt glands have been described in a variety of vertebrates such as the lacrimal gland of turtles (K. Schmidt-Nielsen and Fänge 1958), the sublingual gland of sea snakes (K. Schmidt-Nielsen and Fänge 1958), the rectal gland of elasmobranchs (Burger and Hess 1960), the nasal salt gland in some terrestrial birds (K. Schmidt-Nielsen et al. 1963), the nasal salt glands of various reptiles (W. A. Dunson 1976; K. Schmidt-Nielsen et al. 1963; Templeton 1964), and the dendritic organ of marine catfish (van Lennep and Lanzing 1967). The term "salt gland" is generally used for extrarenal salt-excreting glands in vertebrates (van Lennep and Young 1979). However, glandular organs with a similar osmoregulatory function have also been described in invertebrates, such as the crustacean salt glands (Conte 1984; Copeland 1967).

Physiologisches Institut der Albert-Ludwigs-Universität, 7800 Freiburg, FRG

Advances in Comparative and Environmental Physiology
Vol. 1. Edited by R. Greger

The embryological origin of the above mentioned extrarenal salt-excreting organs varies and also the anatomical localization of these salt glands is quite different. The salt glands of birds and reptiles, for example, are located in the head region and the salt glands of elasmobranchs are attached to the rectum. However, the physiological role of all these organs that have developed quite early in evolution [there are reports on nasal salt glands in dinosaurs (Osmolska 1979)] is very similar: They are capable of producing a hypertonic fluid which, in most instances, is an almost pure NaCl solution. Na^+ concentrations in the secreted fluid can be as high as 1400 mmol l^{-1} as has been measured in the marine lizard *Amblyrhynchus cristatus* (W.A. Dunson 1969). Steep Na^+ gradients with a luminal fluid over a plasma concentration ratio for Na^+ of up to 4.5 in some marine birds can be generated (Schmidt-Nielsen 1960). Some terrestrial birds and reptiles, however, secrete rather a potassium bicarbonate-rich solution and can vary the Na^+/K^+ or Cl^-/HCO_3^- ratios according to their dietary intake (cf. van Lennep and Young 1979). The fluid secretion rate relative to the body weight sometimes exceeds the glomerular filtration rate of mammals (K. Schmidt-Nielsen 1960). This capability of the salt glands to excrete a hyperosmotic fluid enables these nonmammalian vertebrates and some invertebrates to endure temporarily or permanently the osmotic stress of salt loading or water deprivation caused by living in marine, estuarine, or desert environments. The kidneys of most of these animals are generally not developed enough to allow for a sufficient urine concentration and, thus, sufficient osmoregulation during either temporary or permanent salt stress.

A number of comprehensive reviews have been published dealing with exocrine salt glands in nonmammalian vertebrates and also in invertebrates describing the gross anatomy, the ultrastructure, the function, and the hormonal and nervous regulation of salt secretion of the various salt glands (Conte 1984; Holmes 1972, 1975; Holmes and Phillips 1985; Kirschner 1980; van Lennep and Young 1979; Peaker 1979; Peaker and Linzell 1975; K. Schmidt-Nielsen 1960; Simon 1982). This review, therefore, focuses mainly on more recent results on the cellular mechanism of NaCl secretion in these organs and three examples are chosen: the shark rectal gland, the avian rectal gland as in ducks, and the reptilian salt gland as in lizards. The salt gland studied best so far with respect to functional data at the cellular and subcellular level is the rectal gland of the shark. Besides the elasmobranch rectal gland much less functional data are available on the mechanisms of NaCl transport in the avian or even less in the reptilian salt glands. Most of the information from these organs is more pharmacologically oriented and describes humoral and/or nervous control of the secreted fluid and the secretion rate in these glands. The reason for the lack of more functional data on the cellular mechanism of NaCl secretion in the nasal salt glands in probably due to the anatomy of these organs and their location within the skull, which limits the in vivo study at the cellular level. Isolated gland or secretory tubule preparations obviously have also been impossible so far, mainly due to the high degree of connective tissue in these organs. Different in this respect, however, is the rectal gland of elasmobranchs. This organ has proven suitable for the application of such techniques as isolated perfusion of the whole organ (Silva et al. 1977), isolated in vitro perfusion of individual rectal gland tubules (Greger and Schlatter 1984a, b) and patch-clamp studies of single ion channels (Greger et al. 1985). Studies with membrane vesicles prepared from the rectal gland have also been performed (Hannafin et al. 1983). From these studies the cellular mechanism of NaCl secretion and its hormonal regulation is fairly well understood and therefore this organ may serve as a model for other exocrine salt glands. The shark rectal gland may also be a model for NaCl secretion in general and is mechanistically comparable to such organs as the colon, the trachea, the cornea, or the fish operculum (Degnan et al. 1977; Frizzell et al. 1979; Reuss et al. 1983; Welsh 1983; Welsh et al. 1982). We will first describe the cellular mechanisms and the regulation of NaCl secretion in the shark rectal gland and will then compare the information available for the avian and the reptilian salt glands with this model.

2 The Shark Rectal Gland

2.1 Introduction

The shark has a plasma Na^+ concentration of approximately 270 mmol l^{-1}, but it lives in an environment where the NaCl concentration is almost twice as high. It may be for this reason that the shark has a specific gland, the rectal gland, to excrete excess amounts of NaCl. It can be anticipated that this gland is under hormonal control, and that it will respond to volume expansion (Solomon et al. 1985). In fact, the gland can be found in a state where it is nontransporting with the duct system almost clogged with concentrated and precipitated secretion. In this state the oxygen consumption of this organ would be minimal. On another occasion, the gland may secrete large amounts of a secretion which is isotonic or slightly hypertonic with respect to sea water. In such a situation, the oxygen consumption would be correspondingly high.

Needless to mention, an organ of this kind, with the advantage of easy accessibility, must attract the interest of biologists and physiologists. It does not come as a surprise then that this gland has been well studied through several decades, and that, fortunately enough, most refined techniques, developed in epithelial research, have all been applied to this organ.[1] Consequently, we have a detailed picture of the cellular mechanisms involved in the secretory process, and we begin to understand how shark rectal gland secretion is hormonally controlled. The present short section, instead of repeating what has already been published in earlier, more detailed articles and reviews will briefly summarize the current status and will refer to the pertinent reports.

[1] A review of the literature reveals that the world's center of research for the rectal gland of *Squalus acanthias* has been over many years the Mount Desert Island Biological Laboratory in Maine, USA. Many detailed aspects of the function of this organ are therefore exclusively published in the proceedings of this laboratory: The Bulletin of the MDIBL.

2.2 How is NaCl Secreted?

Figure 1 summarizes the current concept of NaCl secretion as it was developed for the rectal gland of *Squalus acanthias,* and as it has been expanded to other secretory epithelia, such as the trachea or the colonic crypt cell. The primary active pump responsible for NaCl secretion is the basolaterally located ($Na^+ + K^+$)-ATPase. In fact, much of what is known about the characteristics of this enzyme in general stems from preparations of this gland (Skou 1975). It is the high concentration of the enzyme per gram of tissue which made it attractive to biochemists. This pump sustains high cytosolic K^+ and correspondingly low Na^+ concentrations (Greger and Schlatter 1984b). The concentration gradient for Na^+ across the basolateral cell membrane is the driving force for the uptake of this ion, in conjunction with K^+ and 2 Cl^-, via a specific carrier system. Again, it was the rectal gland for which the existence of this carrier protein had been postulated (Hannafin et al. 1983) only a few years after it had been discovered in the Ehrlich ascites tumor cell (Geck et al. 1980) and in the thick ascending limb of the loop of Henle (Greger and Schlatter 1981). Today, it is becoming increasingly clear that this carrier system mediates chloride transport in a broad variety of cells (Greger 1985). This carrier system is the target site for loop diuretics such as furosemide (Hannafin et al. 1983; Palfrey et al. 1984; Greger and Schlatter 1984b). It is postulated today that these substances (Schlatter et al. 1983; Greger and Schlatter 1983a) interfere with one of the Cl^--binding sites, and that by doing so, the carrier ceases to translocate the ion complex from the blood side into the cell (Greger 1985). It is predicted that inhibition of the $Na^+ 2 Cl^- K^+$ carrier largely reduces the requirement to eliminate cellular Na^+ via the ($Na^+ + K^+$)-ATPase. Consequently, ouabain inhibition of this pump is devoid of effects when the $Na^+ 2 Cl^- K^+$ carrier is blocked, e.g., by furosemide (Greger

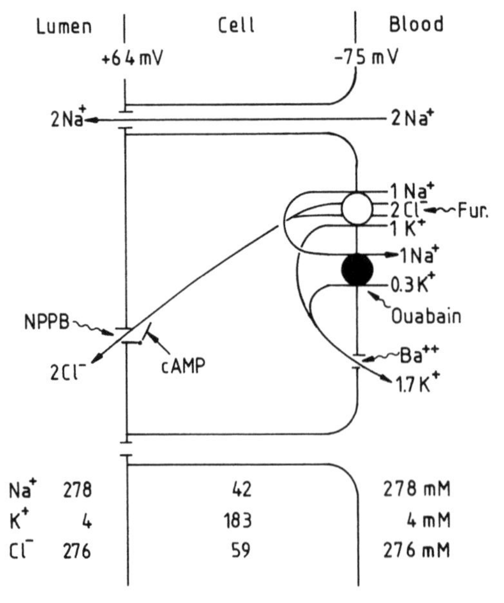

Fig. 1. Cell model for NaCl secretion in the rectal gland of *Squalus acanthias.* ●→ = primary active pump; ○→ = carrier system; ╪→ = ionic conductance; ∿→ = inhibitor.
NPPB 5-nitro-2-(3-phenylpropylamino)-benzoate; *Fur.* furosemide.
The lower half depicts the concentrations of Na^+, K^+, and Cl^- in the respective compartments as measured by ion selective microelectrodes

and Schlatter 1984b). The stoichiometry of the $Na^+ 2 Cl^- K^+$ carrier system has been determined directly by measuring the time constants for the fall in cytosolic Na^+ and Cl^- when furosemide was given instantaneously. It was shown that the initial fall in cytosolic chloride is twice as fast as the fall in cytosolic sodium. This confirms that the stoichiometry is probably 1 Na^+:2 Cl^-. The participation of K^+ in this carrier system has been shown rather directly in isolated membrane preparations (Hannafin et al. 1983). It was confirmed for the intact rectal gland with a rather complicated experimental design. Inspection of Fig. 1 suggests that a sudden removal of K^+ on the blood side would essentially lead to three effects on the membrane voltage: (1) The membrane should hyperpolarize instantly, since a K^+ conductance dominates the properties of this cell membrane, and since the electromotive force for K^+ is hyperpolarized in this experiment. This effect is immediate. (2) The $Na^+ 2 Cl^- K^+$ carrier stops if it was K^+-dependent, and this hyperpolarizes the cell as does the inhibition of this carrier by furosemide. This effect has a time course identical to that of furosemide, i.e., it has a delay of several seconds. (3) Removal of K^+ blocks the ($Na^+ + K^+$) pump and this depolarizes the cell as does inhibition of this pump by ouabain. This depolarization starts rather slowly. All three components could be separated experimentally (Greger and Schlatter 1984), and thus, it was shown for the intact cell that the $Na^+ 2 Cl^- K^+$ carrier requires the presence and participation of the K^+ ion. The apparent affinities of this cotransport system have been determined, and they are very similar to those reported for the same cotransport system of the thick ascending limb of the loop of Henle (Greger 1985).

The K^+ conductance properties of the basolateral membrane have been determined by membrane voltage measurements (Greger and Schlatter 1984a, b) and they have also been examined by the patch-clamp technique (Greger et al. 1987b; Gögelein et al. 1987b). It was shown that these channels have a conductance of 50 – 80 pS, and that they are blocked by Ba^{2+}, Cs^+, Rb^+, tetraethylammonium, quinine, quinidine, and lidocaine. Furthermore, it was found that these channels are present independently of the functional state of the cell, i.e., they are present irrespective of whether chloride secretion is stimulated or not. The calcium dependence of these channels has been tested, and it was found that these channels do not require Ca^{2+} as a gating mechanism (Gögelein et al. 1987b). Therefore, these channels function differently as compared to K^+ channels of other excretory glands like the pancreatic acini (Petersen and Maruyama 1984). For these latter glands, it has been shown that salt secretion is turned on by the Ca^{2+} activation of the respective K^+ channels.

Chloride leaves the rectal gland cell across the apical membrane via a chloride conductance. This conductance has been examined by intracellular microelectrode measurements (Greger et al. 1984), and it was shown that this conductance is normally regulated via cyclic AMP. Within several seconds to a few minutes after exposure to cyclic AMP, the apical membrane chloride conductance increases by one order of magnitude. This effect occurs also in the presence of furosemide which indicates that it is independent of the cellular chloride activity. In fact, it was claimed in these studies that the increase in apical chloride conductance is the first step initiated by cyclic AMP in the cascade of cellular effects. Meanwhile, this chloride conductance has been identified as a 30 – 50 pS chloride channel (Greger et al. 1987b), and it was shown that cyclic AMP increases the density of this channel in apical membranes of rectal glands (Greger et al. 1985). The properties of these channels, in terms of kinetic behavior, have also been examined. The Cl^- channel is also found in other secretory epithelia (cf. M. Welsh, this Vol., on trachea and Hayslett et al. 1987). These channels can be blocked reversibly by a new class of substances: arylamino-benzoates (Wangemann et al. 1986).

The conductive properties of the rectal gland cell, with the K^+ conductance localized in the basolateral membrane, and the chloride conductance in the apical membrane, generate a lumen-negative voltage of some -9 mV. The source of this voltage is the difference in the electromotive forces for potassium and chloride, which amounts to -30 to -50 mV. The fact that the measured value of -9 mV is only one-fourth of the above estimate, indicates that the shunt short-circuits most of the theoretical voltage. Indeed, the shunt has a high conductance of ≈ 30 mS (Greger and Schlatter 1984a). This conductance reflects mostly the permeability to sodium. Therefore, for each Cl^- ion secreted into the lumen, one Na^+ ion follows paracellularly. In terms of energy requirement, this entire process is highly economical, since 2 NaCl are secreted for each Na^+ ion pumped by the $(Na^+ + K^+)$ATPase. Accordingly, the rate of chloride secretion per oxygen consumed has been reported to be as high as 30 mol Cl^-/mol O_2, whereas a ratio of around 15 is predicted for transcellular transport and in fact has been measured, e.g., for the frog skin (Silva et al. 1980).

2.3 Hormonal Control of Secretion

An adequate stimulus for secretion by the rectal gland is volume expansion (Solomon et al. 1984, 1985a). The response of the gland must at least in part be mediated by circulating hormones, since the stimulatory effect can be mediated from an volume-expanded animal to an isolated gland by transfusion. It seems clear today that the secretory response is mediated by the second messenger cyclic AMP. Several findings support this view: (1) Hormones known to increase secretion, do so by increasing cyclic AMP. (2) Phosphodiesterase inhibition increases the response. (3) Forskolin also increases both cyclic AMP production and secretion. (4) Somatostatin, on the other hand, reduces secretion, and also reduces cyclic AMP production (Epstein 1979; Solomon et al. 1984; Silva et al. 1985). (5) Finally, it has been shown (cf. above) that cyclic AMP, added in its membrane permeable form, i.e., dibutyryl cyclic AMP, increases secretion. Also, it is apparent today that the cellular mechanism of stimulus-secretion coupling involves as the first step the "gating" of chloride channels in the apical cell membrane (Greger et al. 1985). It is reasonable to assume that this mechanism of channel gating involves a cyclic AMP-dependent phosphorylation process.

It is by no means clear, however, how the other functional components involved in chloride secretion (cf. Fig. 1) are readjusted once the chloride conductance has opened up. It can be excluded that this readjustment, e.g., of the rate of the $Na^+ 2Cl^- K^+$ carrier is a mere function of the driving force (Greger et al. 1984). Rather, it has to be assumed that it must be controlled by some other mediator.

It should be noted that the basic concept of stimulus-secretion coupling in the rectal gland is entirely different from that, e.g., of the pancreatic acinar cell (Petersen and Maruyama 1984), even though the transport systems involved appear similar. The secretion in the latter preparation is Ca^{2+}-mediated, and the target is the K^+ channel. In the rectal gland, the secretion is cyclic AMP-mediated, and the primary effect is on the chloride channel.

Whereas there is a general consensus with respect to the second messenger mediating chloride secretion in the rectal gland, current research focuses on the hormonal mediation. It has been shown that the rectal gland cell responds with increased secretion when exposed, e.g., to adenosine, vasoactive intestinal peptide, or rectin (a peptide similar, but not identical with vasoactive intestinal peptide) (Shuttleworth 1984). More recent experiments also point to a causal role for atriopeptide. This is a peptide released from the heart atria when they are distended, e.g., by volume expansion. Atriopeptide is a powerful diuretic and saluretic compound in mammalian kidney. It has been shown that the shark atria contain a factor promoting rectal gland chloride secretion (Solomon et al. 1985b), and that the synthetic peptide has a similar effect (Silva et al. 1987). It is worth noting that this peptide, even though it was effective in the whole gland, did not have an effect in isolated glandular tubules. This may be reconciled by the assumption that atriopeptide leads to the local generation or release of some other hormone (e.g., vasointestinal peptide) and that this locally produced hormone is responsible for the increased cyclic AMP production.

3 The Nasal Salt Gland of Birds

3.1 Introduction

The avian kidney combines both mammalian and reptilian type nephrons, i.e., nephrons with highly developed loops of Henle and those with no or only poorly developed loops. Its concentration ability therefore is reduced compared to the mammalian kidney. An exemption, however, are some desert birds, which are able to concentrate their urine enough to conserve water entirely through their kidneys (Emery et al. 1972; Serventy 1971). Marine birds and some birds that migrate between fresh water and salt water habitats or live in estuarine environments, like some ducks, geese, and gulls, however, have in addition nasal salt glands to support the kidneys in the homeostasis of salt and water balance. But in contrast to kidneys, salt glands only secrete fluid when there is a physiological need to produce a hypertonic excretion in excess of the urine. Otherwise they may rest for long periods or even decrease in gland size and cellular membrane area (vide infra). Salt glands are also found in some species of terrestrial birds that live around alkaline inland lakes or in deserts. The avian salt glands are capable of producing a hypertonic fluid which can have an osmolality up to seven times higher than that of the plasma (cf. van Lennep and Young 1979). The salt glands enable the birds either to excrete excess salt ingested with drinking water or food (marine birds or terrestrial birds at alkaline lakes) or to conserve water while excreting salt in areas with little drinking water available (desert birds) to a larger extent than their kidneys are able to do alone. The occurrence of salt glands in birds has been extensively reviewed by Peaker and Linzell (1975) and K. Schmidt-Nielsen (1960). The secreted fluid contains almost exclusively NaCl, similar to the secreted fluid of the elasmobranch rectal gland. In some terrestrial birds a high amount of potassium-bicarbonate is secreted by the salt glands, depending on their dietary salt intake (cf. van Lennep and Young 1979).

The possibilities of studying the mechanism of NaCl secretion experimentally in the avian salt glands unfortunately have been limited. This is mainly caused by the anatomical localization of the gland within the skull and sometimes partially under the beak of the birds and the amount of connective tissue surrounding the salt gland. The salt glands are quite complex with poorly specialized peripheral cells at the blind ends of the primary ducts and principal cells which are progressively more developed along primary and secondary ducts. Several primary ducts join into central canals and secondary ducts and finally into the main collecting duct. So far, the avian salt gland has been inaccessible to in vivo or in vitro micropuncture or microperfusion studies. For a long time in vivo collection of the secreted fluid and its chemical analysis and in vitro incubation of salt gland slices, again with chemical or electron microprobe analysis of cellular ion uptake or loss, have been the only techniques used to study NaCl secretion (reviewed by Butler 1984; Holmes 1972, 1975; Holmes and Phillips 1985; Peaker 1979; Peaker and Linzell 1975; K. Schmidt-Nielsen 1960, Simon 1982). More recently, the use of isolated dispersed salt gland cells (Fisher et al. 1983) and the primary culture of duck salt gland cells have been utilized (Lowy et al. 1985). Whereas in the isolated cell preparation, again only uptake, binding, or metabolism, and protein and DNA synthesis studies have been performed, with the method of cultured cells, the first transepithelial studies were possible. These cells are grown on permeable supports and mounted in an Ussing-type chamber under controlled conditions on both epithelial sides. The regulatory mechanisms for salt secretion of the whole gland are relatively well understood, as well as humoral and nervous control interactions, and morphological changes which occur during salt adaptation and de-adaptation (Butler 1984; Harvey et al. 1985; Holmes and Phillips 1985; Peaker and Linzell 1975; Simon 1982; Stewart et al. 1979; Wilson et al. 1985). Much less is known about the cellular mechanisms, which cellular transport processes take part in the secretory process, and how they are linked together. We will review below the available data for the avian salt gland, especially regarding these cellular aspects, and will compare these data with the model for NaCl secretion proposed for the shark rectal gland (see Sect. 2).

3.2 The $(Na^+ + K^+)$-ATPase

Salt glands in the resting state have much lower $(Na^+ + K^+)$-ATPase levels than those after acute salt loading or especially after salt adaptation (Flechter et al. 1967; Stewart et al. 1976) or after direct methacholine, i.e., cholinergic, stimulation (Hootman and Ernst 1981). The $(Na^+ + K^+)$-ATPase is localized at the basolateral membrane (Barrnett et al. 1983; Ernst and Mills 1977; van Lennep and Young 1979) which has extended basal or lateral infoldings and interdigitations (W. A. Dunson et al. 1976; Kirschner 1977). Earlier proposals of an apically localized ($Na^+ + K^+$)-ATPase or a Cl^- pump as a primary active mechanism (Gilmore et al. 1977; van Rossum 1966; Thesleff and K. Schmidt-Nielsen 1962), which seemed to explain the secretion of NaCl against its electrochemical gradient to concentrations above the plasma level, however, lack any experimental evidence and are no longer considered. The basolateral ($Na^+ + K^+$)-ATPase is sensitive to the glycoside ouabain as was shown first by retrograde infusion of ouabain into the ducts (Thesleff and K. Schmidt-Nielsen 1962). Later, this was also shown in slice preparations (Ernst and van Rossum 1982), in dispersed cells (Hootman and Ernst 1981), and finally also in Ussing-chamber experiments using primary cell cultures that were grown on permeable supports (Lowy et al. 1985). It has been demonstrated that ouabain reduces the cellular ATP levels and oxygen consumption (Ernst and Mills 1977; Ernst and van Rossum 1982; Hootman and Ernst 1981). In Ussing-chamber experiments the increase in transepithelial short-circuit current induced by cholinergic stimulation was inhibited by ouabain (Lowy et al. 1985). The ouabain sensitivity, the oxygen consumption, and the short-circuit current are minimal in resting glands or cells, but are remarkably increased after cholinergic stimulation. Cellular ATP content decreased upon stimulation. This correlates directly with the ($Na^+ + K^+$)-ATPase activity and final NaCl secretion of the gland measured under similar conditions of stimulation or salt adaptation. Hootman and Ernst (1981) showed in the dispersed cell preparation of duck salt glands that cholinergically stimulated cells have an increased ($Na^+ + K^+$)-pump turnover rate, rather than an increased number of pump sites, as measured by [3 H] ouabain binding. Lingham et al. (1980) and Stewart et al. (1976), however, reported that during salt adaptation new catalytic subunits of the ATPase are incorporated and activation of existing pumps may be only secondary (cf. Holmes and Phillips 1985). This increase in ($Na^+ + K^+$)-ATPase activity parallels an increase in basolateral membrane area. In this respect, the avian salt gland has become a model tissue to study membrane incorporation and destruction following salt adaptation or de-adaptation (cf. Merchant et al. 1985).

Thus, the primary active, energy-consuming process in the formation of an isotonic or slightly hypertonic luminal fluid in the secretory tubules of the avian nasal salt gland appears to be the ($Na^+ + K^+$)-ATPase, as in the shark rectal gland. This is a general feature of all NaCl secretory and reabsorptive tissues described so far in mammalian and nonmammalian species (Degnan et al. 1977; Frizzell et al. 1979; Greger and Schlatter 1983, 1984a; Reuss et al. 1983; Shorofsky et al. 1982; Welsh et al. 1982). The activity of the ($Na^+ + K^+$) pump in the basolateral membrane keeps the cellular Na^+ concentration low (Andrews et al. 1983; Marshall et al. 1985). Values reported from chemical measurements of intracellular Na^+ concentrations in slices or dispersed cells range between 23 und 65 $mmol\,l^{-1}$ (cf. Marshall et al. 1985) and those determined by electron probe microanalysis in slices vary between 32 and 47 $mmol\,l^{-1}$ (Andrews et al. 1983; Marshall et al. 1985). These values are comparable with those reported for the stimulated rectal gland of the shark (see Sect. 2; Greger et al. 1984). A low intracellular Na^+ concentration provides energy, i.e., the driving force to move Cl^- coupled to Na^+ across the basolateral membrane against its electrochemical gradient (see Sect. 3.2). Therefore, the transport of Cl^- would be defined as secondary active.

In addition, the activity of the mainly laterally localized $(Na^+ + K^+)$-ATPase (given the long intercellular spaces with their extensive infoldings and interdigitations) could produce a higher sodium concentration in these spaces than in the lumen and/or the plasma. The resulting standing gradient could drive Na^+ passively across the tight junctions into the lumen, possibly favored by an additional lumen-negative potential difference (compared with the shark rectal gland). Alternatively, this gradient could provide the osmotic driving force to reabsorb water from the isotonic luminal fluid that was secreted more proximally. Both facts would increase the hypertonicity of the luminal fluid. These questions, however, especially with respect to the size and polarity of the transepithelial voltage, the water permeabilities of the cellular and paracellular barriers, or the shunt selectivity still await further experimentation.

3.3 The NaCl Cotransport System

Ernst and van Rossum (1982) could demonstrate in slice preparations of salt glands that removal of either Na^+ or Cl^- from the bathing medium inhibits the uptake of the other ion into the cells. This suggests a direct coupling of Na^+ and Cl^- uptake by one transport mechanism also in the avian salt gland. In addition, in vivo studies showed a sensitivity of the NaCl secretion by the avian salt gland to furosemide (Gilmore et al. 1977). The stimulatory effect of metacholine on ouabain binding and oxygen consumption was prevented by furosemide in incubated slices (Ernst and van Rossum 1982; Hootman and Ernst 1981). Furosemide also abolished the response of the short-circuit current to stimulation in the Ussing-chamber experiments of Lowy et al. (1985). This sensitivity to the loop diuretic furosemide is characteristic for the $Na^+ - 2\,Cl^- - K^+$ cotransport system described in the mammalian renal thick ascending limb of Henle's loop (Schlatter et al. 1983). Ernst and van Rossum (1982) also demonstrated that furosemide prevented the K^+ accumulation in the slices seen after methacholine stimulation under control conditions without the diuretic. Other inhibitors of coupled Na^+ or Cl^- cotransport systems such as chlorothiazide (electroneutral NaCl cotransport) or the disulfonic stilbenes SITS or DIDS (Cl^-/HCO_3^- exchange) did not affect NaCl uptake, respiration, cellular ATP levels, or the total salt gland secretion (Ernst and van Rossum 1982; Gilmore et al. 1977).

It is well established for several NaCl secreting as well as reabsorbing epithelia that the uptake of either Na^+ or Cl^- into the cells is coupled and that this coupling also involves K^+ (Geck and Heinz 1986; Greger 1985; Greger and Schlatter 1983). Given the stoichiometry of $1\,Na^+ : 2\,Cl^- : 1\,K^+$ (Geck et al. 1980; Greger and Schlatter 1984; Greger et al. 1983) in other preparations, including the shark rectal gland, this cotransport system is electrically neutral and only driven by the chemical gradients of the participating ions. The above described observations on the coupling of Na^+ and Cl^- and possibly of K^+ uptake into cells or slices and the furosemide sensitivity render this $Na^+ - 2\,Cl^- - K^+$ cotransport system very likely to be responsible for NaCl uptake in the avian salt gland as it is in the shark rectal gland (see Sect. 2). One, however, has to accept that all the above described experiments so far do not directly prove the existence of this particular transport system. On the other hand, systems also discussed in the literature, such as the Na^+/H^+ and the Cl^-/HCO_3^- antiporter (Peaker and Stockley 1974), seem to be highly unlikely.

3.4 Model for NaCl Secretion

There is increasing evidence for the $(Na^+ + K^+)$-ATPase and a coupled $Na^+-2\,Cl-K^+$ cotransport system in the basolateral membrane to drive NaCl secretion in the avian salt gland. These two systems allow for secondary active uptake of Cl^- across the basolateral membrane against its electrochemical gradient. No experimental data, however, are available on the pathway for Cl^- to leave the cells across the apical membrane. Again, since the two transport systems in the basolateral membrane increase the intracellular Cl^- concentration probably above its equilibrium concentration, Cl^- could exit passively through an apical Cl^- conductance as has been demonstrated in the shark rectal gland (Greger and Schlatter 1984a). Intracellular Cl^- concentrations seem to be relatively high (38 – 90 mmol l^{-1}, cf. Marshall et al. 1985) and increase after acute cholinergic stimulation (Ernst and van Rossum 1982). Since there are no stable intracellular voltage measurements available, an exact evaluation of the driving force for the conductive movement of Cl^- across the apical membrane is not possible. Assuming a cell potential of some −60 to −70 mV, the reported cellular Cl^- concentrations would clearly be above equilibrium values. The secondary active Cl^- secretion would generate a lumen-negative transepithelial voltage that could allow for passive Na^+ movement across the tight junctions. Freeze fracture studies have demonstrated tight junctions in the avian salt glands that are structurally simple as in other low resistance epithelia (Ernst et al. 1980; Riddle and Ernst 1979). This model for NaCl secretion has been discussed and proposed for the avian salt gland by Ernst et al. (1980) and it would be virtually identical to the model described for the elasmobranch rectal gland described by Greger and Schlatter (1984a, b, see Sect. 2).

Further questions that are still open and await more specific experimentation are how K^+ is secreted in some instances, whether it recycles across the basolateral membrane as in the rectal gland of the shark, or whether a pathway exists for K^+ exit across the apical membrane. Also, the nature of the observed bicarbonate secretion in those specialized species that produce a fluid rich in potassium-bicarbonate instead of a NaCl solution is completely unknown. There are no data available on the pathways for the movement of water in the primary and secondary ducts or the main collecting duct. The only study that gives some hints on the contribution of the various ducts in the overall production of a highly hypertonic final secretion is the electron microprobe analysis by Marshall et al. (1985). In this study a slightly hypoosmotic fluid was reported for the secretory tubules that did not change at all along the total length of these tubules. The major ion concentrations of this primary secreted fluid also did not change after stimulation. These authors found an increase in the total ion concentration of the luminal fluid only in the main ducts of the salt gland. There, Na^+ and Cl^- concentrations increased, while the K^+ concentration decreased steadily to the values found in the finally secreted fluid. Unfortunately, there are no data available on the transport processes in the cells along the main ducts. If these measurements of ion concentrations by microprobe analysis from slices are correct, the role of the main duct in producing the hypertonic NaCl secretion may be dominating that of the primary ducts and may also be an important site for the regulation of secretion.

3.5 Regulation of NaCl Secretion

When marine or coastal birds are exposed to high salt intake after drinking only fresh water for some time, the salt glands immediately start to secrete a hypertonic NaCl fluid. The stimulatory influence of extracellular osmolality and volume on salt gland control in the duck has been described by Deutsch et al. (1979). Within a few days the salt glands proliferate markedly and protein synthesis increases (Ernst and Ellis 1969; Holmes and Stewart 1968). These processes are fully reversible when the birds readapt to fresh water. In the early phase of the adaptation to high salt intake, membrane synthesis and infolding of the basolateral membrane increase (Ernst and Ellis 1969; Fisher et al. 1983). These long-term adaptations of the salt glands during periods of increased or decreased salt intake are superimposed by a short-term regulation of NaCl secretion. The salt glands do not secrete continuously, but only in times when an excess excretion of salt is necessary to maintain osmotic homeostasis. Simon (1982), Holmes and Phillips (1985), and Peaker and Linzell (1975) reviewed the initial stimulatory signals to increase the activity of the avian salt glands: an increased extracellular volume or increased extracellular osmolality is sensed by receptors most likely situated in the heart and the brain.The majority of studies demonstrates the following cholinergic stimulation of muscarinic receptors in the salt glands (Fänge et al. 1958; Lowy et al. 1985; Peaker and Stockley 1974; cf. Holmes and Phillips 1985; Simon 1982). Stewart et al. (1979) reported an increase in cellular cyclic GMP (but not cAMP) concentrations following cholinergic stimulation which could be a second messenger to prompt phosphorylation of transport proteins. An increase in intracellular Ca^{2+} as well as in inositol phosphates after cholinergic stimulation in this preparation has been recently reported by Snider et al. (1986), indicating that this may be at least one system for intracellular stimulus-response coupling. The adrenal mineralocorticoids also seem to participate in the regulation of the NaCl secretion, since adrenalectomy decreases the secretion and readmission of corticosterone reverses this effect (Harvey and Phillips 1982; Thomas and Phillips 1975). The same authors (Harvey et al. 1985) and Butler (1980, 1984), however, concluded that these adrenocortical effects are not direct effects on the salt glands, but rather indirect effects through changes in the vascular system and/or control of NaCl reabsorption in the gut and kidney. Indirect effects via vascular and hemodynamic changes may also be responsible for the action of angiotensin on salt glands, which reduces secretion, and the converting enzyme inhibitor captopril which increases secretion (Wilson 1984; Wilson et al. 1985). The apparently complex nature of short-term and long-term regulation and adaptation of the avian salt gland to changing ingestive salt loads has been extensively reviewed recently by Holmes and Phillips (1985).

The cellular events in the secretory salt gland cells that follow cholinergic or corticosteroid-induced stimulation and finally lead to an increased NaCl secretion, however, are far less understood. Cholineric stimulation of the salt gland cells seems to increase the cellular Cl^- content (Ernst and van Rossum 1982; B. Schmidt-Nielsen 1976) and that of Na^+ (Ernst and van Rossum 1982) in incubated slices. In other studies no change in the cellular Na^+ concentration during stimulation was seen (Hootman and Ernst 1981; B. Schmidt-Nielsen 1976). More recently with the electron microprobe analysis in slices from normal duck salt glands, an increase in cellular Na^+ and Cl^- content in the primary ducts was demonstrated after salt loading (Marshall et al. 1985). In the same study in salt-adapted ducks, however, intracellular Na^+ and Cl^- concentrations decreased slightly after salt loading. The reason for the differences between the studies may be due to the techniques used to measure intracellular ion concentrations in slices chemically or with X-ray analysis. However, the increase in cellular Na^+ and Cl^- concentrations in slices from normal ducks after acute cholinergic stimulation or salt loading would be compatible with a primarily increased entry step, i.e., the cotransport system in the basolateral membrane. In the other case of long-term salt-adapted ducks, the decrease in intracellular Na^+ and Cl^- concentrations upon salt loading could be explained by primarily increased exit steps for both ions, i.e., a Cl^- conductance in the apical membrane and the (Na^++K^+)-ATPase in the basolateral membrane. The latter situation has been demonstrated for the shark rectal gland (Greger et al. 1984; see Sect. 2), where intracellular ion activities were measured directly using double-barreled ion selective microelectrodes. The influence of cholingeric stimulation or salt loading on intracellular K^+ concentrations is not as clear. Ernst and van Rossum (1982) reported a decrease in slice K^+ content upon stimulation. The electron microprobe analysis of Marshall et al. (1985) revealed a decrease in the cellular K^+ concentration upon acute salt loading only in the peripheral cells of normal, i.e., fresh water-adapted ducks. Peripheral cells of salt water-adapted ducks and principal cells of fresh water-adapted ducks, however, decreased the cellular K^+ content after acute salt loading. In the shark rectal gland intracellular K^+ concentration did not significantly change upon stimulation (see Sect. 2).

4 Salt Glands in Reptilia

A salt gland has been identified in all reptiles studied so far, which represent all major reptilian orders (W.A. Dunson 1979; Taplin and Grigg 1981). The localization and embryological origin of the reptilian salt gland is quite variable with lacrimal glands in turtles and tortoises (Cowan 1974), nasal salt glands in lizards (Bradshaw et al. 1984b), sublingual salt glands in snakes (W.A. Dunson and M.K. Dunson 1973), or salt glands within the tongue in crocodiles (Taplin and

Grigg 1981). The secreted fluid of the reptilian salt glands is more versatile than for example that of elasmobranchs or birds which mainly secrete a NaCl solution. The ratio of Na^+ to K^+ or Cl^- to HCO_3^- concentrations in the final secretion can vary depending on the dietary intake in these animals (cf. van Lennep and Young 1979; Shoemaker et al. 1972). In general, in reptiles the secreted fluid has a high $KHCO_3$ concentration. K^+ concentrations have been found as high as 1400 mmol l^{-1} in the salt gland fluid in some lizards (cf. van Lennep and Young 1979; K. Schmidt-Nielsen et al. 1963). A positive correlation between the secretion rate of the salt gland and the Cl^- concentration of that fluid has been described in sea snakes (M.K. Dunson and W.A. Dunson 1974).

Compared to the information available, especially for the elasmobranch rectal gland and also for the avian nasal salt gland, very little experimental physiological data have been published on the various reptilian salt glands. This includes functional data on the mechanisms responsible for NaCl or $KHCO_3$ secretion as well as the hormonal or nervous control of the secretion. Ellis and Goertemiller (1974) demonstrated in a lizard salt gland the $(Na^+ + K^+)$-ATPase and its localization mostly at the lateral infoldings. Intracellular Na^+ and K^+ concentrations that have been measured chemically in slices of the sea snake salt gland are within the range reported for the avian salt glands (M.K. Dunson and W.A. Dunson 1974). These very few data only prompt the conclusion that salt excretion in reptiles may be driven by the activity of the $(Na^+ + K^+)$-ATPase in the lateral membrane of the secretory cells. This finding would be analogous to the situation in the salt glands of elasmobranchs and birds. Based on the composition of the secreted fluid, the cellular mechanisms and possible pathways for ion transport may be different from the model proposed for the shark rectal gland (see Sect. 2). They may have more in common with the few terrestrial birds that also secrete a $KHCO_3$-rich fluid. Unfortunately, the mechanism responsible for the salt secretion in those birds is also unknown.

Unlike the situation in birds, in reptiles the salt gland obviously does not increase in size upon chronic stimulation. There are, however, indications that the ratio between the secretory striated cells and the seromucous cells changes when the animals and, thus, the salt glands are in an active state compared to the resting state (Saint-Girons et al. 1977). Seasonal variations between the inactive animal during winter and the highly active animal during summer in the corticosteroid binding to the salt gland of a lizard have been shown by Bradshaw et al. (1984b). In contrast to the avian salt glands, the nasal salt glands of lizards do not seem to respond to increased plasma osmolality with increased salt secretion (Shoemaker et al. 1972). These authors proposed that the increased concentration of alkali metals is the signal that is sensed and leads to an increased salt secretion in the nasal salt glands. Bradshaw et al. (1984a, b) reported a high-affinity binding of corticosterone and aldosterone to the salt glands in a lizard and, therefore, postulated a possible, direct regulatory effect of these steroid hormones on the salt secretion. They also found a positive correlation between plasma aldosterone levels and the rate of K^+ excretion, and a negative correlation between the steroid levels and Na^+ excretion. Therefore, corticosteroids may be at least in part responsible for the variation in the composition of the fluid secreted by the reptilian salt glands in response to changing dietary intake.

References

Andrews SB, Mazurkiewicz JE, Kirk GR (1983) The distribution of intracellular ions in the avian salt gland. J Cell Biol 96:1389 – 1399

Barrnett RJ, Mazurkiewicz JE, Addis JS (1983) Avian salt gland: a model for the study of membrane biogenesis. In: Fleischer S, Fleischer B (eds), vol 96. Academic Press, London New York, pp 627 – 659

Bradshaw SD, Lemire M, Vernet R, Grenot CJ (1984a) Aldosterone and the control of secretion by the nasal salt gland of the North African desert lizard, *Uromastix acanthinurus*. Gen Comp Endocrinol 54:314 – 323

Bradshaw SD, Tom JA, Bunn SE (1984b) Corticosteroids and control of nasal salt gland function in the lizard *Tiliqua rugosa*. Gen Comp Endocrinol 54:308 – 313

Burger JW, Hess WN (1960) Function of the rectal gland in the spiny dogfish. Science 131:670 – 671

Butler DG (1980) Functional nasal glands in adrenalectomized domestic ducks (*Anas platyrhynchos*). Gen Comp Endocrinol 40:15 – 26

Butler DG (1984) Endocrine control of the nasal salt glands in birds. J Exp Zool 232:725 – 736

Conte FP (1984) Structure and function of the crustacean larval salt gland. Int Rev Cytol 91:45 – 104

Copeland DE (1967) A study of salt secreting cells in the brine shrimp *Artemia salina*. Protoplasma 63:363 – 384

Cowan FBM (1974) Observations on extrarenal excretion by orbital glands and osmoregulation in *Malaclemys terrapin*. Comp Biochem Physiol A 48:489 – 500

Degnan KJ, Karnaky KJ, Zadunaisky JA (1977) Active chloride transport in the in vitro opercular skin of teleost (*Fundulus heteroclitus*), a gill-like epithelium rich in chloride cells. J Physiol (London) 271:155 – 191

Deutsch H, Hammel HT, Simon E, Simon-Oppermann C (1979) Osmolality and volume factors in salt gand control of Pekin ducks after adaptation to chronic salt loading. J Comp Physiol 129:301 – 308

Dunson MK, Dunson WA (1974) The relation between plasma Na concentration and salt gland Na-K ATPase content in the diamond backed terrapin and the yellow-bellied sea snake. J Comp Physiol 101:89 – 97

Dunson WA (1969) Electrolyte excretion by the salt gland. Am J Physiol 216:995 – 1002

Dunson WA (1976) Salt glands in reptiles. In: Gans C, Dunson WR (eds) Biology of the Reptilia, vol 5. Physiology A, Academic Press, London New York, pp 413 – 445

Dunson WA (1979) Control mechanisms in reptiles. In: Gilles R (ed) Mechanisms of osmoregulation in animals: maintenance of cell volume. Wiley, New York, pp 273 – 322

Dunson WA, Dunson MK (1973) Convergent evolution of sublingual salt glands in the marine file snake and the true sea snakes. J Comp Physiol 86:193 – 208

Dunson WA, Dunson MK, Ohmart RD (1976) Evidence for the presence of nasal salt glands in the roadrunner and the Cortunix quail. J Exp Zool 198:209 – 216

Ellis RA, Goertemiller CC Jr (1974) Cytological effects of salt-stress and localization of transport adenosine triphosphatase in the lateral nasal salt glands of the desert iguana, *Dipsosaurus dorsalis*. Anat Rec 180:285 – 297

Emery N, Poulson TL, Kinter WB (1972) Production of concentrated urine by avian kidneys. Am J Physiol 223:180 – 187

Epstein FH (1979) The shark rectal gland: a model for the active transport of chloride. Yale J Biol Med 52:517 – 523

Ernst SA, Ellis RA (1969) The development of surface specilization in the secretory epithelium of the avian salt gland in response to osmotic stress. J Cell Biol 40:305 – 321

Ernst SA, Mills JW (1977) Basolateral plasma membrane localization of ouabain-sensitive sodium transport sites in the secretory epithelium of the avian salt gland. J Cell Biol 75:74 – 94

Ernst SA, Rossum GDV van (1982) Ions and energy metabolism in duck salt-gland: possible role of furosemide-sensitive co-transport of sodium and chloride. J Physiol (London) 325:333 – 352

Ernst SA, Riddle CV, Karnaky KJ (1980) Relationship between localization of the ($Na^+ + K^+$)-ATPase, cellular fine structure, and reabsorptive and secretory electrolyte transport. In: Boulpaep EL (Ed) Current topics in membranes and transport, vol 13. Academic Press, London New York, pp 335 – 385

Fänge RK, Schmidt-Nielsen K, Robinson M (1958) Control of secretion from the avian salt gland. Am J Physiol 195 (2):321 – 326
Fisher SK, Hootman SR, Hancock AM, Ernst SA, Agranoff BW (1983) Muscarinic stimulation of phospholipid turnover in dissociated salt gland cells. FEBS Lett 155:43 – 46
Flechter GL, Stainer IM, Holmes WN (1967) Sequential changes in adenosine triphosphatase activity and the electrolyte capacity of the nasal-glands of the duck during the period of adaptation to salt-water. J Exp Biol 47:375 – 391
Frizzell RA, Field M, Schultz SG (1979) Sodium-coupled chloride transport by epithelial tissues. Am J Physiol 236:F1 – F8
Geck P, Heinz E (1986) The Na-K-2 Cl cotransport system. J Membr Biol 91:97 – 105
Geck P, Pietryzk C, Burckhardt BC, Pfeiffer B, Heinz E (1980) Electrically silent cotransport of Na^+, K^+ and Cl^- in Ehrlich cells. Biochim Biophys Acta 600:432 – 447
Gilmore JP, Dietz J, Gilmore C, Zucker IH (1977) Evidence for a chloride pump in the salt gland of the goose. Comp Biochem Physiol 56A:121 – 126
Gögelein H, Schlatter E, Greger R (1987 a) The "small" conductance chloride channel in the luminal membran of the rectal gland of the dogfish (*Squalus acanthias*). Pfluegers Arch 409: 122 – 125
Gögelein H, Greger R, Schlatter E (1987 b) Potassium channels in the basolateral membrane of the rectal gland of *Squalus acanthias*. Regulation and inhibitors. Pfluegers Arch 409:107 – 113
Greger R (1985) Ion transport mechanisms in thick ascending limb of Henle's loop of mammalian nephron. Physiol Rev 65:760 – 797
Greger R, Schlatter E (1981) Presence of luminal K^+, a prerequisite for active NaCl transport in the cortical thick ascending limb of Henle's loop of rabbit kidney. Pfluegers Arch 392:92 – 94
Greger R, Schlatter E (1983 a) Properties of the basolateral membrane of the cortical thick ascending limb of Henle's loop of rabbit kidney. A model for secondary active chloride transport. Pfluegers Arch 396:325 – 334
Greger R, Schlatter E (1983 b) Cellular mechanism of the action of loop diuretics on the thick ascending limb of Henle's loop. Klin Wochenschr 61:1019 – 1027
Greger R, Schlatter E (1984 a) Mechanism of NaCl secretion in the rectal gland of spiny dogfish (*Squalus acanthias*). I. Experiments in isolated in vitro perfused rectal gland tubules. Pfluegers Arch 402:63 – 75
Greger R, Schlatter E (1984 b) Mechanism of NaCl secretion in the rectal gland of spiny dogfish (*Squalus acanthias*). II. Effects of inhibitors. Pfluegers Arch 402:364 – 375
Greger R, Schlatter E, Lang F (1983) Evidence for electroneutral sodium chloride cotransport in the cortical thick ascending limb of Henle's loop of rabbit kidney. Pfluegers Arch 396:308 – 314
Greger R, Schlatter E, Wang F, Forrest JN Jr (1984) Mechanism of NaCl secretion in rectal gland tubules of spiny dogfish (*Squalus acanthias*). III. Effects of stimulation of secretion by cAMP. Pfluegers Arch 402:376 – 384
Greger R, Schlatter E, Gögelein H (1985) Cl^- channels in the apical cell membrane of the rectal gland "induced" by cAMP. Pfluegers Arch 403:446 – 448
Greger R, Schlatter E, Gögelein H (1986) Sodium chloride secretion in rectal gland of dogfish (*Squalus acanthias*). New trends in physiology. Science 1:134 – 136
Greger R, Gögelein H, Schlatter E (1987 a) Potassium channels in the basolateral membrane of the rectal gland of the dogfish (*Squalus acanthias*). Pfluegers Arch 409:100 – 106
Greger R, Schlatter E, Gögelein H (1987 b) Chloride channels in the luminal membrane of the rectal gland of the dogfish (*Squalus acanthias*). Properties of the "larger" conductance channel. Pfluegers Arch 409:114 – 121
Hannafin J, Kinne-Saffran E, Frieman D, Kinne R (1983) Presence of a sodium-potassium chloride cotransport system in the rectal gland of *Squalus acanthias*. J Membr Biol 75:73 – 83
Harvey S, Phillips JG (1982) Endocrinology of salt gland function. Comp Biochem Physiol 71A: 537 – 546
Harvey S, Phillips JG, Rees A (1984) Independence of salt gland function and adrenocortical activity in ducks (*Anas platyrhynchos*). Gen Comp Endocrinol 60:210 – 214
Hayslett JP, Gögelein H, Greger R (1987) Characteristics of apical chloride channels in human colon cells (HT_{29}). Pfluegers Arch (in press)
Holmes WN (1972) Regulation of electrolyte balance in marine birds with special reference to the role of the pituitary-adrenal axis in the duck (*Anas platyrhynchos*). Fed Proc 31:1587 – 1598

Holmes WN (1975) Hormones and osmoregulation in marine birds. Gen Comp Endocrinol 25:249 – 258
Holmes WN, Phillips JG (1985) The avian salt gland. Biol Rev 60:213 – 256
Holmes WN, Stewart DJ (1968) Changes in the nucleic acid and protein composition of the nasal glands from the duck (*Anas platyrhynchos*) during the period of adaptation to hypertonic saline. J Exp Biol 48:509 – 519
Hootman SR, Ernst SA (1981) Effect of methacholine on Na^+ pump activity and ion content of dispersed avian salt gland cells. Am J Physiol 241:R77 – R86
Kirschner LB (1977) The sodium chloride excreting cells in marine vertebrates. In: Gupta BL, Moreton RB, Oshman JL, Wall BW (eds) Transport of ions and water in animals. Academic Press, London New York, pp 427 – 452
Kirschner LB (1980) Comparison of vertebrate salt excreting organs. Am J Physiol 238:R219 – R223
Lennep EW van, Lanzing WJR (1967) The ultrastructure of glandular cells in the external dendritic organ of some marine catfish. J Ultrastruct Res 18:333 – 344
Lennep EW van, Young JA (1979) Salt glands. In: Giebisch G (ed) Transport organs. Membrane transport in biology, vol 4B. Springer, Berlin Heidelberg New York, pp 675 – 692
Lingham RB, Stewart DJ, Sen AK (1980) The induction of ($Na^+ + K^+$)-ATPase in salt gland of the duck. Biochim Biophys Acta 601:229 – 234
Lowy RJ, Dawson DC, Ernst SA (1985) Primary culture of duck salt gland. II Neurohormonal stimulation of active transport. Am J Physiol 249:C41 – C47
Marshall AT, Hyatt AD, Phillips JG, Condron RJ (1985) Isosmotic secretion in the avian nasal salt gland: X-ray microanalysis of luminal and intracellular ion distributions. J Comp Physiol B 156:213 – 227
Merchant JL, Papermaster DS, Barnett RJ (1985) Correlation of ($Na^+ + K^+$)-ATPase content and plasma membrane surface area in adapted and de-adapted salt glands of ducklings. J Cell Sci 78:233 – 246
Osmolska H (1979) Nasal salt gland in dinosaurs. Acta Palaeontol Pol 24:205 – 215
Palfrey HC, Silva P, Epstein FH (1984) Sensitivity of cAMP-stimulated salt secretion in shark rectal gland to "loop" diuretics. Am J Physiol 246:C242 – C246
Peaker M (1979) Control mechanisms in birds. In: Gilles R (ed) Mechanisms of osmoregulation in animals. Wiley, New York, pp 323 – 348
Peaker M, Linzell JL (1975) Salt glands in birds and reptiles. Cambridge Univ Press, London
Peaker M, Stockley SJ (1974) The effects of lithium and methacholine on the intracellular ionic composition of goose salt gland slices: relation to sodium and chloride transport. Experientia 30:158 – 159
Petersen OH, Maruyama Y (1984) Calcium-activated potassium channels and their role in secretion. Nature (London) 307:693 – 696
Reuss L, Reinach P, Weinman SA, Grady TP (1983) Intracellular ion activities and Cl^- transport mechanisms in bullfrog corneal epithelium. Am J Physiol 244:C336 – C347
Riddle CV, Ernst SA (1979) Structural simplicity of the zonula occludens in the electrolyte secreting epithelium of the avian salt gland. J Membr Biol 45:21 – 35
Rossum GDV van (1966) Movements of Na^+ and K^+ in slices of herring gull salt gland. Biochim Biophys Acta 126:338 – 349
Saint-Girons H, Lemire M, Bradshaw SD (1977) Structure de la glande nasale externe de *Tiliqua rugosa* (Reptilia, scincidae) et rapports avec sa fonction. Zoomorphologie 88:277 – 288
Schlatter E, Greger R, Weidtke C (1983) Effect of "high ceiling" diuretics on active salt transport in the cortical thick ascending limb of Henle's loop of rabbit kidney. Correlation of chemical structure and inhibitory potency. Pfluegers Arch 396:210 – 217
Schmidt-Nielsen B (1976) Intracellular concentrations of the salt gland of the herring gull *Larus argentatus*. Am J Physiol 230:514 – 521
Schmidt-Nielsen K (1960) The salt secreting gland of marine birds. Circulation 21:955 – 967
Schmidt-Nielsen K, Fänge R (1958) Salt glands in marine reptiles. Nature (London) 182:783 – 785
Schmidt-Nielsen K, Jörgensen CB, Osaki H (1957) Secretion of hypertonic solutions in marine birds. Fed Proc 16:113 – 114
Schmidt-Nielsen K, Jörgensen CB, Osaki H (1958) Extrarenal salt excretion in birds. Am J Physiol 193:101 – 107

Schmidt-Nielsen K, Borut A, Lee P, Crawford E (1963) Nasal salt secretion and the possible function of the cloaca in water conservation. Science 142:1300–1301
Serventy DL (1971) Biology of desert birds. In: Farner DS, King JR (eds) Avian biology, vol 1, Academic Press, London New York, pp 287–339
Shoemaker VH, Nagy KA, Bradshaw SD (1972) Studies on the control of electrolyte excretion by the nasal gland of the lizard *Dipsosaurus dorsalis.* Comp Biochem Physiol A 42:749–757
Shorofsky SR, Field M, Fozzard HA (1982) The cellular mechanism of active chloride secretion in vertebrate epithelia: studies in intestine and trachea. Philos Trans R Soc London Ser B 299:597–607
Shuttleworth TJ, Thorndyke MC (1984) An endogenous peptide stimulates secretory activity in the elasmobranch rectal gland. Science 225:319–321
Silva P, Stoff J, Field M, Fine L, Forrest JN, Epstein FH (1977) Mechanism of active chloride secretion by shark rectal gland: role of Na-K-ATPase in chloride transport. Am J Physiol 233: F298–F306
Silva P, Stoff JS, Solomon RJ, Rosa R, Stevens A, Epstein J (1980) Oxygen cost of chloride transport in perfused rectal gland of *Squalus acanthias.* J Membr Biol 53:215–221
Silva P, Stoff JS, Leone DR, Epstein FH (1985) Mode of action of somatostatin to inhibit secretion by shark rectal gland. Am J Physiol 249:R329–R334
Silva P, Stoff JS, Solomon RJ (1987) Atrial natriuretic peptide stimulates salt secretion by shark rectal gland by releasing VIP. Am J Physiol 252:F99–F103
Simon E (1982) The osmoregulatory system of birds with salt glands. Comp Biochem Physiol 71A:547–556
Skou JC (1975) The ($Na^+ + K^+$)-activated enzyme system and its relationship to transport of sodium and potassium. Q Rev Biophys 7:401–434
Snider RM, Roland RM, Lowy RJ, Agranoff BW, Ernst SA (1968) Muscarinic receptor-stimulated Ca^{2+} signaling and inositol lipid metabolism in avian salt gland cells. Biochim Biophys Acta 889:216–224
Solomon R, Taylor M, Stoff JS, Silva P, Epstein FH (1984) In vivo effect of volume expansion on rectal gland function. I. Humoral factors. Am J Physiol 246:R63–R66
Solomon R, Taylor M, Sheth S, Silva P, Epstein EH (1985a) Primary role of volume expansion in stimulation of rectal gland function. Am J Physiol 248:R638–R640
Solomon R, Taylor M, Dorsey D, Silva P, Epstein FH (1985b) Atriopeptin stimulation of rectal gland function in *Squalus acanthias.* Am J Physiol 249:R348–R354
Stewart DJ, Semple EW, Swart GT, Sen AK (1976) Induction of the catalytic protein of ($Na^+ + K^+$)-ATPase in the salt gland of the duck. Biochim Biophys Acta 419:150–163
Stewart DJ, Sax J, Funk R, Sen AK (1979) Possible role of cyclic GMP in stimulus-secretion coupling by salt gland of the duck. Am J Physiol 237:C200–C204
Taplin LE, Griff GC (1981) Salt glands in the tongue of the estuarine crocodile *Crocodylus porosus.* Science 212:1045–1047
Templeton JR (1964) Nasal salt excretion in terrestrial lizards. Comp Biochem Physiol 11:223–229
Thesleff S, Schmidt-Nielsen K (1962) An electrophysiological study of the salt gland of the herring-gull. Am J Physiol 202:597–600
Thomas DH, Phillips JG (1975) Studies in avian adrenal steroid function. IV Adrenalectomy and the response of the domestic duck (*Anas platyrhynchos*) to hypertonic NaCl loading. Gen Comp Endocrinol 26:412–419
Wangemann P, Wittner M, Di Stefano A, Englert HC, Lang HJ, Schlatter E, Greger R (1986) Cl^--channel blockers in the thick ascending limb of the loop of Henle. Structure activity relationship. Pfluegers Arch 407 (Suppl 2):S128–141
Welsh MJ (1983) Intracellular chloride activities in canine tracheal epithelium. Direct evidence for sodium-coupled intracellular chloride accumulation in a chloride-secreting epithelium. J Clin Invest 71:1392–1401
Welsh MJ, Smith PL, Fromm M, Frizzell RA (1982) Crypts are the site of intestinal fluid and electrolyte secretion. Science 218:1219–1221
Wilson JX (1984) The renin-angiotensin system and the nervous control of blood circulation. Endocrinol Rev 5:45–61
Wilson JX, Pham D van, Tau-Wilson HI (1985) Angiotensin and converting enzyme regulate extrarenal salt excretion in ducks. Endocrinology 117:135–140

Chapter 8

Regulation of NaCl Transport in Tight Epithelia

L. G. Palmer

Contents

Department of Physiology, Cornell University Medical College, New York, NY 10021, USA

Advances in Comparative and Environmental Physiology
Vol. 1. Edited by R. Greger

1 Introduction

Regulation by vertebrates of the amount of salt in the body is accomplished by the controlled transport of Na and Cl ions by various epithelial tissues. The amount of NaCl retained in the body determines, in turn, the volume of extra- and intracellular water, which must be maintained within narrow limits. This review will focus on the mechanisms controlling a specific transport process, the reabsorption of NaCl, by a class of high resistance epithelia represented by the mammalian cortical collecting tubule, the amphibian skin, the colon in mammals, amphibians, reptiles, and birds, and the urinary bladders of mammals, amphibians, and reptiles. These mechanisms involve modulations in the activities of ion channels, ion pumps, and metabolic processes. The major thesis of this chapter will be that different types of control mechanisms govern these processes on different time scales.

2 Sites of Regulation

The cellular mechanisms which effect the reabsorption of Na by this class of epithelia are illustrated in Fig. 1. Na enters the epithelial cells from the mucosal side through Na-selective ion channels and leaves through the Na pump in the basal-lateral membrane. Both of these processes are under the control of hormones and other factors, as discussed in detail below. The Na pump uses ATP as a substrate, transducing the energy of hydrolysis of the nucleotide to move Na ions against their electrochemical activity gradients. The supply of energy to the pump delivered by the mitochondria is also under hormonal control in some cases. Finally, although Na is the actively transported species, the passive reabsorption of an accompanying anion or the exchange for a cation moving in the opposite direction is essential for the net movement of Na. Cl ions can move either across the cell layer through ion channels or between the cells, through the paracellular spaces. Both of these pathways are to some extent under control. Alternatively,

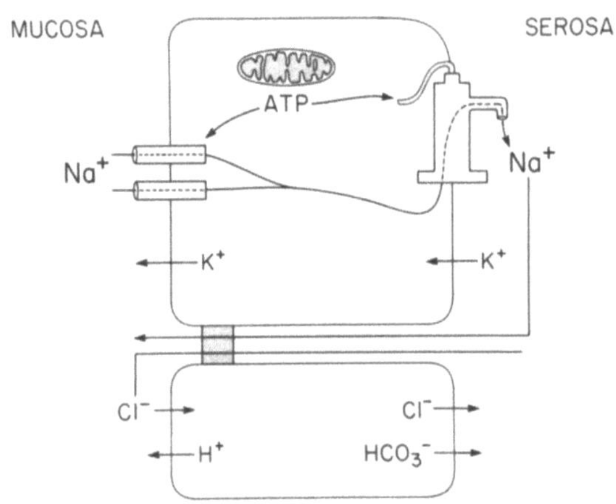

Fig. 1. A Na-reabsorbing epithelium. The three cellular aspects which directly participate in active transport are the Na channels, the Na pump, and the sources of metbolic energy, such as the mitochondria. Factors which indirectly influence the Na transport rate include K^+-selective channels in the basolateral membrane, channels for K^+ and Cl^- in the apical membrane, the paracellular conductance, and the parallel transepithelial pathway for active transport of H^+

Na can be exchanged for either K or H ions, both of which are secreted by some of the Na-reabsorbing epithelia. In any case, promotion of the transport of co- or counterions can, in turn, promote the reabsorption of Na.

Early studies of the regulation of Na reabsorption emphasized the identification of ratelimiting steps in the overall transport process (Sharp and Leaf 1966). More recently, it has become evident that all of these steps are so closely interrelated that it is difficult to affect one and not the others. In many cases, regulation appears to involve the coordinated modulation of several of these processes. For example, aldosterone has been shown to increase the number of conducting Na channels in the apical membrane, to promote the biosynthesis of the Na pump and of the mitochondrial enzyme citrate synthase, to enhance the conductance of K across the apical membrane, to stimulate H-ion secretion, and possibly to regulate the conductance of the paracellular pathway (Table 1). All of these effects will tend to promote Na reabsorption.

The bias of this review, however, is that the most important site of regulation is the conductive pathway for Na ions through apical ion channels. Modulation of these channels by a variety of processes is a dominant force in determining the transport rate, although certainly not the only factor involved. In subsequent sections attention will be focused on these channels, but regulation at the other sites will also be discussed. Because of the many regulatory processes involved, they will be summarized briefly to give an overview of their rich variety. References will be made to articles and reviews which describe the particular processes in more detail.

Table 1. Effects of aldosterone on tight epithelia

Effect	Tissues	Time frame	Reference
↑ Apical Na permeability	All responsive epithelia	1 – 6 h	see Garty (1986)
↑ Na-K-ATPase	ADX rabbit, rat CCT	1 – 3 h[a]	Horster et al. (1980), Petty et al. (1981) El-Mernissa and Doucet (1983)
	Toad bladder	6 – 24 h	Rossier (1984)
	Intact rabbit CCT	1 – 7 days	Garg et al. (1981) El-Mernissa et al. (1983) O'Neil and Hayhurst (1985)
↑ Citrate synthase	Toad bladder	1 – 6 h	Kirsten et al. (1968)
	Rat kidney	1 – 3 h	Law and Edelman (1978)
↑ H^+ secretion	Turtle bladder	1 h	Al-Awqati (1978)
↑ Apical K conductance	Toad bladder	6 – 18 h	Palmer (1986)
	Rabbit CCT	2 – 4 days	Sansom and O'Neil (1985)
↑ Basolateral K conductance	Rabbit CCT	9 days	Sansom and O'Neil (1986)

[a] Effect may depend on concomitant ↑ apical Na permeability.

3 Regulation in the Time Frame of 0 – 10 s

Several factors can result in the very rapid modification of apical membrane Na permeability by affecting Na channels. These factors involve changes in the composition of the mucosal fluid (i.e., the urine).

3.1 Mucosal Na Concentration

In many tight epithelia the rate of Na transport is a saturable function of the mucosal Na concentration, with concentrations of 20 – 40 mM required for half-maximal rates. Since this can be observed even when the driving force for Na entry into the cell continues to increase with increasing mucosal Na concentration, the phenomenon can be viewed as a decrease in apical membrane Na permeability. The decrease in permeability is very rapid, occurring within seconds (Fuchs et al. 1977) and does not depend on an increase in intracellular Na (Lindemann 1984). There are at least two processes which contribute to this phenomenon, which has been termed Na self-inhibition (Lindemann 1984). The rate of transport of Na ions through individual Na channels can saturate as the concentration increases. This is analogous to the saturation of a rate of catalysis by an enzyme as the concentration of substrate increases. Saturation of single-channel conductance has been observed in measurements of currents through individual channels from A6 cells (Olans et al. 1984) and cortical collecting tubules (Palmer and Frindt 1986). The other process involves a decrease in the number of conducting Na channels as the Na concentration is increased (van Driessche and Lindemann 1979), possibly by an allosteric mechanism involving the interaction of Na ions at a site distant from the conduction pathway.

The effect of self-inhibition is to limit the amount of Na entering the cell when the delivery of Na to the channel-containing organ is increased. This may help the cell from being overloaded with Na. The various feedback inhibition processes discussed in the next section also contribute to intracellular homeostasis with respect to Na.

3.2 Mucosal pH

As mucosal pH is decreased from neutral, Na transport in the frog skin and toad urinary bladder increases, reaching an optimum at pH 5.5 to 6.0, and then falls (Leaf et al. 1964; Fanestil et al. 1984). The stimulation by acidification appears to involve an increase in the number of conducting channels (Li and Lindemann 1980). The physiological significance of this effect in unclear.

3.3 Amiloride

The K-sparing diuretic amiloride blocks epithelial Na channels from the mucosal side at micromolar concentrations (Benos 1982). The drug acts rapidly, on the time scale of 100 to 1000 ms, and reversibly. The action of amiloride may be to bind within the outer mouth of the channel, physically obstructing the pore (Palmer 1985a; Warncke and Lindemann 1985). Amiloride itself is a synthetic compound and affects Na transport only when administered pharmacologically. There is a report, however, of the possible existence of a natural amiloridelike compound in the sweat glands of cystic fibrosis patients (Mangos and McSherry 1967). Sweat from CF patients were found to inhibit the reabsorption of Na by normal epithelia, implying that a substance in the secreted fluids could block reabsorption by the sweat ducts and account in part for the high concentrations of Cl in sweat and saliva from these patients. The putative amiloridelike substance has yet to be identified.

3.4 Transmembrane Voltage

The process of Na transport is electrogenic in these tissues, resulting in a transepithelial potential which is negative on the mucosal side relative to the serosal side. An increase in the Na transport rate will increase the transepithelial voltage, and at least part of this change will be manifested across the apical membrane, making the cell less negative (or in some cases more positive) with respect to the mucosal fluid. This instantaneous effect on membrane potential will thus decrease the driving force for Na entry and serve as another powerful self-inhibiting factor. In addition, any other circumstance in which the intracellular potential is made less negative will tend to diminish Na entry and the overall transport rate.

Besides the direct effect on the electrical driving force for Na entry, changes in the apical membrane potential can influence at least three other ion transport processes in tight epithelia. Making the potential less negative by directly clamping the transepithelial voltage to more negative values (mucosal with respect to serosal) activates both a Cl conductance (Hviid Larsen and Kristensen 1978) and a K conductance (Palmer 1986) in the apical membranes of toad skin and bladder. The resulting Cl influx into the cell, or K efflux from the cell, will tend to repolarize the apical membrane and sustain the driving force for Na entry. As a result, either NaCl reabsorption, or Na-K exchange, will be promoted. Finally, the paracellular conductance is also increased by making the mucosal side more negative (Finn and Rogenes 1980). This will enhance the transport of both Cl and K, as well as the backleak of Na, through this pathway.

In contrast to these conductances, the apical Na channels themselves do not appear to be strongly voltage gated. Evidence for this comes from measurements of macroscopic conductance, in which the amiloride-sensitive current-voltage relationship obeys the constant field equation (Fuchs et al. 1977) which would not be expected if the number of open channels were strongly affected by voltage. In addition, direct measurements of single-channel kinetics in A6 cells using the patch-clamp technique revealed only a mild voltage dependence (Hamilton and Eaton 1985).

4 Regulation in the Time Frame of 10 s to 10 min

Included in this time range are processes which appear to involve changes in the intracelluar fluids, especially in the concentrations of Ca^{2+}, H^+, and metabolites. These can be classified as feedback inhibitors; increased entry of Na into the cell will result in alterations which subsequently reduce apical permeability and limit the further entry of the ion (Taylor and Windhager 1979). As was the case for the Na self-inhibition discussed above, this will help to prevent excessive increases in intracellular Na after Na permeability is increased. Also included are effects of osmolarity and hydrostatic pressure across the epithelium.

4.1 Intracellular Ca

Addition of Ca ionophores to Na-transporting epithelia, including toad urinary bladder (Ludens 1978; Wiesmann et al. 1978a) and cortical collecting tubule (Frindt and Windhager 1986) diminishes Na transport. Subsequently, it was shown that Ca can block amiloride-sensitive (i.e., channel-mediated) Na fluxes in apical membrane vesicles from the toad bladder (Chase and Al-Awqati 1983; Garty 1984). The inhibition was observed over concentration ranges of Ca ions expected to be present in the cell (0.1 to 1 μM). On the other hand (Table 2) changes in cytoplasmic-side Ca ion concentrations had no effect on currents through individual Na channels in excised membrane patches from the apical membrane of the rat CCT (Palmer and Frindt 1987).

A separate indirect effect of Ca was reported by Garty and Asher (1985), who incubated permeabilized toad bladder cells in various concentrations of Ca and found that the permeability of apical membrane vesicles made from these cells was inversely related to the Ca concentration in the incubation medium. This effect was independent of the subsequent concentration of Ca present when fluxes were measured. It was suggested that an enzymatic process activated by Ca might be involved. A related phenomenon was observed in the rat CCT, where addition of a Ca ionophore diminished Na channel activity observed in cell-attached patches using the patch-clamp technique (Palmer and Frindt 1987). This process was also apparently an indirect effect of Ca, as it was observed between 3 and 10 min after the addition of the ionophore, while the increase in cell Ca concentration occurred within seconds. In addition, the phenomenon was not reproducible by addition of Ca to excised, inside-out patches.

Although the biochemical bases for these effects of Ca remain unknown, one possible important factor is the concentration of prostaglandins in the cell. Erlij et al. (1986) showed that in the isolated frog skin epithelium Ca ionophores can *stimulate* Na transport and that this effect apparently involves the Ca-dependent production of prostaglandins. Exogenous prostaglandins stimulate Na transport in the frog skin (Els and Helman 1981) and the toad bladder (Lipson and Sharp 1971; Albert and Handler 1974) but inhibit it in the CCT (Stokes and Kokko 1977; Iino and Imai 1978).

Although the regulation of intracellular Ca by these epithelial cells is not completely understood, a Na – Ca exchange mechanism on the basal-lateral membrane appears to be a major route of Ca extrusion from the cell (Grinstein and Erlij 1978; Taylor and Windhager 1979; Chase and Al-Awqati 1981). Thus, an increase in intracellular Na concentration will slow Ca exit, causing the buildup of Ca within the cell and the eventual decrease in Na permeability and Na entry. Consistent with this view are the observations that reduction of serosal Na concentration inhibits Na transport and apical Na entry into the cells (Grinstein and Erlij 1978; Chase and Al-Awqati 1981), and that increasing serosal Ca decreases apical Na permeability in the K-depolarized toad bladder, but only after the cell Na concentration has been raised (Garty and Lindemann 1984; Palmer 1985b).

Table 2. Cytoplasmic factors which may influence epithelial Na channels

Factor	Tissue	Action[a]	Direct effect[b]	Reference
Ca^{2+}	Toad bladder	I	Yes	Chase and Al-Awqati (1983), Garty (1984)
Ca^{2+}	Toad bladder	I	No	Garty and Asher (1985)
	Rat CCT	I	No	Palmer and Frindt (1987)
H^+	Rat CCT	I	Yes	Palmer and Frindt (1987)
cAMP/protein kinase A (ADH)	Toad bladder	S	?	Orloff and Handler (1962)
	Frog skin	S	?	Curran et al. (1963)
	Rabbit, rat CCT	S	?	Frindt and Burg (1972), Reif et al. (1984), Tomita et al. (1985)
Prostaglandins	Toad bladder	S	?	Lipson and Sharp (1971), Albert and Handler (1974)
	Frog skin	S	?	Erlij et al. (1986), Els and Helman (1981)
	Rabbit CCT	I	?	Iino and Imai (1978), Stokes and Kokko (1977)
Phorbol ester/protein kinase C	Frog skin	S	?	Civan et al. (1985)
	A6	I	?	Yanase and Handler (1986)
Methyl donors	A6	S	Yes	Sariban-Schraby et al. (1984)

[a] S = stimulatory, I = inhibitory.
[b] "Direct" effects are those demonstrable in isolated membranes.

4.2 Cell pH

Acidification of the cytoplasm by a rapid increase in CO_2 tension reduces Na transport rates by the frog skin and toad bladder (Ussing and Zerahn 1951; Palmer 1985b). In the latter tissue the apical Na permeability was also shown to decrease. In patch-clamp experiments using excised, inside-out patches from apical membranes of the rat CCT, acidification of the cytoplasmic side of the membrane was shown to decrease the open-state probability of Na channels over the pH range of 6.4 to 7.4 (Palmer and Frindt 1987). On the other hand, in apical membrane vesicles made from toad bladder cells, Na fluxes were independent of pH over the range of 7.0 to 8.0, although they declined when the solution was made more acidic (Garty et al. 1985).

The regulation of pH may be in part analogous to the regulation of Ca, in that a Na – H exchange mechanism could serve to extrude H ions across the basal-lateral membrane, as shown for the rabbit CCT (Chaillet et al. 1985). Thus, the same sort of feedback system thought to involve changes in Ca could also operate via changes in intracellular pH. The relative importance of Ca and pH in the feedback control of Na permeability in Na-transporting epithelia remains to be established.

4.3 Cellular Metabolism

Although alterations in cellular metabolism have obvious potential effects on the rate of Na extrusion by the Na pump, apical Na permeability seems to be even more sensitive to metabolic events. In frog skin, 2-deoxy-glucose was found to reduce transcellular conductance (Hong and Essig 1976). Subsequently, is was shown that this inhibitor reduced apical Na permeability in the toad bladder and that the effect was reversed by addition of pyruvate to bypass the metabolic block (Palmer et al. 1980; Garty et al. 1983). The biochemical basis for the inhibition is unknown. It could involve changes in the concentrations of metabolites, such as a decrease in ATP or an increase in ADP, or in the cell Ca or H-ion concentrations discussed above.

Feedback involving metabolism could be important whenever the metabolic demands of transport exceed the capacity of the cells to provide metabolic energy. This could be the result of either a rapid increase in Na entry into the cell, or to a diminished metabolic capacity as in anoxia or, perhaps in Amphibia, at low temperatures. The resulting reduction in Na entry and hence in the delivery of Na to the energy-consuming Na pump would tend to conserve the metabolic resources of the cell.

4.4 Effects on the Na Pump

The Na pump is also sensitive to changes in cellular Ca, pH and metabolism. In red cell ghosts, intracellular Ca concentrations of 1 to 10 μM reduce the pump-mediated Na efflux (Yingst and Hoffman 1984). In rabbit urinary bladder, acidification of the cytoplasm reduces the rate of Na extrusion by the pump after maximizing the Na entry rate with the ionophore nystatin (Eaton et al. 1984). Finally, reductions in cell ATP concentrations below those required for optimizing the turnover rate of the Na/K-ATPase [about 1 mM (Jorgensen 1986)], will also slow the pump. Note that these effects on the pump would tend to work against the feedback control of intracellular Na, in that an increase in cell Na would, by increasing cell Ca, decreasing cell pH, or draining metabolic resources, tend to block the Na extrusion process.

There appear to be other mechanisms, however, which are capable of stimulating the pump after a primary increase in Na entry. In the *Necturus* urinary bladder, increasing mucosal Na concentrations resulted in an increased rate of Na transport without a detectable rise in intracellular Na (Thomas et al. 1983). This implies either an increase in the pump capacity or an increase in the affinity of the pump for Na. The mechanism by which this change takes place is unknown. However, it takes place within minutes, much too rapidly to be accounted for by changes in the rate of synthesis of pump units. Other possibilities include insertion of pumps into the membrane or an activation or unmasking of existing pumps.

4.5 Osmotic Effects

In the toad urinary bladder, osmotic swelling of the epithelial cells stimulates transport, while shrinkage inhibits it (Lipton 1972; Lewis et al. 1985). Although the effect of hypertonic solutions was very rapid, consistent with a direct effect of cell shrinkage on the transport process, the effects of hypotonic solutions were much slower, not peaking for 30 to 60 min. The mechanisms underlying both of these effects and the sites at which regulation takes place have also not been fully determined. There is evidence, however, that one site of action could be at the basal-lateral membrane K conductance, which appears to increase with cell swelling in the toad bladder (Lewis et al. 1985) as well as in the amphibian intestine (Lau et al. 1984). In line with this suggestion, Wong and Chase (1986) reported that swelling of toad bladder cells is accompanied by an increase in the cytoplasmic Ca^{2+} concentration which could activate K channels. An increased basal-lateral K conductance would hyperpolarize the basal-lateral cell membrane and increase the driving force for Na entry across the apical membrane.

Other studies have implicated changes in paracellular conductance in alterations of NaCl reabsorption in response to changes in external osmolarity. Civan and Dibona (1978) have shown in the toad bladder that under conditions of a dilute mucosal fluid, the paracellular pathway is inhibited by the closing down of the intercellular spaces. This will tend to conserve the low NaCl concentrations in the urine. When the osmolarity of the mucosal fluid is increased by addition of NaCl, the paracellular pathway opens up, permitting the more rapid passive reabsorption of Cl and hence a greater rate of NaCl reabsorption when the delivery to the bladder is high.

4.6 Stretch and Hydrostatic Pressure

In the rabbit urinary bladder, stretching of the tissue or application of hydrostatic pressure increases the Na transport rate rapidly and reversibly (Lewis and de Moura 1982, 1984). The mechanism involves the insertion of additional membrane into the apical surface membrane, with the concomitant insertion of Na channels. The phenomenon may be involved in the adaptation of the bladder to stretch as it fills with urine. The toad urinary bladder also increases its transport rate in response to stretch (Walser 1969), although the mechanisms involved have not been thoroughly studied in that tissue.

5 Regulation in the Time Frame of 10 min to 1 h

Over this time period, the major regulatory pathways that have been studied involve hormones with cell surface receptors (Table 3). These hormones, which include ADH, acetylcholine, and insulin, do not require protein synthesis to exert their effects, at least over this time interval.

Table 3. Hormones which affect Na transport in tight epithelia

Hormone	Time frame	Sites of action	Tissue
Aldosterone	1 h to 1 week	Many (Table 1)	All Na-reabsorbing tight epithelia
Insulin	15 to 90 min	Na permeability Na pump	Amphibian skin and bladder
ADH	5 to 30 min	Na permeability Apical K conductance	Amphibian skin and bladder, CCT
ACh	5 to 10 min	?	Amphibian skin and bladder

5.1 ADH

The effects of antidiuretic hormone (ADH) and its analogs to increase Na transport by frog skin and toad bladder have been known for many years (Leaf and Anderson 1958; Frazier et al. 1962; Curran et al. 1963). The effect is measurable within 5 min after adding the hormone, peaking after 10 to 20 min and declining thereafter with a highly variable time course. In the rabbit CCT the stimulation is transient and is followed by a period of inhibited Na transport (Frindt and Burg 1972; Schuster 1985). In the rat CCT the stimulation of transport is sustained (Reif et al. 1984; Tomita et al. 1985).

The effects of ADH can be mimicked by the addition of exogenous cAMP or its more membrane-permeant analogs, or by theophylline which inhibits the breakdown of cellular cAMP by phosphodiesterase (Orloff and Handler 1962). The major action of ADH on Na transport is an increase in apical membrane Na permeability, which can be accounted for by an increase in the density of conducting Na channels in the apical membrane (Li et al. 1982; Helman et al. 1983). It is not know whether this effect is result of phosphorylation of the Na channels themselves by cAMP-dependent protein kinase, or by a less direct action. Garty and Edelman (1983) suggested, on the basis of sensitivity of channels to proteolysis, that ADH might recruit Na channels to the apical membrane from cytoplasmic stores.

Besides its effects on Na channels, ADH also stimulates the voltage-dependent apical membrane K conductance in the toad bladder (Palmer 1986). As discussed above (Sect. 2) K movement through this pathway could effect an electrically coupled Na – K exchange across the apical membrane and thus enhance Na reabsorption as well as K secretion.

5.2 Insulin

Insulin stimulates Na transport in frog skin and toad urinary bladder (Herrera et al. 1963; Herrera 1965; Siegel and Civan 1976; Cox and Singer 1977). This peptide hormone acts somewhat more slowly than ADH, producing its maximal effects in 1 to 1.5 h. Longer-term effects have also been reported, which unlike the acute effects, apparently require the synthesis of new proteins (Cobb et al. 1981).

At least one effect of insulin may be to stimulate apical Na permeability, as the hormone increases transcellular conductance in toad bladder and frog skin (Siegel and Civan 1976; Schoen and Erlij 1985). Insulin also increases the apparent electromotive driving force of the basal-lateral Na pump (Siegel and Civan 1976). This is consistent with the well-documented effect of the hormone to stimulate pump-mediated Na efflux from skeletal muscle (Creese 1968; Moore 1973) and in adipocytes (Resh et al. 1980).

5.3 Muscarinic Agonists and Phorbol Esters

In toad urinary bladder acetylcholine and its muscarinic agonists inhibit Na transport (Sahib et al. 1978; Weismann et al. 1978b). The effect is complete within 10 min or less and is readily reversible. The second messengers involved in this response have not been identified, although cGMP has been suggested as a possibility (Sahib et al. 1978). The physiological significance of this control mechanism is not clear, nor have the site(s) at which these neurohormones and their second messengers act been established. In other systems muscarinic agonists can increase phosphoinositol turnover and presumably activate protein kinase C (Michell 1975; Berridge and Irvine 1984). Tumor-promoting phorbol esters, which can also activate protein kinase C, reduce Na transport in A6 cells (Yanase and Handler 1986) raising the possibility that protein kinase C could mediate the effects of muscarinic agonists, perhaps by inhibiting apical Na channels. Activators of protein kinase C *stimulate* Na transport in the frog skin (Civan et al. 1985). Acetylcholine can also stimulate Na transport in this tissue under some conditions (Cuthbert and Wilson 1981). These authors showed, however, that the stimulatory effect may be secondary to an increased production of prostaglandins. Thus, the biochemical details of the effects of muscarinic agents, phosphoinositol turnover, protein kinase C and prostaglandins in epithelia remain to be worked out.

6 Regulation in the Time Frame of 1 – 3 h

The major effector of Na transport over this time period is aldosterone. This hormone is probably the most important single determinant of Na transport by the tight epithelia of most vertebrates, and is a major factor in the overall regulation of Na and fluid balance. Aldosterone promotes Na transport in nearly all epithelia which exhibit amiloride-sensitive transport. The mechanism of stimulation involves a classical steroid receptor-mediated sequence of events, and requires the synthesis of new proteins through a selective increase in the transcription rate of certain genes (Edelman 1978; Cox and Geheb 1984; Garty 1986). The effects observed within the first 3 h after a challenge with aldosterone are often referred to as the acute effects of the hormone.

6.1 Effects on Na Channels

Aldosterone increases apical membrane permeability in the toad bladder with a time course that parallels the early increase in Na transport elicited by the mineralocorticoid. After a lag period of about 1 h, transport and permeability increase in parallel over the next several hours (Palmer et al. 1982). It was further shown that the increase can be accounted for by an increase in the number of conducting channels in the membrane.

Although the increased number of channels and the requirement for protein synthesis suggest that the mechanism involves the de novo synthesis of channels, two types of experiments suggest that this is not the case. First, irreversible modification of Na channels by several different agents in the absence of aldosterone results in parallel decreases in basal transport rate and the subsequent response to the hormone. This suggests that the aldosterone-dependent channels were already in the membrane (see Garty 1986 for review). Second, in membrane vesicles made from cultured A6 cells, the additional Na permeability seen in vesicles made from aldosterone-treated tissues can also be elicited in a cell-free environment by addition of methyl-donating substrates (Sariban-Sohraby et al. 1984). The implication is that the hormone-dependent channels were preexistent in the membrane, and were activated by an aldosterone-promoted methylation reaction.

6.2 Effects on Energy Metabolism

A second early event in the stimulation of tight epithelia by aldosterone is the enhancement of the activities of mitochondrial enzymes involved in the TCA cycle (Kirsten et al. 1968; Law and Edelman 1978). At least one of these enzymes, citrate synthase, is synthesized at a higher rate in response to the hormone. As mentioned above, the increase in the metabolic capacity of the cell may be helpful both in providing ATP to the Na pump and in keeping apical Na channels open.

6.3 Effects on the Na Pump

In adrenalectomized rabbits or rats, the Na-K-ATPase (or pump) activity is diminished, and can be restored within 3 h by aldosterone. This effect of the steroid has been demonstrated with in vitro exposure of isolated tubules to the steroid (Horster et al. 1980), injection of aldosterone in vivo, with subsequent isolation of tubules (Petty et al. 1981; El Mernissa et al. 1983), and with incubation of medullary tubular suspensions (Rayson and Lowther 1984). In the mouse CCT this early response of the pump to aldosterone was not detected (Doucet and Katz 1981).

This up-regulation of the pump apparently is secondary to an increase in apical membrane Na transport as the effect could be blocked by amiloride, which abolishes Na entry through Na channels (Petty et al. 1981; Rayson and Gupta 1985). One possibility is that the increase in cell Na which is expected to arise from the primary increase in Na permeability could influence the activation of the pump. However, increased cell Na itself may not be sufficient, as raising cell Na with ouabain resulted in a much slower increase in pump activity (Rayson Gupta 1985).

6.4 Effects on H^+ Transport

Another early effect of aldosterone which has been studied in toad and turtle bladders in the enhancement of the electrogenic secretion of protons (see Al-Awqati 1978 for review). As discussed above, this could indirectly stimulate Na reabsorption by making the transepithelial electrical gradient more favorable, i.e., the mucosal fluid less negative with respect to serosa.

7 Regulation in the Time Frame of 3 – 24 h

These effects include the long-term effects of aldosterone and possibly those of proteolytic enzymes.

7.1 Aldosterone

The effects of aldosterone exerted after about 3 h of exposure to the steroid are in some respects qualitatively different from those observed earlier, particularly in the well-studied toad urinary bladder model. Although apical membrane permeability continues to increase during this period (Palmer and Speez 1986), there is, in addition, an increase in the synthesis of the Na/K pump which, unlike the early response in the rabbit CCT, is not dependent on an increase in the rate of Na entry into the cell (Geering et al. 1982; Rossier 1984). The rate of transcription of mRNA coding for the pump was detectably increased 6 h after addition of the hormone and was doubled after 18 h. There is also electrophysiological data suggesting that the maximal rate of pumping of Na across the basal-lateral membrane is enhanced during this time (Palmer and Speez 1986).

The slower responses to the hormone could be the result of different receptors for aldosterone, or of different rates of expression of genes which are regulated by the same steroid receptor. Rossier and his co-workers have reported that two agents, thyroid hormone and Na butyrate, can selectively inhibit the late response, including the induction of the Na–K ATPase (Geering et al. 1984; Truscello et al. 1983). This implies that the two responses are in some way distinct.

7.2 Proteolytic Degradation of Channels

Recently, a novel mechanism of control of apical Na permeability through changes in degradation rates of the channels has been suggested (Lewis and Alles 1986). It was shown that apical Na channels in rabbit urinary bladder can be degraded by kallikrein in the mucosal solution. A similar effect of another proteolytic enzyme, trypsin, on Na channels in the toad bladder was reported by Garty and Edelman (1983). Since kallikrein is secreted by the kidney and is found in urine, it could be a factor in determining the lifetimes of the channels in vivo. Furthermore, the level of kallikrein in urine is increased during elevation of plasma aldosterone (Margolius 1984), suggesting the intriguing possibility that the rate of channel degradation could be regulated.

8 Regulation in the Time Frame of 24 h

8.1 Chronic Exposure to Mineralocorticoids

Chronic treatment of adrenal-intact rabbits with corticosteroids for a week or more leads to an enhancement in net Na reabsorption by the CCT (O'Neil and Helman 1977; Schwartz and Burg 1978). This stimulation is accompanied by marked changes in the morphology of the Na-transporting of principal cells of the renal CCT (Wade et al. 1979). The area of the basal-lateral membranes of these cells increases by almost twofold. The effect is specific for the basal-lateral membranes, as the apical membrane area does not change, and does not occur in the intercalated cells, which are primarily involved in acid-base rather than Na transport. The increased area results in a greater degree of membrane infolding. It can also be elicited by chronic injection of dexamethasone, presumably a specific glucocorticoid.

Activity of Na–K-ATPase increases more or less in parallel with the increase in membrane area in these cells (Garg et al. 1978; El Mernissa et al. 1983; O'Neil and Hayhurst 1985). The maximal increase requires 2–7 days of treatment with DOCA, a potent mineralocorticoid. Sansom and O'Neil (1985) found that apical membrane Na and K conductances also increased with chronic DOCA treatment. The increase in Na conductance was measurable after 1 day, but that of K required 4 days of hormone treatment. Finally, Sansom and O'Neil (1986) reported that the basolateral K conductance in the CCT was enhanced after DOCA treatment for 8 days.

The relationship of these chronic effects of steroids to those which occur more rapidly is not well established. They could represent a delayed reaction to activation of the same receptors, a reaction to activation of different receptors, or an indirect result of activation of the original receptor, possibly mediated by changes in ion composition (see below).

8.2 Chronic Changes in Cellular Ion Contents

In several cell types chronic inhibition of the Na pump by exposure to ouabain or to low K solution for up to 24 h has been shown to increase the activity of the ATPase and the number of pump sites in the membrane (Vaughn and Cook 1972; Kim et al. 1984; Rayson and Gupta 1985; Pressley et al. 1986). This is generally attributed to an increase in intracellular Na activity. Thus, any chronic increase in the rate of Na entry into the cell, either by stimulation of apical Na permeability or by an increase in the load of Na to the cell, could through this mechanism secondarily elevate the number of pumps in the basal-lateral membrane. This mechanism is distinct from the acute regulation of the pump activity discussed in Section 4.4 as the long-term response involves changes in the rates of synthesis and/or degradation of the pumps. The results, however, are similar, in that the capacity for Na extrusion is increased to balance either a greater rate of Na entry or a partial block of the extrusion mechanism.

9 Summary

Many different pathways exist through which Na transport in tight epithelia can be regulated. The mechanisms range from physical effects on membrane proteins, which can happen very rapidly, to much slower changes in the rates of synthesis and degradation of the transporters. In between are a variety of processes involving chemical alterations in the channels and pumps as well as their translocation from one part of the cell to another. A number of cellular processes bear either directly or indirectly on the rate of Na reabsorption by these organs. These processed are sometimes controlled in a coordinated way, as exemplified by regulation by aldosterone, and can in addition be tightly coupled to each other through various feedback systems. The variety of stimuli and responses which together determine the single process of Na reabsorption illustrate the intricacy of homeostasis in higher organisms.

References

Al-Awqati Q (1978) H^+ transport in urinary epithelia. Am J Physiol 235:F77 – F88

Albert WC, Handler JS (1974) Effects of PGE_1, indomethacin and polyphloretin phosphate in the toad bladder response to ADH. Am J Physiol 226:1382 – 1386

Benos DJ (1982) Amiloride: a molecular probe of sodium transport in tissues and cells. Am J Physiol 242:C131 – C145

Berridge MJ and Irving RF (1984) Inositol triphosphate, a novel second messenger in cellular signal transduction. Nature (London) 312:315 – 321

Chaillet JR, Lopes A, Boron WF (1985) Basolateral Na – H exchange in the rabbit cortical collecting tubule. J Gen Physiol 86:795 – 812
Chase HS Jr, Al-Awqati Q (1981) Regulation of the sodium permeability of the luminal border of toad bladder by intracellular sodium and calcium. J Gen Physiol 77:693 – 712
Chase HS Jr, Al-Awqati Q (1983) Calcium reduces the sodium permeability of luminal membrane vesicles from toad bladder. J Gen Physiol 81:643 – 665
Civan MM, Dibona DR (1978) Pathways for movement of ions and water across toad urinary bladder. III. Physiologic significance of the paracellular pathway. J Membr Biol 38:359 – 386
Civan MM, Rubenstein D, Mauro T, O'Brien TG (1985) Effects of tumor promoters on sodium ion transport across frog skin. Am J Physiol 248:C457 – C465
Cobb MH, Skipski IA, Scott WN (1981) Role of induced proteins in insulin-stimulated sodium transport. Ann N Y Acad Sci 372:247 – 269
Cox M, Geheb M (1984) Aldosterone-induced proteins in renal epithelia. Curr Top Membr Transp 20:271 – 293
Cox M, Singer I (1977) Insulin-mediated Na^+ transport in the toad urinary bladder. Am J Physiol 232:F270 – F277
Creese R (1968) Sodium fluxes in diaphragm muscle and the effects of insulin and serum proteins. J Physiol (London) 197:255 – 278
Curran PF, Herrera FC, Flanigan WJ (1963) The effect of Ca and antidiuretic hormone on Na-transport across the frog skin. II. Sites and mechanisms of action. J Gen Physiol 46: 1011 – 1027
Cuthber AW, Wilson SA (1981) Mechanisms for the effects of acetylcholine on sodium transport in frog skin. J Membr Biol 59:65 – 75
Doucet A, Katz AI (1981) Short-term effect of aldosterone on Na, K-ATPase in single nephron segments. Am J Physiol 241:F273 – F278
Driessche W van, Lindemann B (1979) Concentration-dependence of currents through single sodium-selective pores in frog skin. Nature (London) 282:519 – 520
Eaton DC, Hamilton KL, Johnson KE (1984) Intracellular acidosis blocks the Na^+ pump in rabbit urinary bladder. Am J Physiol 16:F946 – F955
Edelman IS (1978) Candidate mediators in the action of aldosterone on Na^+ transport. In: Hoffman JF (ed) Membrane transport processes, vol 1. Raven, New York, pp 125 – 140
El Mernissa G, Doucet A (1983) Short-term effect of aldosterone on renal sodium transport and Na – K-ATPase in the rat. Pfluegers Arch 399:139 – 146
El Mernissa G, Chabardes D, Doucet A, Hus-Citharel A, Imbert-Teboul M, Le Bouffant F, Montegut M, Siaume S, Morel F (1983) Changes in tubular basolateral membrane markers after chronic DOCA treatment. Am J Physiol 245:F100 – F109
Els WJ, Helman SI (1981) Vasopressin, theophylline, PGE_2 and indomethacin on active Na transport in frog skin: studies with microelectrodes. Am J Physiol 241:F279 – F288
Erlij D, Gersten L, Sterba G, Schoen HF (1986) Role of prostaglandin release in the response of tight epithelia to Ca^{2+} ionophores. Am J Physiol 250:C629 – C636
Fanestil DD, Kessler R, Park CS (1984) Probing molecular characteristics of ion transport proteins. Curr Top Membr Transp 20:259 – 270
Finn AL, Rogenes P (1980) The effects of voltage clamping in tight epithelia. Curr Top Membr Transp 13:245 – 255
Frazier HS, Dempsey EF, Leaf A (1962) Movement of sodium across the mucosal surface of the isolated toad bladder and its modification by vasopressin. J Gen Physiol 45:529 – 543
Frindt G, Burg MB (1972) Effect of vasopressin on sodium transport in renal cortical collecting tubules. Kidney Int 1:224 – 231
Frindt G, Windhager EE (1986) Effect of ionomycin on Na fluxes in perfused cortical collecting tubules (CCT's) of rabbit kidneys. Fed Proc 45:540.
Fuchs W, Hviid Larsen E, Lindemann B (1977) Current voltage curve of sodium channels and concentration dependence of sodium permeability in frog skin. J Physiol (London) 267:137 – 166
Garg L, Knepper M, Burg M (1981) Mineralocorticoid stimulation of Na – K-ATPase in nephron segments. Am J Physiol 240:F536 – F544
Garty H (1984) Amiloride blockable sodium fluxes in toad bladder membrane vesicles. J Membr Biol 82:269 – 280
Garty H (1986) Mechanism of aldosterone action in tight epithelia. J Membr Biol 90:193 – 205

Garty H, Asher C (1985) Ca^{2+}-dependent, temperature-sensitive regulation of Na^+ channels in tight epithelia. A study using membrane vesicles. J Biol Chem 260:8330–8335
Garty H, Edelman IS (1983) Amiloride-sensitive trypsinization of apical sodium channels. Analysis of hormonal regulation of sodium transport in toad bladder. J Gen Physiol 81:785–803
Garty H, Lindemann B (1984) Feedback inhibition of sodium uptake in K-depolarized toad urinary bladders. Biochim Biophys Acta 771:89–98
Garty H, Edelman IS, Lindemann B (1983) Metabolic regulation of apical sodium permeability in toad urinary bladder in the presence and absence of aldosterone. J Membr Biol 74:15–24
Garty H, Civan ED, Civan MM (1985) Effects of internal and external pH on amiloride-blockable Na^+ transport across toad urinary bladder vesicles. J Membr Biol 87:67–75
Geering K, Girardet M, Bron C, Kraehenbuhl JP, Rossier BC (1982) Hormonal regulation of (Na^+,K^+)-ATPase biosynthesis in the toad bladder. Effects of aldosterone and 3,5,3′-triiodo-L-thyronine. J Biol Chem 257:10338–10343
Geering K, Gaeggler HP, Rossier BC (1984) Effects of thyrometric drugs on aldosterone-dependent sodium transport in the toad bladder. J Membr Biol 77:15–23
Grinstein S, Erlij D (1978) Intracellular Ca^{++} and the regulation of Na^+ transport in the frog skin. Proc R Soc Lond Ser B 202:353–360
Hamilton KL, Eaton DC (1985) Single channel recordings from the amiloride-sensitive Na^+ channel. Am J Physiol 249:C200–C207
Helman SI, Cox TC, VanDriessche W (1983) Hormonal control of apical membrane Na transport in epithelia: studies with fluctuation analysis. J Gen Physiol 82:201–220
Herrera FC (1965) Effect of insulin on short-circuit current and sodium transport across toad urinary bladder. Am J Physiol 209:819–824
Herrera FC, Whittembury G, Planchant A (1963) Effect of insulin on short circuit current across isolated frog skin in the presence of calcium and magnesium. Biochim Biophys Acta 66:170–172
Hong CD, Essig A (1976) Effects of 2-deoxy-D-glucose, amiloride, vasopressin and ouabain on active conductance and E_{Na} in the toad bladder. J Membr Biol 28:121–142
Horster M, Schmid H, Schmidt U (1980) Aldosterone in vitro restores nephron Na–K-ATPase of distal segments from adrenalectomized rabbits. Pfluegers Arch 384:203–206
Hviid Larsen E, Kristensen P (1978) Properties of a conductive cellular chloride pathway in the skin of the toad (*Bufo bufo*). Acta Physiol Scand 102:121
Iino Y, Imai M (1978) Effects of prostaglandins on Na transport in isolated collecting tubules. Pfluegers Arch 373:125–132
Jorgensen PL (1986) Structure, function and regulation of Na, K-ATPase in the kidney. Kidney Int 29:10–20
Kim D, Marsh JD, Barr WH, Smith TW (1984) Effects of growth in low potassium medium or ouabain on membrane Na, K-ATPase, cation transport and contractility in cultured chick heart cells. Circ Res 55:39–48
Kirsten E, Kirsten R, Leaf A, Sharp GWG (1968) Increased activity of enzymes of the tricarboxylic acid cycle in response to aldosterone in the toad bladder. Pfluegers Arch 300:213–225
Lau KR, Hudson RL, Schultz SG (1984) Cell swelling increases a barium-inhibitable potassium conductance in the basolateral membrane of *Necturus* small intestine. Proc Natl Acad Sci USA 81:3591–3594
Law P-Y, Edelman IS (1978) Induction of citrate synthase by aldosterone in the rat kidney. J Membr Biol 41:41–64
Leaf A, Anderson P (1958) Active sodium transport in the isolated toad bladder. J Gen Physiol 41:657–668
Leaf A, Keller A, Dempsey EF (1964) Stimulation of sodium transport in toad bladder by acidification of the mucosal medium. Am J Physiol 207:547–552
Lewis SA, Alles WP (1986) Urinary kallikrein: a physiological regulator of epithelial Na^+ absorption. Proc Natl Acad Sci USA 83:5345–5348
Lewis SA, Moura JLC de (1982) Incorporation of cytoplasmic vesicles into apical membrane of mammalian urinary bladder. Nature (London) 297:685–688
Lewis SA, Moura JLC de (1984) Apical membrane area of rabbit urinary bladder increases by fusion of intracellular vesicles: an electrophysiological study. J Membr Biol 82:123–136
Lewis SA, Butt AG, Bowler JM, Leader JP, Macknight ADC (1985) Effects of anions on cellular volume and transepithelial Na^+ transport across toad urinary bladder. J Membr Biol 83:119–137

Li JH-Y, Lindemann B (1980) pH-dependence of apical Na-transport in frog skin. Adv Physiol Sci 3:151 – 155
Li JH-Y, Palmer LG, Edelman IS, Lindemann B (1982) The role of sodium-channel density in the natriferic response of the toad urinary bladder to an antidiuretic hormone. J Membr Biol 64:77 – 89
Lindemann B (1984) Fluctuation analysis of sodium channels in epithelia. Ann Rev Physiol 46:497 – 515
Lipson LC, Sharp GWG (1971) Effect of prostaglandin E_1 on sodium transport and osmotic water flow in the toad bladder. Am J Physiol 220:1046 – 1052
Lipton P (1972) Effect of changes in osmolarity on sodium transport across toad bladder. Am J Physiol 222:821 – 828
Ludens JH (1978) Studies on the inhibition of Na^+ transport in toad bladder by the ionophore A23187. J Pharmacol Exp Ther 206:414 – 422
Mangos JA, McSherry NR (1967) Sodium transport: inhibitory factor in sweet of patients with cystic fibrosis. Science 158:135 – 136
Margolius HS (1984) The kallikrein-kinin system and the kidney. Ann Rev Physiol 46:309 – 326
Michell RH (1975) Inositol phospholipids and cell surface receptor function. Biochim Biophys Acta 415:81 – 147
Moore RD (1973) Effect of insulin upon the sodium pump in frog skeletal muscle. J Physiol (London) 232:23 – 45
Olans L, Sariban-Sohraby S, Benos DJ (1984) Saturation behavior of single, amiloride-sensitive Na^+ channels in planar lipid bilayers. Biophys J 46:831 – 835
O'Neil RG, Hayhurst RA (1985) Sodium-dependent modulation of the renal Na – K-ATPase: influence of mineralocorticoids on the cortical collecting duct. J Membr Biol 85:169 – 179
O'Neil RG, Helman SI (1977) Transport characteristics of renal collecting tubules: influences of DOCA and diet. Am J Physiol 233:F544 – F588
Orloff J, Handler JS (1962) The similarity of effects of vasopressin, adenosine 3′-5′ phosphate (cAMP) and theophylline on the toad bladder. J Clin Invest 41:702 – 709
Palmer LG (1984) Voltage-dependent block by amiloride and other monovalent cations of apical Na channels in the toad urinary bladder. J Membr Biol 80:153 – 165
Palmer LG (1985 a) Interactions of amiloride and other blocking cations with the apical Na channel in the toad urinary bladder. J Membr Biol 87:191 – 199
Palmer LG (1985 b) Modulation of apical Na permeability of the toad urinary bladder by intracellular Na, Ca and H. J Membr Biol 83:57 – 69
Palmer LG (1986) Apical membrane K conductance in the toad urinary bladder. J Membr Biol 92:217 – 226
Palmer LG, Frindt G (1986) Amiloride-sensitive Na channels from the apical membrane of the rat cortical collecting tubule. Proc Natl Acad Sci USA 83:2767 – 2770
Palmer LG, Frindt G (1987) Effects of cell Ca and pH on Na channels from rat cortical collecting tubule. Am J Physiol 253:F333 – F339
Palmer LG, Speez N (1986) Stimulation of apical Na permeability and basolateral Na pump of toad urinary bladder by aldosterone. Am J Physiol 250:F273 – F281
Palmer LG, Edelman IS, Lindemann B (1980) Current-voltage analysis of apical sodium transport in toad urinary bladder: effects of inhibitors of transport and metabolism. J Membr Biol 57:59 – 71
Palmer LG, Li JH-Y, Lindemann B, Edelman IS (1982) Aldosterone control of the density of sodium channels in the toad urinary bladder. J Membr Biol 64:91 – 102
Petty KJ, Kokko JP, Marver DC (1981) Secondary effect of aldosterone on Na – K-ATPase activity in the rabbit cortical collecting tubule. J Clin Invest 68:1514 – 1521
Pressley TA, Haber RS, Loeb JN, Edelman IS, Ismail-Beigi F (1986) Stimulation of Na,K-activated adenosine triphosphatase and active transport by low external K^+ in a rat liver cell line. J Gen Physiol 87:591 – 606
Rayson BM, Gupta RK (1985) Steroids, intracellular sodium levels, and Na^+/K^+-ATPase regulation. J Biol Chem 260:12740 – 12743
Rayson BM, Lowther SO (1984) Steroid regulation of Na^+ K^+-ATPase: differential sensitivities along the nephron. Am J Physiol 246:F656 – F662
Reif MC, Troutman SL, Schafer JA (1984) Sustained response to vasopressin in isolated rat cortical collecting tubule. Kidney Int 26:725 – 732

Resh MD, Memenoff RA, Guidotti G (1980) Insulin stimulation of (Na^+,K^+)-adenosine triphosphatase-dependent $^{86}Rb^+$ uptake in rat adipocytes. J Biol Chem 255:10938 – 10945
Rossier BC (1984) Biosynthesis of (Na^+, K^+) ATPase in amphibian epithelial cells. Curr Top Membr Transp 20:125 – 145
Sahib MK, Schwarz JH, Handler JS (1978) Inhibition of toad urinary bladder sodium transport by carbamylcholine: possible role of cyclic GMP. Am J Physiol 235:F586 – F591
Sansom SC, O'Neil RG (1985) Mineralocorticoid regulation of apical cell membrane Na^+ and K^+ transport of cortical collecting duct. Am J Physiol 248:F858 – F868
Sansom SC, O'Neil RG (1986) Effects of mineralocorticoids on transport properties of cortical collecting duct basolateral membrane. Am J Physiol 251:F743 – F757
Sariban-Sohraby S, Burg M, Wiesmann WP, Chiang PK, Johnson JP (1984) Methylation increases sodium transport into A6 apical membrane vesicles: possible mode of aldosterone action. Science 225:745 – 746
Schoen HF, Erlij D (1985) Basolateral membrane responses to transport modifiers in the frog skin epithelium. Pfluegers Arch 405:S33 – S38
Schuster VL (1985) Mechanism of bradykinin, ADH and cAMP interaction in rabbit cortical collecting duct. Am J Physiol 249:F645 – F653
Schwartz GJ, Burg MB (1978) Mineralocorticoid effects on cation transport by cortical collecting tubules in vitro. Am J Physiol 233:576 – 585
Sharp GWG, Leaf A (1966) Mechanism of action of aldosterone. Physiol Rev 46:593 – 633
Siegel B, Civan MM (1976) Aldosterone and insulin effects on driving force of Na^+ pump in toad bladder. Am J Physiol 230:1603 – 1608
Stokes JB, Kokko JP (1977) Inhibition of sodium transport by prostaglandin E_2 across isolated, perfused rabbit collecting tubule. J Clin Invest 59:1099 – 1104
Taylor A, Windhager EE (1979) Possible role of cytosoic calcium and Na – Ca exchange in regulation of transepithelial sodium transport. Am J Physiol 236:F505 – F512
Thomas SR, Suzuki Y, Schultz SG (1983) The electrophysiology of *Necturus* urinary bladder. I. "Instantaneous" current-voltage relations in the presence of varying sodium concentrations. J Membr Biol 73:157 – 175
Tomita K, Pisano JJ, Knepper MA (1985) Control of sodium and potassium transport in the cortical collecting duct of the rat. J Clin Invest 76:132 – 136
Truscello A, Geering K, Gaeggler HP, Rossier BC (1983) Effects of butyrate and histone deacylation on aldosterone-dependent Na^+ transport in the toad bladder. J Biol Chem 258:3388 – 3395
Ussing HH, Zerahn K (1951) Active transport of sodium as the source of electric current in the short-circuited frog skin. Acta Physiol Scand 23:110 – 127
Vaughn Gl, Cook JS (1972) Regulation of cation transport capacity in HeLa cell membrane after specific blockade by ouabain. Proc Natl Acad Sci USA 69: 2627 – 2631
Wade JB, O'Neil RG, Pryor JL, Boupaep EL (1979) Modulation of cell membrane area in renal collecting tubules by corticosteroid hormones. J Cell Biol 81:439 – 445
Walser M (1969) Reversible stimulation of sodium transport in the toad bladder by stretch. J Clin Invest 48:1714 – 1723
Warncke J, Lindemann B (1985) Voltage dependence of Na channel blockage by amiloride: relaxation effects in admittance spectra. J Membr Biol 86:255 – 265
Wiesmann W, Sinha S, Klahr S (1977) Effects of insulin, ADH, and cyclic AMP on sodium transport in the toad bladder. Am J Physiol 232:307 – 314
Wiesmann W, Sinha S, Klahr S (1978a) Effects of ionophore A23187 on base-line and vasopressin-stimulated sodium transport in the toad bladder. J Clin Invest 59:418 – 425
Wiesmann W, Sinha S, Yates J, Klahr S (1978b) Cholinergic agents inhibit sodium transport across the isolated toad bladder. Am J Physiol 235:F564 – F569
Wong SME, Chase HS Jr (1986) Role of intracellular calcium in cellular volume regulation. Am J Physiol 250:C841 – C852
Yanase M, Handler JS (1986) Activators of protein kinase C inhibit sodium transport in A6 epithelia. Am J Physiol 250:C517 – C522
Yingst DR, Hoffman JF (1984) Ca-induced K transport in human red blood cell ghosts containing arsenazo III. Transmembrane interactions of Na, K and Ca and the relationship to the functioning Na – K pump. J Gen Physiol 83:19 – 45

Subject Index

MIX
Papier aus verantwortungsvollen Quellen
Paper from responsible sources
FSC® C105338

If you have any concerns about our products,
you can contact us on
ProductSafety@springernature.com

In case Publisher is established outside the EU,
the EU authorized representative is:
Springer Nature Customer Service Center GmbH
Europaplatz 3, 69115 Heidelberg, Germany

Printed by Libri Plureos GmbH
in Hamburg, Germany